Decision Theory

This edition first published 2009
© 2009 John Wiley & Sons, Ltd.

Registered office
John Wiley & Sons Ltd, The Atrium, Southern Gate, Chichester, West Sussex,
PO19 8SQ, United Kingdom

For details of our global editorial offices, for customer services and for information
about how to apply for permission to reuse the copyright material in this book
please see our website at www.wiley.com.

Library of Congress Cataloging-in-Publication Data

Parmigiani, G. (Giovanni)
  Decision theory : principles and approaches / Giovanni Parmigiani, Lurdes Inoue.
    p.   cm.
  Includes bibliographical references and index.
  ISBN 978-0-471-49657-1 (cloth)
  1. Statistical decision.   2. Axiomatic set theory.   3. Experimental design.
  I. Inoue, Lurdes.   II. Title.
  QA279.4.P37 2009
  519.5′42—dc22

                                                             2009008345

A catalogue record for this book is available from the British Library.

ISBN 978-0-471-49657-1

Set in 10/12pt Times by Integra Software Services Pvt. Ltd, Pondicherry, India

To our advisors: Don Berry, Morrie De Groot, and Jay Kadane;
and to their advisors: Jay Kadane, Jimmy Savage, and Herman Chernoff

Contents

Preface

Goals

The goal of this book is to give an overview of fundamental ideas and results about rational decision making under uncertainty, highlighting the implications of these results for the philosophy and practice of statistics. The book grew from lecture notes from graduate courses taught at the Institute of Statistics and Decision Sciences at Duke University, at the Johns Hopkins University, and at the University of Washington. It is designed primarily for graduate students in statistics and biostatistics, both at the Masters and PhD level. However, the interdisciplinary nature of the material should make it interesting to students and researchers in economics (choice theory, econometrics), engineering (signal processing, risk analysis), computer science (pattern recognition, artificial intelligence), and scientists who are interested in the general principles of experimental design and analysis.

Rational decision making has been a chief area of investigation in a number of disciplines, in some cases for centuries. Several of the contributions and viewpoints are relevant to both the education of a well-rounded statistician and to the development of sound statistical practices. Because of the wealth of important ideas, and the pressure from competing needs in current statistical curricula, our first course in decision theory aims for breadth rather than depth. We paid special attention to two aspects: bridging the gaps among the different fields that have contributed to rational decision making; and presenting ideas in a unified framework and notation while respecting and highlighting the different and sometimes conflicting perspectives.

With this in mind, we felt that a standard textbook format would be too constraining for us and not sufficiently stimulating for the students. So our approach has been to write a "tour guide" to some of the ideas and papers that have contributed to making decision theory so fascinating and important. We selected a set of exciting papers and book chapters, and developed a self-contained lecture around each one. Some lectures are close to the source, while others stray far from their original inspiration. Naturally, many important articles have been left out of the tour. Our goal was to select a set that would work well together in conveying an overall view of the fields and controversies.

We decided to cover three areas: the axiomatic foundations of decision theory; statistical decision theory; and optimal design of experiments. At many universities, these are the subject of separate courses, often taught in different departments and schools. Current curricula in statistics and biostatistics are increasingly emphasizing interdisciplinary training, reflecting similar trends in research. Our plan reflects this need. We also hope to contribute to increased interaction among the disciplines by training students to appreciate the differences and similarities among the approaches.

We designed our tour of decision-theoretic ideas so that students might emerge with their own overall philosophy of decision making and statistics. Ideally that philosophy will be the result of contact with some of the key ideas and controversies in the different fields. We attempted to put the contributions of each article into some historical perspective and to highlight developments that followed. We also developed a consistent unified notation for the entire material and emphasized the relationships among different disciplines and points of view. Most lectures include current-day materials, methods, and results, and try at the same time to preserve the viewpoint and flavor of the original contributions.

With few exceptions, the mathematical level of the book is basic. Advanced calculus and intermediate statistical inference are useful prerequisites, but an enterprising student can profit from most of the the book even without this background. The challenging aspect of the book lies in the swift pace at which each lecture introduces new and different concepts and points of view.

Some lectures have grown beyond the size that can be delivered during a $1\frac{1}{2}$ hour session. Some others merge materials that were often taught as two separate lectures. But for the most part, the lecture–session correspondence should work reasonably well. The style is also closer to that of transcribed lecture notes than that of a treatise. Each lecture is completed by worked examples and exercises that have been helpful to us in teaching this material. Many proofs, easy and hard, are left to the student.

Acknowledgments

We have intellectual debt to more people than we can list, but a special place in this list is occupied by courses we took and lecture notes we read. Giovanni's course at Duke was initially developed from two main sources. The first is the lectures from Teddy Seidenfeld's course on the Foundations of Statistics. Giovanni only took it five times—he will firmly hold he did not choose the stopping rule: left to his own devices he would have taken that class forever. The second is the lectures from Schervish's course on Advanced Statistics, from which his book on the theory of statistics would ultimately develop. We also had access to a very insightful bootleg of Charles Stein's lecture notes at Stanford, from an edition of the course taught by Persi Diaconis.

Dennis Lindley reviewed an early draft and gave very constructive comments and encouragement. Other anonymous reviewers gave helpful feedback. Bruno Sansó used our notes to teach his class at the University of California at Santa Cruz, and gave us detailed comments. We have used our notes in teaching for over a decade. Many students braved earlier drafts, gave useful feedback through questions, conversations, solutions to problems, and sometimes highly informative puzzled looks. Martin McIntosh shared his precious correspondence with Herman Chernoff.

Both of us are grateful to Hedibert Lopes, with whom our long journey to writing this book had started back in the mid 1990s. His notes from Giovanni's classes were used extensively in early versions of this book, and some figures, problems, and examples still carry his hallmark.

Lurdes is thankful to Sergio Wechsler who opened the door to new ways of thinking about statistics and introducing her to decision theory. She thanks Giovanni for inviting her on this journey, which through bumps and laughter has been a lifetime experience. She cannot wait for the next one (well, give and take some time off for her recovery from the thrill!). She wishes to thank the loving support from her brothers Roberto, Carlos, and Silvio and from her uncles Masao, Olinda, and Tadazumi. Finally, her loving gratitude goes to her parents, Satie and Kasuo, and her grandmother Matta, for the inspiring memories and lessons that guide Lurdes.

Giovanni still has mixed feelings about the day Marco Scarsini handed him a copy of Wald's book on decision functions, with the assignment of reporting on it to an undergraduate discussion group. Later Michele Cifarelli, Guido Consonni, Morrie DeGroot, Jay Kadane, Teddy Seidenfeld, Mark Schervish, Nick Polson, Don Berry, Pietro Muliere, Peter Müller, and David Matchar fueled his intellectual passion for rational decision making. Giovanni's wife Francesca is a statistician who, despite her impact on national policy making at various levels, is a bit bored by the kind of decision theory her husband favors, and perhaps baffled by the scant influence all the talking about rationality has had on his personal behavior. Nevertheless, she has been fully supportive of this never-ending project, in more ways than one can list. Giovanni thinks working with Lurdes has been absolutely fantastic. He has not told her yet, but he is already thinking about notation changes for the second edition . . .

Giovanni Parmigiani, Lurdes Y. T. Inoue

Acknowledgments

1. Page 15; Extracted from Savage LJ (1981a). A panel discussion of personal probability, The writings of Leonard Jimmie Savage – A memorial Selection, American Statistical Association, Alexandria, VA, pp. 508–513, American Statistical Association.
2. Pages 16, 22; Extracted from de Finetti, B. (1937). Foresight: its logical laws, its subjective sources, in H. E. Kyburg and H. E. Smokler (eds.), Studies in Subjective Probability, Krieger, New York, pp. 55–118.
3. Pages 24–25; Extracted from Goldstein, M. (1983). The prevision of a prevision, Journal of the American Statistical Association 78: 817–819, American Statistical Association.
4. Pages 36–37, 48, 82–83, 88, 115–116; Extracted from Savage, L. J. (1954). The foundations of statistics, John Wiley & Sons, Inc., New York.
5. Page 38; Extracted from Bonferroni, C. (1924). La media esponenziale in matematica finanziaria, Annuario del Regio Istituto Superiore di Scienze Economiche e Commerciali di Bari, Vol. 23–24, pp. 1–14, Regio Istituto Superiore di Scienze Economiche e Commerciali di Bari.
6. Pages 61, 63; Extracted from Pratt, J. (1964). Risk aversion in the small and in the large, Econometrica 32: 122–136, The Econometric Society.
7. Page 93; Extracted from Ellsberg, D. (1961). Risk, ambiguity and the savage axioms, Quarterly Journal of Economics 75: 643–669, The MIT Press.
8. Page 93; Extracted from Gärdenfors, P. and Sahlin, N.-E. (1988). Decision, Probability and Utility, Cambridge University Press, Cambridge.
9. Pages 97–98; Extracted from Savage, L. J. (1981a). A panel discussion of personal probability, The writings of Leonard Jimmie Savage – A memorial selection, American Statistical Association, Alexandria, VA, pp. 508–513, American Statistical Association.
10. Page 99; Extracted from Anscombe, F. J. and Aumann, R. J. (1963). A definition of subjective probability, Annals of Mathematical Statistics 34: 199–205, Institute of Mathematical Statistics.
11. Page 103; Extracted from Schervish, M. J., Seidenfeld, T. and Kadane, J. B. (1990). State-dependent utilities, Journal of the American Statistical Association 85: 840–847, American Statistical Association.
12. Page 111; Extracted from Edgeworth, F. Y. (1887). The method of measuring probability and utility, Mind 12 (47): 484–488, Oxford University Press.

13. Page 113; Extracted from Wald, A. (1939). Contributions to the theory of statistical estimation and testing hypotheses, Annals of Mathematical Statistics 10: 299–326, Institute of Mathematical Statistics.

14. Page 115; Extracted from Chernoff, H. (1954). Rational selection of a decision function, Econometrica 22: 422–443, The Econometric Society.

15. Page 133; Extracted from Neyman, J. and Pearson, E. S. (1933). On the problem of the most efficient test of statistical hypotheses, Philosophical Transaction of the Royal Society (Series A) 231: 286–337, Springer-Verlag, New York.

16. Pages 150–151; Extracted from Lindley, D. V. (1968b). Decision making, The Statistician 18: 313–326, Blackwell Publishing.

17. Pages 153–154; Extracted from Birnbaum, A. (1962). On the foundations of statistical inference (Com: P307-326), Journal of the American Statistical Association 57: 269–306. (From Pratt's comments to Birnbaum 1962, pp. 314–315), American Statistical Association.

18. Pages 114, 283; Extracted from Savage, L. J. (1951). The theory of statistical decision, Journal of the American Statistical Association 46: 55–67, American Statistical Association.

19. Page 171; Extracted from Schervish, M. J. (1995). Theory of Statistics, Springer-Verlag, American Statistical Association.

20. Page 175; Extracted from Robert, C. P. (1994). The Bayesian Choice, Springer-Verlag.

21. Pages 191–192; Extracted from Brier, G. (1950). Verification of forecasts expressed in terms of probability, Monthly Weather Review 78: 1–3, American Meterological Society.

22. Page 199; Extracted from Good, I. J. (1952). Rational decisions, Journal of the Royal Statistical Society, Series B, Methodological 14: 107–114, Blackwell Publishing.

23. Page 200; Extracted from DeGroot, M. H. and Fienberg, S. E. (1983). The comparison and evaluation of forecasters, The Statistician 32: 12–22, Blackwell Publishing.

24. Page 207; Extracted from Dawid, A. (1982). The well-calibrated Bayesian, Journal of the American Statistical Association 77: 605–613, American Statistical Association.

25. Page 224; Extracted from Nemhauser, G. (1966). Introduction to Dynamic Programming, John Wiley & Sons, Inc., New York.

26. Page 225; Extracted from Lindley, D. (1961). Dynamic programming and decision, Applied Statistics, Blackwell Publishing.

27. Pages 226, 228–229; Extracted from Lindley, D. V. (1985). Making Decisions, second ed., John Wiley & Sons, Ltd, Chichester.

28. Pages 251–253; Extracted from French, S. (1988). Decision theory: an introduction to the mathematics of rationality, Ellis Horwood, Chichester, Horwood Publishing Ltd.

29. Page 291; Extracted from Grundy, P., Healy, M. and Rees, D. (1956). Economic choice of the amount of experimentation, Journal of The Royal Statistical Society. Series B. 18: 32–55, Blackwell Publishing.

30. Pages 297–298; Extracted from Lee, S. and Zelen, M. (2000). Clinical trials and sample size considerations: another perspective, Statistical Science 15: 95–110, Institute of Mathematical Statistics.

31. Page 324–325; Extracted from Wald, A. (1945). Sequential tests of statistical hypotheses, Annals of Mathematical Statistics 16: 117–186, Institute of Mathematical Statistics.

32. Page 324; Extracted from Wallis,W. A. (1980). The Statistical Research Group, 1942-1945 (C/R: p331–335), Journal of the American Statistical Association 75: 320–330, American Statistical Association.

33. Page 339; Extracted from Berger, J. and Berry, D. (1988). The relevance of stopping rules in statistical inference (with discussion), in J. Berger and S. Gupta (eds.), Statistical Decision Theory and Related Topics IV, Vol. 1, Springer-Verlag, New York, pp. 29–72.

34. Page 341; Extracted from Kadane, J. B., Schervish,M. J. and Seidenfeld, T. (1996). Reasoning to a foregone conclusion, Journal of the American Statistical Association 91: 1228–1235, American Statistical Association.

1

Introduction

We statisticians, with our specific concern for uncertainty, are even more liable than other practical men to encounter philosophy, whether we like it or not.

(Savage 1981a)

1.1 Controversies

Statistics is a mature field but within it remain important controversies. These controversies stem from profoundly different perspectives on the meaning of learning and scientific inquiry, and often result in widely different ways of interpreting the same empirical observations.

For example, a controversy that is still very much alive involves how to evaluate the reliability of a prediction or guess. This is, of course, a fundamental issue for statistics, and has implications across a variety of practical activities. Many are captured by a case study on the evaluation of evidence from clinical trials (Ware 1989). We introduce the controversy with an example. You have to guess a secret number. You know it is an integer. You can perform an experiment that would yield either the number before it or the number after it, with equal probability. You know there is no ambiguity about the experimental result or about the experimental answer. You perform this type of experiment twice and get numbers 41 and 43. What is the secret number? Easy, it is 42. Now, how good an answer do you think this is? Are you tempted to say "It is a perfect answer, the secret number has to be 42"? It turns

Decision Theory: Principles and Approaches G. Parmigiani, L. Y. T. Inoue
© 2009 John Wiley & Sons, Ltd

out that not all statisticians think this is so easy. There are at least two opposed perspectives on how to go about figuring out how good our answer is:

Judge an answer by what it says. *Compare the answer to other possible answers, in the light of the experimental evidence you have collected, which can now be taken as a given.*

versus

Judge an answer by how it was obtained. *Specify the rule that led you to give the answer you gave. Compare your answer to the answers that your rule would have produced when faced with all possible alternative experimental results. How far your rule is from the truth in this collection of hypothetical answers will inform you about how good your rule is. Indirectly, this will tell you how well you can trust your specific answer.*

Let us go back to the secret number. From the first perspective you would compare the answer "42" to all possible alternative answers, realize that it is the only answer that is not ruled out by the observed data, and conclude that there is no ambiguity about the answer being right. From the second perspective, you ask how the answer was obtained. Let us consider a reasonable recipe for producing answers: average the two experimental results. This approach gets the correct answer half the time (when the two experimental results differ) and is 1 unit off the remainder of the time (when the two experiments yield the same number). Most measures of error that consider this entire collection of potential outcomes will result in a conclusion that will attribute some uncertainty to your reported answer. This is in sharp contrast with the conclusion reached following from the first perspective. For example, the standard error of your answer is $1/\sqrt{2}$. By this principle, you would write a paper reporting your discovery that "the secret number is 42 (s.e. 0.7)" irrespective of whether your data are 41 and 43, or 43 and 43. You can think of other recipes, but if they are to give you a single guess, they are all prone to making mistakes when the two experimental results are the same, and so the story will have the same flavor.

The reasons why this controversy exists are complicated and fascinating. When things are not as clear cut as in our example, and multiple answers are compatible with the experimental evidence, the first perspective requires weighing them in some way—a step that often involves judgment calls. On the other hand the second perspective only requires knowing the probabilities involved in describing how the experiments relate to the secret number. For this reason, the second approach is perceived by many to be more objective, and more appropriate for scientific inquiry. Objectivity, its essence, worthiness, and achievability, have been among the most divisive issues in statistics. In an extreme simplification the controversy can be captured by two views of probability:

Probability lives in the world. *Probability is a physical property like mass or wavelength. We can use it to describe stochastic experimental* *mechanisms, generally repeatable ones, like the assignments of exper-imental units to different conditions, or the measurement error of a device. These are the only sorts of probabilistic considerations that should enter scientific investigations.*

versus

Probability lives in the mind. *Probability, like most conceptual con-structs in science, lives in the context of the system of values and theories of an individual scientist. There is no reason why its use should be restricted to repeatable physical events. Probability can for example be applied to scientific hypotheses, or the prediction of one-time events.*

Ramsey (1926) prefaced his fundamental paper on subjective probability with a quote from poet William Blake: "Truth can never be told so as to be understood, and not be believed."

These attitudes define a coordinate in the space of statisticians' personal philoso-phies and opinions, just like the poles of the previous controversy did. These two coordinates are not the same. For example, there are approaches to the secret num-ber problem that give different answers depending on whether data are 41 and 43, or 43 and 43, but do not make use of "subjective" probability to weigh alternative answers. Conversely, it is common practice to evaluate answers obtained from sub-jective approaches, by considering how the same approaches would have fared in other experiments.

A key aspect that both these dimensions have in common is the use of a stochas-tic model as the basis for learning from data. In the secret number story, for example, the starting point was that the experimental results would fall to the left or right of the secret number *with equal probability*. The origin of the role of probability in interpreting experimental results is sampling. The archetype of many statistical the-ories is that experimental units are sampled from a larger population, and the goal of statistical inference is to draw conclusions about the whole population. A *statistical model* describes the stochastic mechanism based on which samples are selected from the population. Sometimes this is literally the case, but more often samples and pop-ulations are only metaphors to guide the construction of statistical procedures. While this has been the model of operation postulated in most statistical theory, in practical applications it is only one pole of yet another important controversy:

Learning requires models. *To rigorously interpret data we need to understand and specify the stochastic mechanism that generated them. The archetype of statistical inference is the sample-population situation.*

versus

Learning requires algorithms. *To efficiently learn from data, it is crit-ical to have practical tools for exploring, summarizing, visualizing,* *clustering, classifying. These tools can be built with or without explicit consideration of a stochastic data-generating model.*

The model-based approach has ancient roots. One of the relatively recent landmarks is Fisher's definition of the likelihood function (Fisher 1925). The algorithmic approach also goes back a long way in history: for example, most measures of dependence, such as the correlation coefficient, were born as descriptive, not inferential tools (Galton 1888). The increasing size and complexity of data, and the interface with computing, have stimulated much exploratory data analysis (Tukey 1977, Chambers *et al.* 1983) and statistical work at the interface with artificial intelligence (Nakhaeizadeh and Taylor 1997, Hastie *et al.* 2003). This controversy is well summarized in an article by Breiman (2001).

The premise of this book is that it is useful to think about these controversies, as well as others that are more technical in statistics, from first principles. The principles we will bring to bear are principles of rationality in action. Of course, this idea is in itself controversial. With this regard, the views of many statisticians distribute along another important dimension of controversy:

> **Statisticians produce knowledge**. The scope of statistics is to rigorously interpret experimental results, and present experimental evidence in an unbiased way to scientists, policy makers, the public, or whoever may be in charge of drawing conclusions or making decisions.

versus

> **Statisticians produce solutions to problems**. Understanding data requires placing them in the context of scientific theories, which allow us to sort important from ancillary information. One cannot answer the question "what is important?" without first considering the question "important for what?"

Naturally, producing knowledge helps solving problems, so these two positions are not in contrast from this standpoint. The controversy is on the extent to which the goals of an experiment should affect the learning approaches, and more broadly whether they should be part of our definition of learning.

The best known incarnation of this controversy is the debate between Fisher and Neyman about the meaning of hypothesis tests (Fienberg 1992). The Neyman–Fisher controversy is broader, but one of the key divides is that the Neyman and Pearson theory of hypothesis testing considers both the hypothesis of interest and at least one alternative, and involves an explicit quantification of the consequences of rejecting or accepting the hypothesis based on the data: the type I and type II errors. Ultimately, Neyman and Pearson's theory of hypothesis testing will be one of the key elements in the development of formal approaches to rationality-based statistical analysis. On the other hand Fisher's theory of significance test does not require considering an alternative and incarnates a view of science in which hypotheses represent working approximations to natural laws, that serve to guide experimentation, until they are refuted with sufficient strength that new theories evolve.

A simple example challenges theories of inference that are based solely on the evidence provided by observation, regardless of scope and context of the theory. Let x_1 and x_2 be two Bernoulli trials. Suppose the experimenter's probabilities are such that $P(x_1 = 0) = P(x_2 = 0) = 0.5$ and $P(x_1 + x_2 = 0) = 0.05$. Then, $P(x_1 + x_2 = 1) = 0.9$ and $P(x_1 + x_2 = 2) = 0.05$. Let e be the new evidence that $x_1 = 1$, let h_1 be the hypothesis that $x_1 + x_2 = 2$, and h_2 be the hypothesis that $x_2 = 1$. Given e, the two hypotheses are equivalent. Yet, probability-wise, h_1 is corroborated by the data, whereas h_2 is not. So if one is to consider the change in probability as a measure of support for a theory, one would be left with either an inconsistent measure of evidence, or the need to defend the position that the two hypotheses are in some sense different even when faced with evidence that proves that they are the same. This and other similar examples seriously question the idea that inductive practice can be adequately represented by probabilities alone, without relation to their rational use in action.

There can be disagreements of principle about whether consideration of consequences and beliefs belongs to scientific inquiry. In reality, though, it is our observation that the vast majority of statistical inference approaches have an implicit or explicit set of goals and values that guide the various steps of the construction. When making a decision as simple as summarizing a set of numbers by their median (as opposed to, say, their mean) one is making judgments about the relative importance of the possible oversimplifications involved. These could be made formal, and in fact there are decision problems for which each of the two summaries is optimal. Our view is that scientific discussion is more productive when goals are laid out in the open, and perhaps formalized, than when they are hidden or unappreciated. As the old saying goes, "there are two types of statisticians: those who know what decision problem they are solving and those who don't."

Despite the draconian simplifications we have made in defining the dimensions along which these four controversies unfold, one would be hard pressed to find two statisticians that live on the same point in this four-dimensional space. One of the goals of this book is to help students find their own spot in a way that reflects their personal intellectual values, and serves them best in approaching the theoretical and applied problems that are important to them.

We definitely lean on the side of "judging answers by what they say" and believing that "probabilities live in the mind." We may be some distance away along the models versus algorithm dimension—at least judging by our approaches in applications. But we are both enthusiastic about the value of thinking about rationality as a guide, though sometimes admittedly a rough guide, to science, policy, and individual action. This guidance comes at two levels: it tells us how to formally connect the tools of an analysis with the goals of that analysis; and it tells us how to use rationality-based criteria to evaluate alternative statistical tools, approaches, and philosophies.

Overall, our book is an invitation to Bayesian decision-theoretic ideas. While we do not think they necessarily provide a solution to every statistical problem, we find much to think about in this comment from Herman Chernoff (from a personal communication to Martin McIntosh):

> Frankly, I am not a Bayesian. I go under the following principle. If you don't understand a problem from a Bayesian decision theory point of view, you don't understand the problem and trying to solve it is like shooting at a target in the dark. Once you understand the problem, it is not necessary to attack it from a Bayesian point of view. Bayesian methods always had the difficulty that our approximations to our subjective beliefs could carry a lot more information than we thought or felt willing to assume.

1.2   A guided tour of decision theory

We think of this book as a "tour guide" to the key ideas in decision theory. The book grew out of graduate courses where we selected a set of exciting papers and book chapters, and developed a self-contained lecture around each one. We make no attempt at being comprehensive: our goal is to give you a tour that conveys an overall view of the fields and its controversies, and whets your appetite for more. Like the small pictures of great paintings that are distributed on leaflets at the entrance of a museum, our chapters may do little justice to the masterpiece but will hopefully entice you to enter, and could guide you to the good places.

As you read it, keep in mind this thought from R. A. Fisher:

> The History of Science has suffered greatly from the use by teachers of second-hand material, and the consequent obliteration of the circumstances and the intellectual atmosphere in which the great discoveries of the past were made. A first-hand study is always useful, and often . . . full of surprises. (Fisher 1965)

Our tour includes three parts: foundations (axioms of rationality); optimal data analysis (statistical decision theory); and optimal experimental design.

Coherence. We start with de Finetti's "Dutch Book Theorem" (de Finetti 1937) which provides a justification for the axioms of probability that is based on a simple and appealing rationality requirement called coherence. This work is the mathematical foundation of the "probabilities live in the mind" perspective. One of the implications is that new information is merged with the old via Bayes' formula, which gets promoted to the role of a universal inference rule—or Bayesian inference.

Utility. We introduce the axiomatic theory of utility, a theory on how to choose among actions whose consequences are uncertain. A rational decision maker proceeds by assigning numerical utilities to consequences, and scoring actions by their expected utility. We first visit the birthplace of quantitative utility: Daniel Bernoulli's St. Petersburg paradox (Bernoulli 1738). We then present in detail von Neumann and

Morgenstern's utility theory (von Neumann and Morgenstern 1944) and look at a criticism by Allais (1953).

Utility in action. We make a quick detour to take a look at practical matters of implementation of rational decision making in applied situations, and talk about how to measure the utility of money (Pratt 1964) and the utility of being in good health. For applications in health, we examine a general article (Torrance *et al.* 1972) and a medical article that has pioneered the utility approach in health, and set the standard for many similar analyses (McNeil *et al.* 1981).

Ramsey and Savage. We give a brief digest of the beautiful and imposing axiomatic system developed by Savage. We begin by tracing its roots to the work of Ramsey (1931) and then cover Chapters 2, 3, and 5 from Savage's *Foundations of statistics* (Savage 1954). Savage's theory integrates the coherence story with the utility story, to create a more general theory of individual decision making. When applied to statistical practice, this theory is the foundation of the "statisticians find solutions to problems" perspective. The general solution is to maximize expected utility, and expectations are computed by assigning personal probabilities to all unknowns. A corollary is that "answers are judged by what they say."

State independence. Savage's theory relies on the ability to separate judgment of probability from judgment of utility in evaluating the worthiness of actions. Here we study an alternative axiomatic justification of the use of subjective expected utility in decision making, due to Anscombe and Anmann (1963). Their theory highlights very nicely the conditions for this separation to take place. This is the last chapter on foundations.

Decision functions. We visit the birthplace of statistical decision theory: Wald's definition of a general statistical decision function (Wald 1949). Wald proposed a unifying framework for much of the existing statistical theory, based on treating statistical inference as a special case of game theory, in which the decision maker faces nature in a zero-sum game. This leads to maximizing the smallest utility, rather than a subjective expectation of utility. The contrast of these two perspectives will continue through the next two chapters.

Admissibility. Admissibility is the most basic and influential rationality requirement of Wald's classical statistical decision theory. A nice surprise for Savage's fans is that maximizing expected utility is a safe way, and often, at least approximately, the only way, to build admissible statistical decision rules. Nice. In this chapter we also reinterpret one of the milestones of statistical theory, the Neyman–Pearson lemma (Neyman and Pearson 1933), in the light of the far-reaching theory this lemma sparked.

Shrinkage. The second major surprise from the study of admissibility is the fact that $\bar{x}$—the motherhood and apple pie of the statistical world—is inadmissible in estimating the mean of a multidimensional normal vector of observations. Stein (1955) was the first to realize this. We explore some of the important research directions that stemmed from Stein's paper, including shrinkage estimation, empirical Bayes estimation, and hierarchical modeling.

Scoring rules. We change our focus to prediction and explore the implications of holding forecasters accountable for their predictions. We study the incentive systems that must be set in place for the forecasters to reveal their information/beliefs rather than using them to game the system. This leads to the study of proper scoring rules (Brier 1950). We also define and investigate calibration and refinement of forecasters.

Choosing models We try to understand whether statistical decision theory can be applied successfully to the much more elusive tasks of constructing and assessing statistical models. The jury is still out. On this puzzling note we close our tour of statistical decision theory and move to experimental design.

Dynamic programming. We describe a general approach for making decisions dynamically, so that we can both learn from accruing knowledge and plan ahead to account for how present decisions will affect future decisions and future knowledge. This approach, called dynamic programming, was developed by Bellman (1957). We will try to understand why the problem is so hard (the "curse of dimensionality").

Changes in utility as information. In decision theory, the value of the information carried by a data set depends on what we intend to do with the data once we have collected them. We use decision trees to quantify this value (DeGroot 1984). We also explore in more detail a specific way of measuring the information in a data set, which tries to capture "generic learning" rather than specific usefulness in a given problem (Lindley 1956).

Sample size. We finally come to terms with the single most common decision statisticians make in their daily activities: how big should a data set be? We try to understand how all the machinery we have been setting in place can help us and give some examples. Our discussion is based on the first complete formalization of Bayesian decision-theoretic approaches to sample size determination (Raiffa and Schlaifer 1961).

Stopping. Lastly, we apply dynamic programming to sequential data collection, where we have the option to stop an experiment after each observation. We discuss the stopping rule principle, which states that within the expected utility paradigm, the rule used to arrive at the decision to stop at a certain stage is not informative about parameters controlling the data-generating mechanism. We also study whether it is possible to design stopping rules that will stop experimentation only when one's favorite conclusion is reached.

A terrific preparatory reading for this book is Lindley (2000) who lays out the philosophy of Bayesian statistics in simple, concise, and compelling terms. As you progress through the book you will find, generally in each chapter's preamble, alternative texts that dwell on individual topics in greater depth than we do. Some are also listed next. A large number of textbooks overlap with ours and we make no attempt at being comprehensive. An early treatment of statistical decision theory is Raiffa and Schlaifer

(1961), a text that contributed enormously to defining practical Bayesian statistics and decision making in the earliest days of the field. Their book was exploring new territory on almost every page and, even in describing the simplest practical ideas, is full of deep insight. Ferguson (1967) is one of the early influential texts on statistical decision theory, Bayes and frequentist. DeGroot (1970) has a more restricted coverage of Part Two (no admissibility) but a more extensive discussion of Part Three and an in-depth discussion of foundations, which gives a quite independent treatment of the material compared to the classical papers discussed in our book. A mainstay statistical decision theory book is Berger (1985) which covers topics throughout our tour. Several statistics books have good chapters on decision-theoretic topics. Excellent examples are Schervish (1995) and Robert (1994), both very rigorous and rich in insightful examples. Bernardo and Smith (1994) is also rich in foundational discussions presented in the context of both statistical inference and decision theory. French (1988), Smith (1987), and Bather (2000) cover decision-analytic topics very well. Kreps (1988) is an accessible and very insightful discussion of foundations, covered in good technical detail. A large number of texts in decision analysis, medical decision making, microeconomics, operations research, statistics, machine learning, and stochastic processes cover individual topics.

Part One
Foundations

2

Coherence

Decision making under uncertainty is about making choices whose consequences are not completely predictable, because events will happen in the future that will affect the consequences of actions taken now. For example, when deciding whether to play a lottery, the consequences of the decision will depend on the number drawn, which is unknown at the time when the decision is made. When deciding treatment for a patient, consequences may depend on future events, such as the patient's response to that treatment. Political decisions may depend on whether a war will begin or end within the next month. In this chapter we discuss de Finetti's justification, the first of its kind, for using the calculus of probability as a quantification of uncertainty in decision making.

In the lottery example, uncertainty can be captured simply by the chance of a win, thought of, at least approximately, as the long-term frequency of wins over many identical replications of the same type of draw. This definition of probability is generally referred to as frequentist. When making a prognosis for a medical patient, chances based on relative frequencies are still useful: for example, we would probably be interested in knowing the frequency of response to therapy within a population of similar patients. However, in the lottery example, we could work out properties of the relevant frequencies on the basis of plausible approximations of the physical properties of the draw, such as independence and equal chance. With patients, that is not so straightforward, and we would have to rely on observed populations. Patients in a population are more different from one another than repeated games of the lottery, and differ in ways we may not understand. To trust the applicability of observed relative frequencies to our decision we have to introduce an element of judgment about the comparability of the patients in the population. Finally, events like "Canada will go to war within a month" are prohibitive from a relative frequency standpoint. Here the element of judgment has to be preponderant, because it is not easy to assemble, or even imagine, a collection of similar events of which the event in question is a typical representative.

The theory of subjective probability, developed in the 1930s by Ramsey and de Finetti, is an attempt to develop a formalism that can handle quantification of uncertainty in a wide spectrum of decision-making and prediction situations. In contrast to frequentist theories, which interpret probability as a property of physical phenomena, de Finetti suggests that it is more useful and general to define probability as a property of decision makers. An intelligent decision maker may recognize and use probabilistic properties of physical phenomena, but can also go beyond. Somewhat provocatively, de Finetti often said that *probability does not exist*—meaning that it is not somewhere out there to be discovered, irrespective of the person, or scientific community, trying to discover it. De Finetti posed the question of whether there could be a calculus for these more general subjective probabilities. He proposed that the axioms of probability commonly motivated by the frequency definition could alternatively be justified by a single rationality requirement now known as coherence. Coherence amounts to avoiding loss in those situations in which the probabilities are used to set betting odds. De Finetti's proposal for subjective probability was originally published in 1931 (de Finetti 1931b) and in English, in condensed form, in 1937 (de Finetti 1937).

In Section 2.1.1 we introduce the simple betting game that motivates the notion of coherence—the fundamental rationality principle underlying the theory. In Section 2.1 we present the so-called *Dutch Book argument*, which shows that an incoherent probability assessor can be made a sure loser, and establishes a connection between coherence and the axioms of probability. In Section 2.1.3, we will also show how to derive conditional probability, the multiplication rule, and Bayes' theorem from coherence conditions. In Section 2.2 we present a temporal coherence theory (Goldstein 1985) that extends de Finetti's to situations where personal or subjective beliefs can be revised over time. De Finetti's 1937 paper is important in statistics for other reasons as well. For example, it was a key paper in the development of the notion of exchangeability.

Featured article:

de Finetti, B. (1937). Foresight: Its logical laws, its subjective sources, *in* H. E. Kyburg and H. E. Smokler (eds.), *Studies in Subjective Probability*, Krieger, New York, pp. 55–118.

Useful general readings are de Finetti (1974) and Lindley (2000). For a comprehensive overview of de Finetti's contributions to statistics see Cifarelli and Regazzini (1996).

2.1   The "Dutch Book" theorem

2.1.1   Betting odds

De Finetti's theory of subjective probability is usually described using the metaphor of betting on the outcome of an as yet unknown event, say a sports event. This is a stylized situation, but it is representative of many simple decision situations, such as setting insurance premiums. The odds offered by a sports bookmaker, or the premiums set by an insurance company, reflect their judgment about the probabilities of events. So this seems like a natural place to start thinking about quantification of uncertainty.

Savage worries about a point that matters a great deal to philosophers and surprisingly less so to statisticians writing on foundations:

> The idea of facts known is implicit in the use of the preference theory. For one thing, the person must know what acts are available to him. If, for example, I ask what odds you give that the fourth toss of this coin will result in heads if the first three do, it is normally implicitly not only that you know I will keep my part of the bargain if we bet, but also that you will know three heads if you see them. The statistician is forever talking about what reaction would be appropriate to this or that set of data, or givens. Yet, the data never are quite given, because there is always some doubt about what we have actually seen. Of course, in any applications, the doubt can be pushed further along. We can replace the event of three heads by the less immediate one of three tallies-for-head recorded, and then take into our analysis the possibility that not every tally is correct. Nonetheless, not only universals, but the most concrete and individual propositions are never really quite beyond doubt. Is there, then, some avoidable lack of clarity and rigor in our allusion to known facts? It has been argued that since indeed there is no absolute certainty, we should understand by "certainty" only strong relative certainty. This counsel is provocative, but does seem more to point up, than to answer, the present question. (Savage 1981a, p. 512)

Filing away this concern under "philosophical aches and pains," as Savage would put it, let us continue with de Finetti's plan.

Because bookmakers (and insurance companies!) make a profit, we will, at least for now, dissect the problem so that only the probabilistic component is left. So we will look at a situation where bookmakers are willing to buy and sell bets at the same odds. To get rid of considerations that come up when one bets very large sums of money, we will assume, like de Finetti, that we are in a range of bets that involves enough money for the decision maker to take things seriously, but not big enough that aversion to potentially large losses may interfere. In the next several chapters we will discuss how to replace monetary amounts with a more abstract and general measure, built upon ideas of Ramsey and others, that captures the utility to a decision maker of owning that money. For now, though, we will keep things simple and follow de Finetti's assumptions. With regard to this issue, de Finetti (1937) notes that:

> Such a formulation could better, like Ramsey's, deal with expected *utilities*; I did not know of Ramsey's work before 1937, but I was aware of the difficulty of money bets. I preferred to get around it by considering sufficiently small stakes, rather than to build up a complex theory to deal with it. (de Finetti 1937, p. 140)

So de Finetti separates the derivation of probability from consideration of utility, although rationality, more broadly understood, is part of his argument. In later writings (de Finetti 1952, de Finetti 1964b), he discussed explicitly the option of deriving both utilities and probability from a single set of preferences, and seemed to consider it the most appropriate way to proceed in decision problems, but maintained that the separation is preferable in general, giving two reasons:

> First, the notion of probability, purified from the factors that affect utility, belongs to a logical level that I would call "superior". Second, constructing the calculus of probability in its entirety requires vast developments concerning probability alone. (de Finetti 1952, p. 698, our translation)

These "vast developments" begin with the notion of coherent probability assessments. Suppose we are interested in predicting the result of an upcoming tennis match, say between Fisher and Neyman. Bookmakers are generally knowledgeable about this, so we are going to examine the bets they offer as a possible way of quantifying uncertainty about who will win. Bookmakers post odds. If the posted odds in favor of Fisher are, say, 1:2, one can bet one dollar, and win two dollars if Fisher wins and nothing if he does not. In sports you often see the reciprocal of the odds, also called odds "against," and encounter expressions like "Fisher is given 2:1" to convey the odds against.

To make this more formal, let θ be the indicator of the event "Fisher wins the game." We say, equivalently, that θ occurred or θ is true or that the true value of θ is 1. A bet is a ticket that will be worth a stake S if θ occurs and nothing if θ does not occur. A bookmaker generally sells bets at a price $\pi_\theta S$. The price is expressed in units of the stake; when there is no ambiguity we will simply use π. The ratio $\pi : (1 - \pi)$ is the betting odds in favor of the event θ. In our previous example, where odds in favor are 1:2, the stake S is three dollars, the price πS is one dollar, and π is $1/3$.

The action of betting on θ, or buying the ticket, will be denoted by $a_{S,\theta}$. This action can be taken by either a client or the bookmaker, although in real life it is more often the client's. What are the consequences of this action? The buyer will have a net gain of $(1 - \pi)S$, that is the stake S less the price πS, if $\theta = 1$, or a net gain of $-\pi S$, if $\theta = 0$. These net gains are summarized in Table 2.1. We are also going to allow for negative stakes. The action $a_{-S,\theta}$, whose consequences are also shown

Table 2.1   Payoffs for the actions corresponding to buying and selling bets at stake S on event θ at odds $\pi : (1 - \pi)$.

Action		States of the world	
		$\theta = 1$	$\theta = 0$
Buy bet on θ	$a_{S,\theta}$	$(1 - \pi)S$	$-\pi S$
Sell bet on θ	$a_{-S,\theta}$	$-(1 - \pi)S$	πS

in Table 2.1, reverses the role of the buyer and seller compared to $a_{S,\theta}$. A stake of zero will represent abstaining from the bet. Also, action $a_{-S,\theta}$ (selling the bet) has by definition the same payoff as buying a bet on the event $\theta = 0$ at stake S and price $\pi_{1-\theta} = 1 - \pi_\theta$.

We will work in the stylized context of a bookmaker who posts odds $\pi : (1 - \pi)$ and is willing to buy or sell bets at those odds, for any stake. In other words, once the odds are posted, the bookmaker is indifferent between buying and selling bets on θ, or abstaining. The expression we will use for this is that the odds are *fair* from the bookmaker's perspective. It is implicitly assumed that the bookmaker can assess his or her willingness to buy and sell directly. Assessing odds is therefore assumed to be a primitive, as opposed to derivative, way of expressing preferences among actions involving bets. We will need a notation for binary comparisons among bets: $a_1 \sim a_2$ indicates indifference between two bets. For example, a bookmaker who considers odds on θ to be fair is indifferent between $a_{S,\theta}$ and $a_{-S,\theta}$, that is $a_{S,\theta} \sim a_{-S,\theta}$. Also, the symbol $\succ$ indicates a strict preference relation. For example, if odds in favor of θ are considered too high by the bookmaker, then $a_{S,\theta} \succ a_{-S,\theta}$.

2.1.2   Coherence and the axioms of probability

Before proceeding with the formal development, let us illustrate the main idea using a simple example. Lindley (1985) has a similar one, although he should not be blamed for the choice of tennis players. Let us imagine that you know a bookmaker who is willing to take bets on the outcome of the match between Fisher and Neyman. Say the prices posted by the bookmaker are 0.2 (1:4 odds in favor) for bets on the event θ: "Fisher wins," and 0.7 (7:3 odds in favor) for bets on the event "Neyman wins." In the setting of the previous section, this means that this bookmaker is willing to buy or sell bets on θ for a stake S at those prices. If you bet on θ, the bookmaker cashes $0.2S$ and then gives you back S (a net gain to him of $-0.8S$) if Fisher wins, and nothing (a net gain of $0.2S$) if Fisher loses. In tennis there are no ties, so the event "Neyman wins" is the same as "Fisher loses" or $\theta = 0$. The bookmaker has posted separate odds on "Neyman wins," and those imply that you can also bet on that, in which case the bookmaker cashes $0.7S$ and returns S (a net gain of $-0.3S$) if Neyman wins and nothing (a net gain of $0.7S$) if Neyman loses. Let us now see what happens if you place both bets. Your gains are:

	Fisher wins	Neyman wins
Bet 1	$0.8S$	$-0.2S$
Bet 2	$-0.7S$	$0.3S$
Both bets (total)	$0.1S$	$0.1S$

So by placing both bets you can make the bookmaker lose money irrespective of whether Fisher or Neyman will win! If the stake is 10 dollars, you win one dollar either way. There is an internal inconsistency in the prices posted by the bookmaker that can be exploited to an economic advantage. In some bygone era, this used to be called "making Dutch Book against the bookmaker." It is a true embarrassment, even aside from financial considerations. So the obvious question is "how to avoid Dutch Book?" Are there conditions that we can impose on the prices so that a bookmaker posting those prices cannot be made a sure loser? The answer is quite simple: prices π have to satisfy the axioms of probability. This argument is known as the "Dutch Book theorem" and is worth exploring in some detail.

Avoiding Dutch Book is the rationality requirement that de Finetti had in mind when introducing *coherence*.

Definition 2.1 (Coherence)   *A bookmaker's betting odds are coherent if a client cannot place a bet or a combination of bets such that no matter what outcome occurs, the bookmaker will lose money.*

The next theorem, due to de Finetti (1937), formalizes the claim that coherence requires prices that satisfy the axioms of probabilities. For a more formal development see Shimony (1955). The conditions of the theorem, in the simple two-event version, are as follows. Consider disjoint events θ_1 and θ_2, and assume a bookmaker posts odds (and associated prices) on all the events in the algebra induced by these events, that is on $1 - \Theta, \theta_1, \theta_2, (1 - \theta_1)(1 - \theta_2), \theta_1 + \theta_2, 1 - \theta_1, 1 - \theta_2, \Theta$, where Θ is the indicator of the sure event. As we discussed, there are two structural assumptions being made:

DBT1: The odds are fair to the bookmaker, that is the bookmaker is willing to both sell and buy bets on any of the events posted.

DBT2: There is no restriction about the number of bets that clients can buy or sell, as long as this is finite.

The first condition is required to guarantee that the odds reflect the bookmaker's knowledge about the relevant uncertainties, rather than desire to make a profit. The second condition is used, in de Finetti's words, to "purify" the notion of probability from the factors that affect utility. It is strong: it implies, for example, that the bookmaker values the next dollar just as much as if it were his or her last dollar. Even with this caveat, this is a very interesting set of conditions for an initial study of the rational underpinnings of probability.

Theorem 2.1 (Dutch Book theorem)   *If DBT1 and DBT2 hold, a necessary condition for a set of prices to be coherent is to satisfy Kolmogorov's axioms, that is:*

Axiom 1 $0 \leq \pi_\theta \leq 1$, for every θ.

Axiom 2 $\pi_\Theta = 1$.

Axiom 3   If θ_1 and θ_2 are such that $\theta_1\theta_2 = 0$, then $\pi_{\theta_1} + \pi_{\theta_2} = \pi_{\theta_1 + \theta_2}$.

Proof: We assume that the odds are fair to the bookmaker, and consider the gain g_θ made by a client buying bets from the bookmaker on event θ. If the gain is strictly positive for every θ in a partition, then the bookmaker can be made a sure loser and is incoherent.

Axiom 1: Suppose, by contradiction, that $\pi_\theta > 1$. When $S < 0$, the gain g_θ to a client is

$$g_\theta = \begin{cases} (1 - \pi_\theta)S & \text{if } \theta = 1 \\ -\pi_\theta S & \text{if } \theta = 0 \end{cases}$$

which is strictly positive for both values of θ. Similarly, $\pi_\theta < 0$ and $S > 0$ also imply a sure loss.

Axiom 2: Let us assume that Axiom 1 holds. Say, for a contradiction, that $0 \leq \pi_\Theta < 1$. For any $S > 0$ the gain g_Θ is $(1 - \pi_\Theta)S > 0$, if $\Theta = 1$. Because $\Theta = 1$ by definition, this implies a sure loss.

Axiom 3: Let us consider separate bets on θ_1, θ_2, and $\theta_3 = \theta_1 + \theta_2 - \theta_1\theta_2 = \theta_1 + \theta_2$. θ_3 is the indicator of the union of the two events represented by θ_1 and θ_2. Say stakes are $S_{\theta_1}, S_{\theta_2}$, and S_{θ_3}, respectively. Consider the partition given by θ_1, θ_2, and $(1 - \theta_1)(1 - \theta_2)$. The net gains to the client in each of those cases are

$$g_{\theta_1} = S_{\theta_1} + S_{\theta_3} - (\pi_{\theta_1}S_{\theta_1} + \pi_{\theta_2}S_{\theta_2} + \pi_{\theta_3}S_{\theta_3})$$
$$g_{\theta_2} = S_{\theta_2} + S_{\theta_3} - (\pi_{\theta_1}S_{\theta_1} + \pi_{\theta_2}S_{\theta_2} + \pi_{\theta_3}S_{\theta_3})$$
$$g_{(1-\theta_1)(1-\theta_2)} = -(\pi_{\theta_1}S_{\theta_1} + \pi_{\theta_2}S_{\theta_2} + \pi_{\theta_3}S_{\theta_3}).$$

These three equations can be rewritten in matrix notation as $\boldsymbol{Rs} = \boldsymbol{g}$, where

$$\boldsymbol{g} = (g_{\theta_1}, g_{\theta_2}, g_{(1-\theta_1)(1-\theta_2)})'$$
$$\boldsymbol{s} - (S_{\theta_1}, S_{\theta_2}, S_{\theta_3})'$$
$$\boldsymbol{R} = \begin{pmatrix} 1 - \pi_{\theta_1} & -\pi_{\theta_2} & 1 - \pi_{\theta_3} \\ -\pi_{\theta_1} & 1 - \pi_{\theta_2} & 1 - \pi_{\theta_3} \\ -\pi_{\theta_1} & -\pi_{\theta_2} & -\pi_{\theta_3} \end{pmatrix}.$$

If the matrix $\boldsymbol{R}$ is invertible, the system can be solved to get $\boldsymbol{s} = \boldsymbol{R}^{-1}\boldsymbol{g}$. This means that a client can set $\boldsymbol{g}$ to be a vector of positive values, corresponding to losses for the bookmaker. Thus coherence requires that the matrix $\boldsymbol{R}$ be singular, that is $|\boldsymbol{R}| = 0$, which in turn implies, after a little bit of algebra, that $\pi_{\theta_1} + \pi_{\theta_2} - \pi_{\theta_3} = 0$. $\square$

We stated and proved this theorem in the simple case of two disjoint events. This same argument can be extended to an arbitrary finite set of disjoint events, as done in de Finetti (1937). From a mathematical standpoint it is also possible, and standard, to define axioms in terms of countable additivity, that is additivity over a denumerable partition. This also permits us to talk about coherent probability distributions on both discrete and continuous random variables.

The extension of coherence results from finite to denumerable partitions is controversial, as some find it objectionable to state a rationality requirement in terms of a circumstance as abstract as taking on an infinite number of bets. In truth, the theory as stated allows for any finite number of bets, so this number can easily be made to be large enough to be ridiculously unrealistic anyway. But there are other reasons as well why finite-bets theory is fun to explore. Seidenfeld (2001) reviews some differences between the countably additive theory of probability and the alternative theory built solely using finitely additive probability.

A related issue concerns events of probability zero more generally. Shimony (1955), for example, has criticized the coherence condition we discussed in this chapter as too weak, and prefers a stricter version that would not consider it rational to choose a bet whose return is never positive and sometimes negative. This version implies that no possible event can have probability zero—a requirement sometimes referred to a "Cromwell's Rule" (Lindley 1982a).

2.1.3   Coherent conditional probabilities

In this section we present a second Dutch Book theorem that applies to coherence of conditional probabilities. The first step in de Finetti's development is to define conditional statements. These are more general logical statements based on a three-valued logic: statement A conditional on B can be either true, if both are true, or false, if A is false and B is true, or void if B is false. In betting terminology this idea is operationalized by the so-called "called-off" bets. A bet on θ_1, with stake S at a price π, called off if θ_2 does not occur, means buying at price πS a ticket worth the following. If θ_2 does not occur the price πS will be returned. If θ_2 occurs the ticket will be worth S if θ_1 occurs and nothing if θ_1 does not occur, as usual. We denote by $\pi_{\theta_1|\theta_2}$ the price of this bet. The payoff is then described by Table 2.2. Under the same structural conditions of Theorem 2.1, we have that:

Theorem 2.2 (Multiplication rule)   *A necessary condition for coherence of prices of called-off bets is that* $\pi_{\theta_1|\theta_2}\pi_{\theta_2} = \pi_{\theta_1\theta_2}$.

Table 2.2   Payoffs corresponding to buying a bet on $\theta_1 = 1$, called off it $\theta_2 = 0$.

Action	States of the world		
	$\theta_1\theta_2 = 1$	$(1 - \theta_1)\theta_2 = 1$	$\theta_2 = 0$
bet on θ_1, called off if $\theta_2 = 0$	$(1 - \pi_{\theta_1|\theta_2})S$	$-\pi_{\theta_1|\theta_2}S$	0

Proof: Consider bets on θ_2, $\theta_1\theta_2$, and $\theta_1|\theta_2$ with stakes S_{θ_2}, $S_{\theta_1\theta_2}$, and $S_{\theta_1|\theta_2}$, respectively, and the partition $\theta_1\theta_2, (1-\theta_1)\theta_2, (1-\theta_2)$. The net gains are

$$g_{\theta_1\theta_2} = S_{\theta_2} + S_{\theta_1\theta_2} + S_{\theta_1|\theta_2} - (\pi_{\theta_2}S_{\theta_2} + \pi_{\theta_1\theta_2}S_{\theta_1\theta_2} + \pi_{\theta_1|\theta_2}S_{\theta_1|\theta_2})$$
$$g_{(1-\theta_1)\theta_2} = S_{\theta_2} - (\pi_{\theta_2}S_{\theta_2} + \pi_{\theta_1\theta_2}S_{\theta_1\theta_2} + \pi_{\theta_1|\theta_2}S_{\theta_1|\theta_2})$$
$$g_{1-\theta_2} = -(\pi_{\theta_2}S_{\theta_2} + \pi_{\theta_1\theta_2}S_{\theta_1\theta_2} + 0).$$

These three equations can be rewritten in matrix notation as $\boldsymbol{Rs} = \boldsymbol{g}$, where

$$\boldsymbol{g} = (g_{\theta_1\theta_2}, g_{(1-\theta_1)\theta_2}, g_{1-\theta_2})'$$
$$\boldsymbol{s} = (S_{\theta_2}, S_{\theta_1\theta_2}, S_{\theta_1|\theta_2})'$$
$$\boldsymbol{R} = \begin{pmatrix} 1-\pi_{\theta_2} & 1-\pi_{\theta_1\theta_2} & 1-\pi_{\theta_1|\theta_2} \\ 1-\pi_{\theta_2} & -\pi_{\theta_1\theta_2} & -\pi_{\theta_1|\theta_2} \\ -\pi_{\theta_2} & -\pi_{\theta_1\theta_2} & 0 \end{pmatrix}.$$

The requirement of coherence implies that $|\boldsymbol{R}| = 0$, which in turn implies that

$$\pi_{\theta_1|\theta_2}\pi_{\theta_2} - \pi_{\theta_1\theta_2} = 0. \qquad \square$$

At this point we have at our disposal all the machinery of probability calculus. For example, a corollary of the law of total probability and the conditioning rule is Bayes' rule. Therefore, coherent probability assessment must also obey the Bayes rule.

If we accept countable additivity, we can use continuous random variables and their properties. One can define a parallel set of axioms in terms of expectations of random variables that falls back on the case we studied if the random variables are binary. An important case is that of conditional expectations. If θ is any continuous random variable and θ_2 is an event, then the "conditional random variable" θ given θ_2 can be defined as

$$\theta|\theta_2 = \theta\theta_2 + (1-\theta_2)E[\theta|\theta_2]$$

where θ is observed if θ_2 occurs and not otherwise. Taking expectations,

$$E[\theta|\theta_2] = E[\theta\theta_2] + (1-\pi_{\theta_2})E[\theta|\theta_2]$$

and solving gives

$$E[\theta|\theta_2] = \frac{E[\theta\theta_2]}{\pi_{\theta_2}}. \qquad (2.1)$$

We will use this relationship in Section 2.2.

2.1.4   The implications of Dutch Book theorems

What questions have we answered so far? We definitely answered de Finetti's original one, that is: assuming that we desire to use probability to represent an individual's knowledge about unknowns, is there a justification for embracing the axioms of probability, as stated a few years earlier by Kolmogorov? The answer is: yes, there is. If the axioms are not satisfied the probabilities are "intrinsically contradictory" and lead to losing in a hypothetical game in which they are put to the test by allowing others to bet at the implied odds.

The laws of probability are, in de Finetti's words, "conditions which characterize coherent opinions (that is, opinions admissible in their own right) and which distinguish them from others that are intrinsically contradictory." Within those constraints, a probability assessor is entitled to any opinion. De Finetti continues:

> a complete class of incompatible events $\theta_1, \theta_2, \ldots, \theta_n$ being given, all the assignments of probability that attribute to $\pi_1, \pi_2, \ldots, \pi_n$ any values whatever, which are non-negative and have a sum equal to unity, are admissible assignments: each of these evaluations corresponds to a coherent opinion, to an opinion legitimate in itself, and every individual is free to adopt that one of these opinions which he prefers, or, to put in more plainly, that which he *feels*. The best example is that of a championship where the spectator attributes to each team a greater or smaller probability of winning according to his own judgment; the theory cannot reject *a priori* any of these judgments unless the sum of the probabilities attributed to each team is not equal to unity. This arbitrariness, which any one would admit in the above case, exists also, according to the conception which we are maintaining, in all other domains, including those more or less vaguely defined domains in which the various objective conceptions are asserted to be valid. (de Finetti 1937, pp. 139–140)

The Dutch Book argument provides a calculus for using subjective probability in the quantification of uncertainty and gives decision makers great latitude in establishing fair odds based on formal or informal processing of knowledge. With this freedom comes two important constraints. One is that probability assessors be ready, at least hypothetically, to "put their money where their mouth is." Unless ready to lie about their knowledge (we will return to this in Chapter 10), the probability assessor does not have an incentive to post capricious odds. The other is implicit in de Finetti's definition of event as a statement whose truth will become known to the bettors. That truth of events can be known and agreed upon by many individuals in all relevant scientific contexts is somewhat optimistic, and reveals the influence of the positivist philosophical school on de Finetti's thought. From a statistical standpoint, though, it is healthy to focus controversies on observable events, rather than theoretical entities that may not be ultimately measured, such as model parameters. The latter, however, are important in a number of scientific settings. De Finetti's theory of exchangeability, also covered in his 1937 article, is a formidable contribution to grounding parametric inference in statements about observables. Covering it here would takes us too far astray. A good entry point to the extensive literature is Cifarelli and Regazzini (1996).

A second related question is the following: is asking someone for their fair betting odds a good way to find out their probability of an event? Or, stated more technically, is the betting mechanism a practical elicitation tool for measuring subjective probability? The Dutch Book argument does not directly mention this, but nonetheless this is an interesting possibility. Discussions are in de Finetti (1974), Kadane and Winkler (1988), Seidenfeld *et al.* (1990a) and Garthwaite *et al.* (2005) who connect statistical considerations to the results of psychological research about "how people represent uncertain information cognitively, and how they respond to questions about that information."

A third question is perhaps the most important: do all rational decision makers facing a decision under uncertainty (say the betting problem) have to act as though they represented their uncertainty using a coherent subjective probability distribution? There is a little bit of extra work that needs to be done before we can answer that, and we will postpone it to our discussion of Ramsey's ideas in Chapter 5.

Lastly, we need to consider the question of temporal coherence. We have seen that the Bayes rule and conditional probabilities are derived in terms of called-off bets, which are assessed before the conditioning events are observed. As such they are static constraints among probabilities of events, all of which are in the future. Much of statistical thinking is about what can be said about unknowns *after* some data are observed. Ramsey (1926, p. 180) first pointed out that the two are not the same. Hacking (1976) draws a distinction between conditional probability and a posteriori probability, the latter being the statement made after the conditioning event is observed. The dominant view among Bayesian statistician has been that the two can be equated without resorting to any additional principle. For example, Howson and Urbach (1989) argue that unless relevant additional background knowledge accrues between the time the conditional probability is stated and the time the conditioning event occurs, it is legitimate to equate conditional probability and a posteriori probability. And one can often make provisions for this background knowledge by incorporating it explicitly in the algebra being considered.

Others, however, have argued that the leap to using the Bayes rule for *a posteriori probability* is not justified by the Dutch Book theorem. Goldstein writes:

> As no coherence principles are used to justify the equivalence of conditional and a posteriori probabilities, this assumption is an arbitrary imposition on the subjective theory. As Bayesians rarely make a simple updating of actual prior probabilities to the corresponding conditional probabilities, this assumption misrepresents Bayesian practice. Thus Bayesian statements are often unclear.... The practical implication is that Bayesian theory does not appear to be very helpful in considering the kind of question that we have raised about the expert and his changing judgments. (Goldstein 1985, p. 232)

There are at least a couple of options for addressing this issue. One is to add to the coherence principle the separate principle, taken at face value, that conditional and a posteriori probabilities are the same. This is sometimes referred to as the conditionality principle. For example, Pratt *et al.* (1964) hold that conditional probability before and after the conditioning event are two different behavioral principles, though in their view, equating the two is a perfectly reasonable additional requirement. We will revisit this in Chapter 5. Another, which we briefly examine next, is to formalize a more general notion of coherence that would apply to the dynamic nature of updating.

2.2   Temporal coherence

Let us start with a simple example. You are about to go to work on a cloudy morning and your current degree of belief about the event θ that it will rain in the afternoon is 0.9. If you ask yourself the same question at lunchtime you may state a different belief perhaps because you hear the weather forecast on the radio, or because you see a familiar weather pattern develop. We will denote by π_θ^0 the present assessment, and by π_θ^T the lunchtime assessment. The quantity π_θ^T is unknown at the present time and, therefore, it can be thought of as a random quantity.

Goldstein describes the dynamic nature of beliefs. Your beliefs, he says, are:

> *temporally coherent* at a particular moment if your current assessments are coherent and you also believe that at each future time point your new current assessments will be coherent. (Goldstein 1985, p. 232)

To make this more formal, one needs to think about degree of belief about future probability assessments. In our example, we would need to consider beliefs about our future probability π_θ^T. We will denote the expected value, computed at time 0 of this probability, by $E^0[\pi_\theta^T]$. Goldstein (1983) proposes that in order for one to be considered coherent over time, his or her expectation for an event's prevision ought to be his or her current probability of that event.

Definition 2.2 (Temporal coherence)   *Probability assessments on event θ at two time points 0 and T are temporally coherent iff*

$$E^0[\pi_\theta^T] = \pi_\theta^0. \tag{2.2}$$

This condition establishes a relation between one's current assessments and those to be asserted at a future time. Also, this relation assures that one's change in prevision, that is $\pi_\theta^T - \pi_\theta^0$, cannot be systematically predicted from current beliefs. In Goldstein's own words:

> I have beliefs. I may not be able to describe "the precise reasons" why I hold these beliefs. Even so, the rules implied by coherence provide logical guidance for my expression of these beliefs. My beliefs will change. I may not now be able to describe precise reasons how or why these changes will occur. Even so, just as with my beliefs, the rules implied

by coherence provide logical guidance for my expression of belief as to how these beliefs will change. (Goldstein 1983, p. 819)

The following theorem, due to Goldstein (1985), extends the previous result to conditional previsions.

Theorem 2.3   *If you are temporally coherent, then your previsions must satisfy*

$$E^0[\pi^T_{\theta_1|\theta_2}|\theta_2] = \pi^0_{\theta_1|\theta_2}$$

where $\pi^T_{\theta_1|\theta_2}$ is the revised value for $\pi^0_{\theta_1|\theta_2}$ declared at time T, still before θ_2 is obtained.

Proof: Consider the definition of conditional expectation given in equation (2.1). If we choose θ to be the random $\pi^T_{\theta_1|\theta_2}$, then

$$E^0[\pi^T_{\theta_1|\theta_2}|\theta_2] = E^0[\pi^T_{\theta_1|\theta_2}\theta_2]/\pi^0_{\theta_2}.$$

Next, applying the definition of temporal coherence (2.2) with $\theta = \pi^T_{\theta_1|\theta_2}\theta_2$, and substituting, we get

$$E^0[\pi^T_{\theta_1|\theta_2}|\theta_2] = E^0[\pi^T_{\pi^T_{\theta_1|\theta_2}\theta_2}]/\pi^0_{\theta_2}.$$

At time T, $\pi^T_{\theta_1|\theta_2}$ is constant, so

$$E^0[\pi^T_{\theta_1|\theta_2}|\theta_2] = E^0[\pi^T_{\theta_1|\theta_2}\pi^T_{\theta_2}]/\pi^0_{\theta_2}$$
$$= E^0[\pi^T_{\theta_1\theta_2}]/\pi^0_{\theta_2}$$

from the definition of conditional probability. Finally, from temporal coherence, $E^0[\pi^T_{\theta_1\theta_2}] = \pi^0_{\theta_1\theta_2}$, so

$$E^0[\pi^T_{\theta_1|\theta_2}|\theta_2] = \pi^0_{\theta_1\theta_2}/\pi^0_{\theta_2} = \pi^0_{\theta_1|\theta_2}. \qquad \square$$

Consider events $\theta_i, i = 1, \ldots, k$, forming a partition of Θ; that is, one and only one of them will occur. Formally $\sum_{i=1}^k \theta_i = 1$ and $\theta_i\theta_j = 0$, for all $i \neq j$. Also, choose θ to be any event in Θ, not necessarily one of the θ_i above. A useful consequence of Theorem 2.3 above is that

$$\pi^0_{\theta|\Theta} = \sum_i \theta_i \pi^0_{\theta|\theta_i}.$$

Similarly, at a future time T,

$$\pi^T_{\theta|\Theta} = \sum_i \theta_i \pi^T_{\theta|\theta_i}.$$

Our next goal is to investigate the relationship between $\pi^0_{\theta|\Theta}$ and $\pi^T_{\theta|\Theta}$. The next theorem establishes that $\pi^0_{\theta|\Theta}$ cannot systematically predict the change $Q = \pi^T_{\theta|\Theta} - \pi^0_{\theta|\Theta}$ in the prevision of conditional beliefs.

Theorem 2.4   *If an agent is temporally coherent, then*

(i) $\pi_Q^0 = 0$,

(ii) Q and $\pi_{\theta|\Theta}^0$ *are uncorrelated,*

(iii) $\pi_{Q|\Theta}^0$ *is identically zero.*

Proof: Following Goldstein (1985) we will prove (iii) by showing that $\pi_{Q|\theta_i}^0 = 0$ for every i:

$$\pi_{Q|\theta_i}^0 = E^0[\pi_{\theta|\Theta}^T - \pi_{\theta|\Theta}^0 | \theta_i]$$

$$= E^0 \left[\sum_j \theta_j \pi_{\theta|\theta_j}^T - \sum_j \theta_j \pi_{\theta|\theta_j}^0 | \theta_i \right]$$

$$= E^0 \left[\sum_j (\pi_{\theta|\theta_j}^T - \pi_{\theta|\theta_j}^0) \theta_j | \theta_i \right]$$

$$= \frac{E^0[(\pi_{\theta|\theta_i}^T - \pi_{\theta|\theta_i}^0) \theta_i]}{\pi_{\theta_i}^0}$$

$$= E^0[\pi_{\theta|\theta_i}^T - \pi_{\theta|\theta_i}^0 | \theta_i]$$

$$= 0.$$

Summing over i gives the desired result.   □

Although the discussion in this section utilizes previsions of event indicators (and, therefore, probabilities of events), the arguments hold for more general bounded random variables. In this general framework, one can directly assess the prevision or expectation of the random variable, which is only done indirectly here, via the event probabilities.

2.3 Scoring rules and the axioms of probabilities

In Chapter 10 we will study measures used for the evaluation of forecasters, after the events that were being predicted have actually taken place. These are called scoring rules and typically involve the computation of a summary value that reflects the correspondence between the probability forecast and the observation of what actually occurred.

Consider the case where a forecaster must announce probabilities $\pi = (\pi_{\theta_1}, \pi_{\theta_2}, \ldots, \pi_{\theta_k})$ for the events $\theta_1, \ldots, \theta_k$, which form a partition of the possible states of nature. A popular choice of scoring rule is the negative of the sum of the squared differences between predictions and events:

$$s(\theta_j, \pi) = - \sum_{i=1}^{k} (\pi_{\theta_i} - \theta_j)^2. \qquad (2.3)$$

It turns out that the score s above of an incoherent forecaster can be improved upon regardless of the outcome. This can provide an alternative justification for using coherent probabilities, without relying on the Dutch Book theorem and the betting scenario. This argument is in fact used in de Finetti's treatise on probability (de Finetti 1974) to justify the axioms of probability. In Section 2.4 we show how to derive Kolmogorov's axioms when the forecaster is scored on the basis of such a quadratic scoring rule.

2.4 Exercises

Problem 2.1 Consider a probability assessor being evaluated according to the scoring rule (2.3). If the assessor's probabilities π violate any of the following two conditions,

1. $0 \le \pi_{\theta_j} \le 1$ for all $j = 1, \ldots, k$,

2. $\sum_{j=1}^{k} \pi_{\theta_j} = 1$,

then there is a vector π', satisfying conditions 1 and 2, and such that $s(\theta_j, \pi) \le s(\theta_j, \pi')$ for all $j = 1, \ldots, k$ and $s(\theta_j, \pi) < s(\theta_j, \pi')$ for at least one j.

We begin by building intuition about this result in low-dimensional cases. When $k = 2$, in the $(\pi_{\theta_1}, \pi_{\theta_2})$ plane the quantities $s(\theta_1, \pi)$ and $s(\theta_2, \pi)$ represent the negative of the squared distances between the point $\pi = (\pi_{\theta_1}, \pi_{\theta_2})$ and the points $e_1 = (1, 0)$ and $e_2 = (0, 1)$, respectively. In Figure 2.1, points C and D satisfy condition 2 but violate condition 1, while point B violates condition 2 and satisfies condition 1. Can we find values which satisfy both conditions and have smaller distances to both canonical vectors e_1 and e_2? In the cases of C and D, e_1 and e_2 do the job. For B, we can find a point b that does the job by looking at the projection of B on the $\pi_{\theta_1} + \pi_{\theta_2} = 1$ line. The scores of B are hypotenuses of the Be_1b and Be_2b triangles. The scores of the projection are the sides of the same triangles, that is the line segments e_1b and e_2b. The point E violates both conditions. If you are not yet convinced by the argument, try to find a point that does better than E in both dimensions.

Figure 2.2 illustrates a similar point for $k = 3$. In that case, to satisfy the axioms, we have to choose points in the shaded area (the simplex on $\mathfrak{R}^3$). This result can be generalized to $\mathfrak{R}^k$ by using the triangle inequality in k dimensions. We now make this more formal.

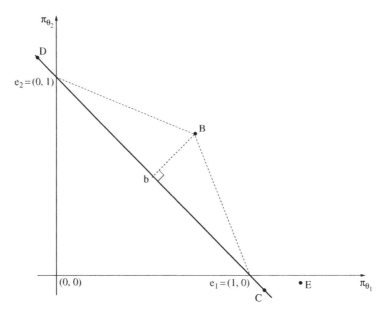

Figure 2.1   Quadratic scoring rule for $k = 2$. *Point B satisfies the first axiom of probability and violates the second one, while points C and D violate the first axiom and satisfy the second. Point E violates them both, while points* b, e_1, *and* e_2 *satisfy both axioms.*

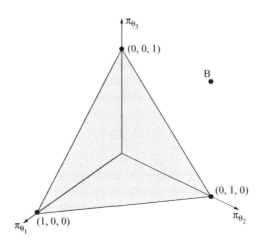

Figure 2.2   Quadratic scoring rule for $k = 3$.

We begin by considering a violation of condition 1. Suppose that the first component of π is negative, that is $\pi_{\theta_1} < 0$, and $0 \le \pi_{\theta_j} \le 1$ for $j = 2, \ldots, k$. Consider now π' constructed by setting $\pi'_{\theta_1} = 0$ and $\pi'_{\theta_j} = \pi_{\theta_j}$ for $j = 2, \ldots, k$. π' satisfies condition 1. If event θ_1 occurs,

$$s(\theta_1, \pi) - s(\theta_1, \pi') = -\sum_{i=1}^{k}\left[(\pi_{\theta_i} - \theta_i)^2 - (\pi'_{\theta_i} - \theta_i)^2\right]$$

$$= -\left[(\pi_{\theta_1} - 1)^2 - (\pi'_{\theta_1} - 1)^2\right] - \sum_{i=2}^{k}(\pi_{\theta_i}^2 - \pi_{\theta_i}'^2)$$

$$= -[(\pi_{\theta_1} - 1)^2 - 1] < 0$$

since $\pi_{\theta_1} < 0$. Furthermore, for $j = 2, \ldots, k$,

$$s(\theta_j, \pi) - s(\theta_j, \pi') = -(\pi_{\theta_1}^2 - \pi'^{2}_{\theta_1}) - \sum_{i=2; i \ne j}^{k}(\pi_{\theta_i}^2 - \pi'^{2}_{\theta_i})$$

$$- \left[(\pi_{\theta_j} - 1)^2 - (\pi'_{\theta_j} - 1)^2\right]$$

$$= -\pi_{\theta_1}^2 < 0.$$

Therefore, if $\pi_{\theta_1} < 0$, $s(\theta_j, \pi) < s(\theta_j, \pi')$ for every $j = 1, \ldots, k$.

We now move to the case where the predictions violate condition 2, that is $0 \le \pi_{\theta_i} \le 1$ for $i = 1, \ldots, k$, but $\sum_{i=1}^{k} \pi_{\theta_i} \ne 1$. Take π' to be the orthogonal projection of π on the plane defined to satisfy both conditions 1 and 2. For any j we have

$$s(\theta_j, \pi) - s(\theta_j, \pi') = -\left[\sum_{i=1, i \ne j}^{k} \pi_{\theta_i}^2 + (\pi_{\theta_j} - 1)^2\right] + \left[\sum_{i=1, i \ne j}^{k} \pi_{\theta_i}'^2 + (\pi'_{\theta_j} - 1)^2\right].$$

The term

$$\left[\sum_{i=1, i \ne j}^{k} \pi_{\theta_i}^2 + (\pi_{\theta_j} - 1)^2\right]$$

corresponds to the squared Euclidean distance between π and e_j, the k-dimensional canonical point with value 1 in its jth coordinate, and 0 elsewhere. Similarly, the term

$$\left[\sum_{i=1, i \ne j}^{k} \pi_{\theta_i}'^2 + (\pi'_{\theta_j} - 1)^2\right]$$

is the squared Euclidean distance between π' and e_j. Since π' is an orthogonal projection of π and satisfies conditions 1 and 2, it follows that $||\pi, \pi'|| + ||\pi', e_j|| = ||\pi, e_j||$, for any $j = 1, \ldots, k$. Here $||\pi_1, \pi_2||$ denotes the squared Euclidean distance between π_1 and π_2. As $||\pi, \pi'|| > 0$, with $||\pi, \pi'|| = 0$ if, and only if, $\pi = \pi'$, we conclude that $||\pi', e_j|| < ||\pi, e_j||$. Therefore $s(\theta_j, \pi) - s(\theta_j, \pi') < 0$ for any $j = 1, \ldots, k$.

Problem 2.2   We can extend this result to conditional probabilities by a slight modification of the scoring rule. We will do this next, following de Finetti (1972, chapter 2) (English translation of de Finetti 1964). Suppose that a forecaster must announce a probability $\pi_{\theta_1|\theta_2}$ for an event θ_1 conditional on the occurrence of the event θ_2, and is scored by the rule $-(\pi_{\theta_1|\theta_2} - \theta_1)^2\theta_2$. This implies that a penalty of $-(\pi_{\theta_1} - \theta_1)^2$ occurs if θ_2 is true, but the score is 0 otherwise. Suppose the forecaster is also required to announce probabilities $\pi_{\theta_1\theta_2}$ and π_{θ_2} for the events $\theta_1\theta_2$ and θ_2 subject to quadratic scoring rules $-(\pi_{\theta_1\theta_2} - \theta_1\theta_2)^2$ and $-(\pi_{\theta_2} - \theta_2)^2$, respectively. Overall, by announcing $\pi_{\theta_1|\theta_2}$, $\pi_{\theta_1\theta_2}$, and π_{θ_2} the forecaster is subject to the penalty $s(\theta_1|\theta_2, \theta_1\theta_2, \theta_2, \pi_{\theta_1|\theta_2}, \pi_{\theta_1\theta_2}, \pi_{\theta_2}) = -(\pi_{\theta_1|\theta_2} - \theta_1)^2\theta_2 - (\pi_{\theta_1\theta_2} - \theta_1\theta_2)^2 - (\pi_{\theta_2} - \theta_2)^2$.

Under these conditions $\pi_{\theta_1|\theta_2}$, $\pi_{\theta_1\theta_2}$, and π_{θ_2} must satisfy

$$\pi_{\theta_1\theta_2} = \pi_{\theta_1|\theta_2}\pi_{\theta_2}$$

or else the forecaster can again be outscored irrespective of the event that occurs.

Let x, y, and z be the values taken by s under the occurrence of $\theta_1\theta_2$, $(1 - \theta_1)\theta_2$, and $(1 - \theta_2)$, respectively:

$$
\begin{aligned}
x &= -(\pi_{\theta_1|\theta_2} - 1)^2 - (\pi_{\theta_1\theta_2} - 1)^2 - (\pi_{\theta_2} - 1)^2 \\
y &= \quad\quad -\pi_{\theta_1|\theta_2}^2 - \quad\quad \pi_{\theta_1\theta_2}^2 - (\pi_{\theta_2} - 1)^2 \\
z &= \quad\quad\quad - \quad\quad \pi_{\theta_1\theta_2}^2 - \quad\quad \pi_{\theta_2}^2.
\end{aligned}
$$

To prove this, let $(\pi_{\theta_1|\theta_2}, \pi_{\theta_1\theta_2}, \pi_{\theta_2})$ be a point at which the gradients of x, y, and z are not in the same plane, meaning that the Jacobian

$$\frac{\partial(x, y, z)}{\partial(\pi_{\theta_1|\theta_2}, \pi_{\theta_1\theta_2}, \pi_{\theta_2})}$$

is not zero. Then, it is possible to make x, y, and z smaller by moving $(\pi_{\theta_1|\theta_2}, \pi_{\theta_1\theta_2}, \pi_{\theta_2})$. For a geometrical interpretation of this idea, look at de Finetti (1972). Therefore, the Jacobian, $8(\pi_{\theta_1|\theta_2}\pi_{\theta_2} - \pi_{\theta_1\theta_2})$, has to be equal to zero, which implies $\pi_{\theta_1\theta_2} = \pi_{\theta_1|\theta_2}\pi_{\theta_2}$. □

Problem 2.3   Next week the Duke soccer team plays American University and Akron. Call θ_1 the event *Duke beats Akron* and θ_2 the event *Duke beats American*. Suppose a bookmaker is posting the following betting odds and is willing to take stakes of either sign on any combination of the events:

Event	Odds
θ_1	4:1
θ_2	3:2
$\theta_1 + \theta_2 - \theta_1\theta_2$	9:1
$\theta_1\theta_2$	2:3

Find stakes that will enable you to win money from the bookmaker no matter what the results of the games are.

Problem 2.4 (Bayes' theorem) Let θ_1 and θ_2 be two events. A corollary of Theorem 2.2 is the following. Provided that $\pi_{\theta_2} \neq 0$, a necessary condition for coherence is that $\pi_{\theta_1|\theta_2} = \pi_{\theta_2|\theta_1}\pi_{\theta_1}/\pi_{\theta_2}$. Prove this corollary without using Theorem 2.2.

Problem 2.5 An insurance company, represented here by you, provides insurance policies against hurricane damage. One of your clients is especially concerned about tree damage (θ_1) and flood damage (θ_2). In order to state the policies he is asked to elicit (personal) *rates* for the following events: $\theta_1, \theta_2, \theta_1 \mid \theta_2$, and $\theta_2 \mid \theta_1$. Suppose his probabilities are as follows:

Event	Rates
θ_1	0.2
θ_2	0.3
$\theta_1 \mid \theta_2$	0.8
$\theta_2 \mid \theta_1$	0.9

Show that your client can be made a sure loser.

3

Utility

Broadly defined, the goal of decision theory is to help choose among actions whose consequences cannot be completely anticipated, typically because they depend on some future or unknown state of the world. Expected utility theory handles this choice by assigning a quantitative utility to each consequence, a probability to each state of the world, and then selecting an action that maximizes the expected value of the resulting utility. This simple and powerful idea has proven to be a widely applicable description of rational behavior.

In this chapter, we begin our exploration of the relationship between *acting rationally* and *ranking actions based on their expected utility*. For now, probabilities of unknown states of the world will be fixed. In the chapter about coherence we began to examine the relationship between *acting rationally* and *reasoning about uncertainty using the laws of probability*, but utilities were fixed and relegated to the background. So, in this chapter, we will take a complementary perspective.

The pivotal contribution to this chapter is the work of von Neumann and Morgenstern. The earliest version of their theory (here NM theory) is included in the appendix of their book *Theory of Games and Economic Behavior*. The goal of the book was the construction of a theory of games that would serve as the foundation for studying the economic behavior of individuals. But the theory of utility that they constructed in the process is a formidable contribution in its own right. We will examine it here outside the context of game theory.

Featured articles:

Bernoulli, D. (1954). Exposition of a new theory on the measurement of risk, *Econometrica* **22**: 23–36.

Jensen, N. E. (1967). An introduction to Bernoullian utility theory: I Utility functions, *Swedish Journal of Economics* **69**: 163–183.

The Bernoulli reference is a translation of his 1738 essay, discussing the celebrated St. Petersburg's paradox. Jensen (1967) presents a classic version of the axiomatic theory of utility originally developed by von Neumann and Morgenstern (1944). The original is a tersely described development, while Jensen's account crystallized the axioms of the theory in the form that was adopted by the foundational debate that followed. Useful general readings are Kreps (1988) and Fishburn (1970).

3.1   St. Petersburg paradox

The notion that mathematical expectation should guide rational choice under uncertainty was formulated and discussed as early as the seventeenth century. An issue of debate was how to find what would now be called a certainty equivalent: that is, the fixed amount of money that one is willing to trade against an uncertain prospect, as when paying an insurance premium or buying a lottery ticket. Huygens (1657) is one of the early authors who used mathematical expectation to evaluate the fair price of a lottery. In his time, the prevailing thought was that, in modern terminology, a rational individual should value a game of chance based on the expected payoff.

The ideas underlying modern utility theory arose during the Age of Enlightenment, as the evolution of the initial notion that the fair value is the expected value. The birthplace of utility theory is usually considered to be St. Petersburg. In 1738, Daniel Bernoulli, a Swiss mathematician who held a chair at the local Academy of Science, published a very influential paper on decision making (Bernoulli 1738). Bernoulli analyzed the behavior of rational individuals in the face of uncertainty from a Newtonian perspective, viewing science as an operational model of the human mind. His empirical observation, that thoughtful individuals do not necessarily take the financial actions that maximize their expected monetary return, led him to investigate a formal model of individual choices based on the direct quantification of value, and to develop a prototypical utility function for wealth. A major contribution has been to stress that:

> no valid measurement of the value of risk can be given without consideration of its utility, that is the utility of whatever gain accrues to the individual. . . . To make this clear it is perhaps advisable to consider the following example. Somehow a very poor fellow obtains a lottery ticket that will yield with equal probability either nothing or twenty thousand ducats. Would he not be ill advised to sell this lottery ticket for nine thousand ducats? To me it seems that the answer is in the negative. On the other hand I am inclined to believe that a rich man would be ill advised to refuse to buy the lottery ticket for nine thousand ducats. (Bernoulli 1954, p. 24, an English translation of Bernoulli 1738).

Bernoulli's development is best known in connection with the notorious St. Petersburg game, a gambling game that works, in Bernoulli's own words, as follows:

> Perhaps I am mistaken, but I believe that I have solved the extraordinary problem which you submitted to M. *de Montmort*, in your letter of September 9, 1713 (problem 5, page 402). For the sake of simplicity I shall assume that A tosses a coin into the air and B commits himself to give A 1 ducat if, at the first throw, the coin falls with its cross upward; 2 if it falls thus only at the second throw, 4 if at the third throw, 8 if at the fourth throw, etc. (Bernoulli 1954, p. 33)

What is the fair price of this game? The expected payoff is

$$1 \times \frac{1}{2} + 2 \times \frac{1}{4} + 4 \times \frac{1}{8} + \ldots,$$

that is infinitely many ducats. So a ticket that costs 20 ducats up front, and yields the outcome of this game, has a positive expected payoff. But so does a ticket that costs a thousand ducats, or any finite amount. If expectation determines the fair price, no price is too large. Is it rational to be willing to spend an arbitrarily large sum of money to participate in this game? Bernoulli's view was that it is not. He continued:

> The paradox consists in the infinite sum which calculation yields as the equivalent which A must pay to B. This seems absurd since no reasonable man would be willing to pay 20 ducats as equivalent. You ask for an explanation of the discrepancy between mathematical calculation and the vulgar evaluation. I believe that it results from the fact that, *in their theory*, mathematicians evaluate money in proportion to its quantity while, *in practice*, people with common sense evaluate money in proportion to the utility they can obtain from it. (Bernoulli 1954, p. 33)

Using this notion of utility, Bernoulli offered an alternative approach: he suggested that one should not act based on the expected reward, but on a different kind of expectation, which he called *moral expectation*. His starting point was that any gain brings a utility inversely proportional to the whole wealth of the individual. Analytically, Bernoulli's utility u of a gain Δ in wealth, relative to the current wealth z, can be represented as

$$u(z + \Delta) - u(z) = c\frac{\Delta}{z}$$

where c is some positive constant. For Δ sufficiently small we obtain

$$du(z) = c\frac{dz}{z}.$$

Integrating this differential equation gives

$$u(z) = c \log(z) - \log(z_0), \tag{3.1}$$

where z_0 is the constant of integration and can be interpreted here as the wealth necessary to get a utility of zero. The worth of the game, said Bernoulli, is the expected value of the gain in wealth based on (3.1). For example, if the initial wealth is 10 ducats, the worth is

$$c \sum_{n=1}^{\infty} \frac{\log(n + 10)}{2^n} - \log(z_0).$$

Bernoulli computed the value of the game for various values of the initial wealth. If you own 10 ducats, the game is worth approximately 3 ducats, 6 ducats if you own 1000.

Bernoulli moved rational behavior away from linear payoffs in wealth, but still needed to face the obstacle that the logarithmic utility is unbounded. Savage (1954) comments on an exchange between Gabriel Cramer and Daniel Bernoulli's uncle Nicholas Bernoulli on this issue:

> Daniel Bernoulli's paper reproduces portions of a letter from Gabriel Cramer to Nicholas Bernoulli, which establishes Cramer's chronological priority to the idea of utility and most of the other main ideas of Bernoulli's paper.... Cramer pointed out in his aforementioned letter, the logarithm has a serious disadvantage; for, if the logarithm were the utility of wealth, the St. Petersburg paradox could be amended to produce a random act with an infinite expected utility (i.e., an infinite expected logarithm of income) that, again, no one would really prefer to the status quo.... Cramer therefore concluded, and I think rightly, that the utility of cash must be bounded, at least from above. (Savage 1954, pp. 91–95)

Bernoulli acknowledged this restriction and commented:

> The mathematical expectation is rendered infinite by the enormous amount which I can win if the coin does not fall with its cross upward until rather late, perhaps at the hundredth or thousandth throw. Now, as a matter of fact, if I reason as a sensible man, this sum is worth no more to me, causes me no more pleasure and influences me no more to accept the game than does a sum amounting only ten or twenty million ducats. Let us suppose, therefore, that any amount above 10 millions, or (for the sake of simplicity) above $2^{24} = 166,777,216$ ducats be deemed by me equal in value to 2^{24} ducats or, better yet, that I can never win more than

that amount, no matter how long it takes before the coin falls with its cross upward. In this case, my expectation is

$$\frac{1}{2}1 + \frac{1}{4}2 + \frac{1}{8}4 + \cdots + \frac{1}{2^{25}}2^{24} + \frac{1}{2^{25}}2^{24} + \frac{1}{2^{26}}2^{24} + \cdots =$$
$$\frac{1}{2} + \frac{1}{2} + \frac{1}{2} + \cdots (24 \text{ times}) \cdots + \frac{1}{2} + \frac{1}{4} + \frac{1}{8} + \cdots =$$
$$12 + 1 = 13.$$

Thus, my moral expectation is reduced in value to 13 ducats and the equivalent to be paid for it is similarly reduced – a result which seems much more reasonable than does rendering it infinite. (Bernoulli 1954, p. 34)

We will return to the issue of bounded utility when discussing Savage's theory. In the rest of this chapter we will focus on problems with a finite number of outcomes, and we will not worry about unbounded utilities. For the purposes of our discussion the most important points in Bernoulli's contribution are: (a) the distinction between the outcome ensuing from an action (in this case the number of ducats) and its value; and (b) the notion that rational behavior may be explained and possibly better guided by quantifying this value. A nice historical account of Bernoulli's work, including related work from Laplace to Allais, is given by Jorland (1987). For additional comments see also Savage (1954, Sec. 5.6), Berger (1985) and French (1988).

3.2 Expected utility theory and the theory of means

3.2.1 Utility and means

The expected utility score attached to an action can be considered as a real-valued summary of the worthiness of the outcomes that may result from it. In this sense it is a type of mean. From this point of view, expected utility theory is close to theories concerned with the most appropriate way of computing a mean. This similarity suggests that, before delving into the details of utility theory, it may be interesting to explore some of the historical contributions to the theory of means. The discussion in this section in based on Muliere and Parmigiani (1993), who expand on this theme.

As we have seen while discussing Bernoulli, the notion that mathematical expectation should guide rational choice under uncertainty was formulated and discussed as early as the seventeenth century. After Bernoulli, the debate on moral expectation was important throughout the eighteenth century. Laplace dedicated to it an entire chapter of his celebrated treatise on probability (Laplace 1812).

Interestingly, Laplace emphasized that the appropriateness of moral expectation relies on individual preferences for relative rather than absolute gains:

> [With D. Bernoulli], the concept of moral expectation was no longer a substitute but a complement to the concept of mathematical expectation, their difference stemming from the distinction between the absolute and the relative values of goods. The former being independent of, the latter increasing with the needs and desires for these goods. (Laplace 1812, pp. 189–190, translation by Jorland 1987)

It seems fair to say that choosing the appropriate type of expectation of an uncertain monetary payoff was seen by Laplace as the core of what is today identified as rational decision making.

3.2.2   Associative means

After a long hiatus, this trend reemerged in the late 1920s and the 1930s in several parts of the scientific community, including mathematics (functional equations, inequalities), actuarial sciences, statistics, economics, philosophy, and probability. Not coincidentally, this is also the period when both subjective and axiomatic probability theories were born. Bonferroni proposed a unifying formula for the calculation of a variety of different means from diverse fields of application. He wrote:

> The most important means used in mathematical and statistical applications consist of determining the number $\bar{z}$ that relative to the quantities $z_1, \ldots, z_n$ with weights $w_1, \ldots, w_n$, is in the following relation with respect to a function ψ:
>
> $$\psi(\bar{z}) = \frac{w_1 \psi(z_1) + \cdots + w_n \psi(z_n)}{w_1 + \cdots + w_n} \qquad (3.2)$$
>
> I will take $z_1, \ldots, z_n$ to be distinct and the weights to be positive. (Bonferroni 1924, p. 103; our translation and modified notation for consistency with the remainder of the chapter)

Here ψ is a continuous and strictly monotone function. Various choices of ψ yield commonly used means: $\psi(z) = z$ gives the arithmetic mean, $\psi(z) = 1/z$, $z > 0$, the harmonic mean, $\psi(z) = z^k$ the power mean (for $k \neq 0$ and z in some real interval I where ψ is strictly monotone), $\psi(z) = \log z$, $z > 0$, the geometric mean, $\psi(z) = e^z$ the exponential mean, and so forth.

To illustrate the type of problem behind the development of this formalism, let us consider one of the standard motivating examples from actuarial sciences, one of the earliest fields to be concerned with optimal decision making under uncertainty. Consider the example (Bonferroni 1924) of an insurance company offering life insurance to a group of N individuals of which w_1 are of age z_1, w_2 are of age z_2, $\ldots$, and w_n are of age z_n. The company may be interested in determining the mean age $\bar{z}$ of the group so that the probability of complete survival of the group after a number y of years is the same as that of a group of N individuals of equal age $\bar{z}$. If these individuals share the same survival function Q, and deaths are independent, the mean $\bar{z}$ satisfies the relationship

$$\left(\frac{Q(\bar{z}+y)}{Q(\bar{z})} \right)^N = \left(\frac{Q(z_1+y)}{Q(z_1)} \right)^{w_1} \times \cdots \times \left(\frac{Q(z_n+y)}{Q(z_n)} \right)^{w_n}$$

which is of the form (3.2) with $\psi(z) = \log Q(z+y) - \log Q(z)$.

Bonferroni's work stimulated activity on characterizations of (3.2); that is, on finding a set of desirable properties of means that would be satisfied if and only if a mean of the form (3.2) is used. Nagumo (1930) and Kolmogorov (1930), independently, characterized (3.2) (for $w_i = 1$) in terms of these four requirements:

1. continuity and strict monotonicity of the mean in the z_i;

2. reflexivity (when all the z_i are equal to the same value, that value is the mean);

3. symmetry (that is, invariance to labeling of the z_i); and

4. associativity (invariance of the overall mean to the replacement of a subset of the values with their partial mean).

3.2.3   Functional means

A complementary approach stemmed from the problem-driven nature of Bonferroni's solution. This approach is usually associated with Chisini (1929), who suggests that one may want to identify a critical aspect of a set of data, and compute the mean so that, while variability is lost, the critical aspect of interest is maintained. For example, when computing the mean of two speeds, he would argue, one can be interested in doing so while keeping fixed the total traveling time, leading to the harmonic mean, or the total fuel consumption, leading to an expression depending, for example, on a deterministic relationship between speed and fuel consumption.

Chisini's proposal was formally developed and generalized by de Finetti (1931a), who later termed it *functional*. De Finetti regarded this approach as the appropriate way for a subject to determine the certainty equivalent of a distribution function. In this framework, he reinterpreted the axioms of Nagumo and Kolmogorov as natural requirements for such choice. He also extended the characterization theorem to more general spaces of distribution functions.

In current decision-theoretic terms, determining the certainty equivalent of a distribution of uncertain gains according to (3.2) is formally equivalent to computing an expected utility score where ψ plays the role of the utility function. De Finetti commented on this after the early developments of utility theory (de Finetti 1952).

In current terms, his point of view is that the Nagumo–Kolmogorov characterization of means of the form (3.2) amounts to translating the expected utility principle into more basic axioms about the comparison of probability distributions.

It is interesting to compare this later line of thinking to the treatment of subjective probability that de Finetti was developing in the early 1930s (de Finetti 1931b, de Finetti 1937), and that we discussed in Chapter 2. There, subjective probability is derived based on an agent's fair betting odds for events. Determining the fair betting odds for an event amounts to declaring a fixed price at which the agent is willing to buy or sell a ticket giving a gain of S if the event occurs and a gain of 0 otherwise. Again, the fundamental notion is that of certainty equivalent. However, in the problem of means, the probability distribution is fixed, and the existence of a well-behaved ψ (that can be thought of as a utility) is derived from the axioms. In the foundation of subjective probability, the utility function for money is fixed at the outset (it is actually linear), and the fact that fair betting odds behave like probabilities is derived from the coherence requirement.

3.3 The expected utility principle

We are now ready to formalize the problem of choosing among actions whose consequences are not completely known. We start by defining a set of outcomes (also called consequences, and sometimes rewards), which we call $\mathcal{Z}$, with generic outcome z. Outcomes are potentially complex and detailed descriptions of all the circumstances that may be relevant in the decision problem at hand. Examples of outcomes include schedules of revenues over time, consequences of correctly or incorrectly rejecting a scientific hypothesis, health states following a treatment or an intervention, consequences of marketing a drug, change in the exposure to a toxic agent that may result from a regulatory change, knowledge gained from a study, and so forth. Throughout the chapter we will assume that only finitely many different outcomes need to be considered.

The consequences of an action depend on the unknown state of the world: an action yields a given outcome in $\mathcal{Z}$ corresponding to each state of the world θ in some set Θ. We will use this correspondence as the defining feature of an action; that is, an action will be defined as a function a from Θ to $\mathcal{Z}$. The set of all actions will be denoted here by $\mathcal{A}$. A simple action is illustrated in Table 3.1.

In expected utility theory, the basis for choosing among actions is a quantitative evaluation of both the utility of the outcomes and the probability that each outcome will occur. Utilities of consequences are measured by a real-valued function u on $\mathcal{Z}$, while probabilities of states of the world are represented by a probability distribution π on Θ. The weighted average of the utility with respect to the probability is then used as the choice criterion.

Throughout this chapter, the focus is on the utility aspect of the expected utility principle. Probabilities are fixed; they are regarded as the description of a well-understood chance mechanism.

Table 3.1   A simple decision problem with three possible actions, four possible outcomes, expressed as cash flows in pounds, and three possible states of the world. You can revisit it in Problem 3.3.

		States of nature		
		$\theta = 1$	$\theta = 2$	$\theta = 3$
Actions	a_1	£100	£110	£120
	a_2	£90	£100	£120
	a_3	£100	£110	£100

An alternative way of describing an action in the NM theory is to think of it as a probability distribution over the set $\mathcal{Z}$ of possible outcomes. For example, if action a is taken, the probability of outcome z is

$$p(z) = \int_{\theta:a(\theta)=z} \pi(\theta)d\theta, \tag{3.3}$$

that is the probability of the set of states of the world for which the outcome is z if a is chosen. If Θ is a subset of the real line, say $(0, 1)$, and π a continuous distribution, by varying $a(\theta)$ we can generate any distribution over $\mathcal{Z}$, as there are only finitely many elements in it. We could, for example, assume without losing generality that π is uniform, and that the "well-understood chance mechanism" is a random draw from $(0, 1)$. We will, however, maintain the notation based on a generic π, to facilitate comparison with Savage's theory, discussed in Chapter 5.

In this setting the expected utility principle consists of choosing the action a that maximizes the expected value

$$\mathcal{U}_\pi(a) = \int_\Theta u(a(\theta))\pi(\theta)d\theta \tag{3.4}$$

of the utility. Equivalently, in terms of the outcome probabilities,

$$\mathcal{U}_\pi(a) = \sum_{z \in \mathcal{Z}} p(z)u(z). \tag{3.5}$$

An optimal action a^*, sometimes called a *Bayes action*, is one that maximizes the expected utility, that is

$$a^* = \text{argmax } \mathcal{U}_\pi(a). \tag{3.6}$$

Utility theory deals with the foundations of this quantitative representation of individual choices.

3.4   The von Neumann–Morgenstern representation theorem

3.4.1   Axioms

Von Neumann & Morgenstern (1944) showed how the expected utility principle captured by (3.4) can be derived from conditions on ordinal relationships among all actions. In particular, they provided necessary and sufficient conditions for preferences over a set of actions to be representable by a utility function of the form (3.4). These conditions are often thought of as basic rationality requirements, an interpretation which would equate acting rationally to making choices based on expected utility. We examine them in detail here.

The set of possible outcomes $\mathcal{Z}$ has n elements $\{z_1, z_2, \ldots, z_n\}$. As we have seen in Section 3.3, an action a implies a probability distribution over $\mathcal{Z}$, which is also called lottery, or gamble. We will also use the notation p_i for $p(z_i)$, and $\boldsymbol{p} = (p_1, \ldots, p_n)$. For example, if $\mathcal{Z} = \{z_1, z_2, z_3\}$, the lotteries corresponding to actions a and a', depicted in Figure 3.1, can be equivalently denoted as $\boldsymbol{p} = (1/2, \ 1/4, \ 1/4)$ and $\boldsymbol{p}' = (0, \ 0, \ 1)$.

The space of possible actions $\mathcal{A}$ is the set of all functions from Θ to $\mathcal{Z}$. A lottery does not identify uniquely an element of $\mathcal{A}$, but all functions that lead to the same probability distribution can be considered equivalent for the purpose of the NM theory (we will meet an exception in Chapter 6 where we consider state-dependent utilities).

As in Chapter 2, the notation $\prec$ is used to indicate a binary preference relation on $\mathcal{A}$. The notation $a \prec a'$ indicates that action a' is strictly preferred to action a.

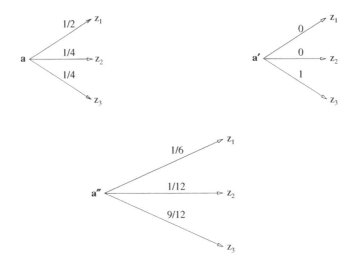

Figure 3.1   Two lotteries associated with a and a', and the lottery associated with compound action a'' with $\alpha = 1/3$.

Indifference between two outcomes (neither $a \prec a'$ nor $a' \prec a$) is indicated by $a \sim a'$. The notation $a \precsim a'$ indicates that a' is preferred or indifferent to a. The relation $\prec$ over the action space $\mathcal{A}$ is called a *preference relation* if it satisfies these two properties:

Completeness: for any two a and a' in $\mathcal{A}$, one and only one of these three relationships must be satisfied:

$a \prec a'$, or
$a \succ a'$, or
neither of the above.

Transitivity: for any a, a'', and a'' in $\mathcal{A}$, the two relationships $a \precsim a'$ and $a' \precsim a''$ together imply that $a \precsim a''$.

Completeness says that, in comparing two actions, "I don't know" is not allowed as an answer. Considering how big $\mathcal{A}$ is, this may seem like a completely unreasonable requirement, but the NM theory will not really require the decision maker to actually go through and make all possible pairwise comparisons: it just requires that the decision maker will find the axioms plausible *no matter which* pairwise comparison is being considered. The "and only one" part of the definition builds in a property sometimes called *asymmetry*.

Transitivity is a very important assumption, and is likely to break down in problems where outcomes have multiple dimensions. Think of a sport analogue. "Team A is likely to beat team B" is a binary relation between teams. It could be that team A beats team B most of the time, that team B beats team C most of the time, and yet that team C beats team A more often than not. Maybe team C has offensive strategies that are more effective against A than B, or has a player that can erase A's star player from the game. The point is that teams are complex and multidimensional and this can generate cyclical, or non transitive, binary relations. Transitivity will go a long way towards turning our problem into a one-dimensional one and paving the way for a real-valued utility function.

A critical notion in what follows is that of a *compound action*. For any two actions a and a', and for $\alpha \in [0, 1]$, a compound action $a'' = \alpha a + (1 - \alpha)a'$ denotes the action that assigns probability $\alpha p(z) + (1 - \alpha)p'(z)$ to outcome z. For example, in Figure 3.1, $a'' = (1/3)a + (2/3)a'$ which implies that $p'' = (1/6, 1/12, 9/12)$. The notation $\alpha a + (1 - \alpha)a'$ is a shorthand for pointwise averaging of the probability distributions implied by actions a and a', and does not indicate that the outcomes themselves are averaged. In fact, in most cases the summation operator will not be defined on $\mathcal{Z}$.

The axioms of the NM utility theory, in the format given by Jensen (1967) (see also Fishburn 1981, Fishburn 1982, Kreps 1988), are

NM1 $\prec$ is complete and transitive.

NM2 *Independence*: for every a, a', and a'' in $\mathcal{A}$ and $\alpha \in (0, 1]$, we have

$$a \succ a' \quad \text{implies} \quad (1 - \alpha)a'' + \alpha a \succ (1 - \alpha)a'' + \alpha a'.$$

NM3 *Archimedean*: for every a, a', and a'' in $\mathcal{A}$ such that $a \succ a' \succ a''$ we can find $\alpha, \beta \in (0, 1)$ such that

$$\alpha a + (1 - \alpha)a'' \succ a' \succ \beta a + (1 - \beta)a''.$$

The independence axiom requires that two composite actions should be compared solely based on components that may be different. In economics this is a controversial axiom from both the normative and the descriptive viewpoints. We will return later on to some of the implications of this axiom and the criticisms it has drawn. An important argument in favor of this axiom as a normative axiom in statistical decision theory is put forth by Seidenfeld (1988), who argues that violating this axiom amounts to a sequential form of incoherence.

The Archimedean axiom requires that, when a is preferred to a', it is not preferred so strongly that mixing a with a'' cannot lead to a reversal of preference. So a cannot be incommensurably better than a'. Likewise, a'' cannot be incommensurably worse than a'. A simple example of a violation of this axiom is lexicographic preferences: that is, preferences that use some dimensions of the outcomes as a primary way of establishing preference, and use other dimensions only as tie breakers if the previous dimensions are not sufficient to establish a preference. Look at the worked Problem 3.1 for more details.

In spite of the different context and structural assumptions, the axioms of the Nagumo–Kolmogorov characterization and the axioms of von Neumann and Morgenstern offer striking similarities. For example, there is a parallel between the associativity condition (substituting observations with their partial mean has no effect on the global mean) and the independence condition (mixing with the same weight an option to two other options will not change the preferences).

3.4.2   Representation of preferences via expected utility

Axioms NM1, NM2 and NM3 hold if and only if there is a real valued utility function u such that the preferences for the options in $\mathcal{A}$ can be represented as in expression (3.4). A given set of preferences identifies a utility function u only up to a linear transformation with positive slope. This result is formalized in the von Neumann–Morgenstern representation theorem below.

Theorem 3.1   *Axioms NM1, NM2, and NM3 are true if and only if there exists a function u such that for every pair of actions a and a'*

$$a \succ a' \quad \Longleftrightarrow \quad \sum_{z \in \mathcal{Z}} p(z)u(z) > \sum_{z \in \mathcal{Z}} p'(z)u(z) \tag{3.7}$$

and u is unique up to a positive linear transformation.

An equivalent representation of expression (3.7) is

$$a \succ a' \quad \Longleftrightarrow \quad \int_{\Theta} u(a(\theta))\pi(\theta)d\theta > \int_{\Theta} u(a'(\theta))\pi(\theta)d\theta. \tag{3.8}$$

Understanding the proof of the von Neumann–Morgenstern representation theorem is a bit of a time investment but it is generally worthwhile, as it will help in understanding the implications and the role of the axioms as well as the meaning of utility and its relation to probability. It will also lead us to an intuitive elicitation approach for individual utilities.

To prove the theorem we will start out with a lemma, the proof of which is left as an exercise.

Lemma 1   *If the binary relation $\succ$ satisfies Axioms NM1, NM2, and NM3, then:*

(a) *If $a \succ a'$ then*
$$\beta a + (1 - \beta)a' \succ \alpha a + (1 - \alpha)a' \quad \text{if and only if} \quad 0 \le \alpha < \beta \le 1.$$

(b) *$a \succeq a' \succeq a'', a \succ a''$ imply that there is a unique $\alpha^* \in [0, 1]$ such that $a' \sim \alpha^* a + (1 - \alpha^*)a''$.*

(c) *$a \sim a'$, $\alpha \in [0, 1]$ imply that*
$\alpha a + (1 - \alpha)a'' \sim \alpha a' + (1 - \alpha)a'', \forall\, a'' \in \mathcal{A}.$

Taking this lemma for granted we will proceed to prove the von Neumann–Morgenstern theorem. Parts (a) and (c) of the lemma are not too surprising. The heavyweight is (b). Most of the weight in proving (b) is carried by the Archimedean axiom. If you want to understand more look at Problem 3.6.

We will prove the theorem in the $\Rightarrow$ direction. The reverse is easier and is left as an exercise. There are two important steps in the proof of the main theorem. First we will show that the preferences can be represented by some real-valued function φ; that is, that there is a φ such

$$a \succ a' \quad \Leftrightarrow \quad \varphi(a) > \varphi(a').$$

This means that the problem can be captured by a one-dimensional "score" representing the worthiness of an action. Considering how complicated an action can be, this is no small feat. The second step is to prove that this φ must be of the form of an expected utility.

Let us start by defining χ_z as the action that gives outcome z for every θ. The implied p assigns probability 1 to z (see Figure 3.2).

A fact that we will state without detailed proof is that, if preferences satisfy Axioms NM1, NM2, and NM3, then there exists z_0 (worst outcome) and z^0 (best outcome) in $\mathcal{Z}$ such that

$$\chi_{z^0} \succeq a \succeq \chi_{z_0}$$

for every $a \in \mathcal{A}$. One way to prove this is by induction on the number of outcomes. To get a flavor of the argument, try the case $n = 2$.

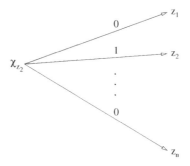

Figure 3.2   The lottery implied by χ_{z_2}, which gives outcome z_2 with certainty.

We are going to rule out the trivial case where the decision maker is indifferent to everything, so that we can assume that z^0 and z_0 are also such that $\chi_{z^0} \succ \chi_{z_0}$. Lemma 1, part (b) guarantees that there exists a unique $\alpha \in [0, 1]$ such that

$$a \sim \alpha \chi_{z^0} + (1 - \alpha)\chi_{z_0}.$$

We define $\varphi(a)$ as the value of α that satisfies the condition above. Consider now another action $a' \in \mathcal{A}$. We have

$$a \sim \varphi(a)\chi_{z^0} + (1 - \varphi(a))\chi_{z_0} \tag{3.9}$$

$$a' \sim \varphi(a')\chi_{z^0} + (1 - \varphi(a'))\chi_{z_0}; \tag{3.10}$$

from Lemma 1, part (a) $\varphi(a) > \varphi(a')$ if and only if

$$\varphi(a)\chi_{z^0} + (1 - \varphi(a))\chi_{z_0} \succ \varphi(a')\chi_{z^0} + (1 - \varphi(a'))\chi_{z_0},$$

which in turn holds if and only if

$$a \succ a'.$$

So we have proved that $\varphi(.)$, the weight at which we are indifferent between the given action and a combination of the best and worst with that weight, provides a numerical representation of the preferences.

We now move to proving that $\varphi(.)$ takes the form of an expected utility. Now, from applying Lemma 1, part (c) to the indifference relations (3.9) and (3.10) we get

$$\alpha a + (1 - \alpha)a' \sim \alpha[\varphi(a)\chi_{z^0} + (1 - \varphi(a))\chi_{z_0}] + (1 - \alpha)[\varphi(a')\chi_{z^0} + (1 - \varphi(a'))\chi_{z_0}].$$

We can reorganize this expression and obtain

$$\alpha a + (1 - \alpha)a' \sim$$
$$[\alpha\varphi(a) + (1 - \alpha)\varphi(a')]\chi_{z^0} + [\alpha(1 - \varphi(a)) + (1 - \alpha)(1 - \varphi(a'))]\chi_{z_0}. \tag{3.11}$$

By definition $\varphi(a)$ is such that

$$a \sim \varphi(a)\chi_{z^0} + (1 - \varphi(a))\chi_{z_0}.$$

Therefore $\varphi(\alpha a + (1 - \alpha)a')$ is such that

$$\alpha a + (1 - \alpha)a' \sim \varphi(\alpha a + (1 - \alpha)a')\chi_{z^0} + (1 - \varphi(\alpha a + (1 - \alpha)a'))\chi_{z_0}.$$

Combining the expression above with (3.11) we can conclude that

$$\varphi(\alpha a + (1 - \alpha)a') = \alpha\varphi(a) + (1 - \alpha)\varphi(a'),$$

proving that $\varphi(.)$ is an affine function.

We now define $u(z) = \varphi(\chi_z)$. Again, by definition of φ,

$$\chi_z \sim u(z)\chi_{z^0} + (1 - u(z))\chi_{z_0}.$$

Eventually we would like to show that $\varphi(a) = \sum_{z \in \mathcal{Z}} u(z)p(z)$, where the p are the probabilities implied by a. We are not far from it. Consider $\mathcal{Z} = \{z_1, \ldots, z_n\}$. The proof is by induction on n.

(i) If $\mathcal{Z} = \{z_1\}$ take $a = \chi_{z_1}$. Then, χ_{z^0} and χ_{z_0} are equal to χ_{z_1} and therefore, $\varphi(a) = u(z_1)p(z_1)$.

(ii) Now assume that the representation holds for actions giving positive mass to $\mathcal{Z} = \{z_1, \ldots, z_{m-1}\}$ only and consider the larger outcome space $\mathcal{Z} = \{z_1, \ldots, z_{m-1}, z_m\}$. Say z_m is such that $p(z_m) > 0$; that is, z_m is in the support of p. (If $p(z_m) = 0$ we are covered by the previous case.) We define

$$p'(z) = \begin{cases} 0, & \text{if } z = z_m \\ \dfrac{p(z)}{1 - p(z_m)}, & \text{if } z \neq z_m. \end{cases}$$

Here, p' has support of size $(m - 1)$. We can reexpress the definition above as

$$p(z) = p(z_m)I_{z_m} + (1 - p(z_m))p'(z),$$

where I is the indicator function, so that

$$a \sim p(z_m)\chi_{z_m} + (1 - p(z_m))a'.$$

Applying φ on both sides we obtain

$$\varphi(a) = p(z_m)\varphi(\chi_{z_m}) + (1 - p(z_m))\varphi(a') = p(z_m)u(z_m) + (1 - p(z_m))\varphi(a').$$

We can now apply the induction hypothesis on a', since the support of p' is of size $(m - 1)$. Therefore,

$$\varphi(a) = p(z_m)u(z_m) + (1 - p(z_m)) \sum_{z \neq z_m} u(z)p'(z).$$

Using the definition of p' it follows that

$$\varphi(a) = p(z_m)u(z_m) + \sum_{z \neq z_m} p(z)u(z) = \sum_{i=1}^{m} p(z_i)u(z_i),$$

which completes the proof.   □

3.5   Allais' criticism

Because of the centrality of the expected utility paradigm, the von Neumann–Morgenstern axiomatization and its derivatives have been deeply scrutinized and criticized from both descriptive and normative perspectives. Empirically, it is well documented that individuals may willingly violate the independence axiom (Allais 1953, Kreps 1988, Kahneman *et al.* 1982, Shoemaker 1982). Normative questions have also been raised about the weak ordering assumption (Seidenfeld 1988, Seidenfeld *et al.* 1995).

In a landmark paper, published in 1953, Maurice Allais proposed an example that challenged both the normative and descriptive validities of expected utility theory. The example is based on comparing the hypothetical decision situations depicted in Figure 3.3. In one situation, the decision maker is asked to choose between lotteries a and a'. In the next, the decision maker is asked to choose between lotteries b and b'. Consider these two choices before you continuing reading. What would you do?

In Savage's words:

> Many people prefer gamble a to gamble a', because, speaking qualitatively, they do not find the chance of winning a *very* large fortune in place of receiving a large fortune outright adequate compensation for even a small risk of being left in the status quo. Many of the same people prefer gamble b' to gamble b; because, speaking qualitatively, the chance of winning is nearly the same in both gambles, so the one with the much larger prize seems preferable. But the intuitively acceptable pair of preferences, gamble a preferred to gamble a' and gamble b' to gamble b, is not compatible with the utility concept. (Savage 1954, p. 102, with a change of notation)

Why is it that these preferences are not compatible with the utility concept? The pair of preferences implies that any utility function must satisfy

$$u(5) > 0.1u(25) + 0.89u(5) + 0.01u(0)$$

$$0.1u(25) + 0.9u(0) > 0.11u(5) + 0.89u(0).$$

Here arguments are in units of $\$100\,000$. You can easily check that one cannot have both inequalities at once. So there is no utility function that is consistent with gamble a being preferred to gamble a' and gamble b' to gamble b.

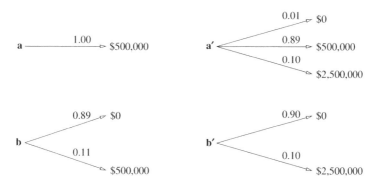

Figure 3.3   The lotteries involved in the so-called Allais paradox. Decision makers often prefer a to a′ and b′ to b—a violation of the independence axiom in the NM theory.

Specifically, which elements of the expected utility theory are being criticized here? Lotteries a and a' can be viewed as compound lotteries involving another lottery a^*, which is depicted in Figure 3.4. In particular,

$$a = 0.11a + 0.89a$$
$$a' = 0.11a^* + 0.89a.$$

By the independence axiom (Axiom NM2 in the NM theory) we should have

$$a \succ a' \text{ if and only if } a \succ a^*.$$

Observe, on the other hand, that

$$b = 0.89\chi_0 + 0.11\chi_5 = 0.89\chi_0 + 0.11a$$
$$\succ 0.89\chi_0 + 0.11a^*$$
$$= 0.89\chi_0 + 0.11\left(\frac{0.01}{0.11}\chi_0 + \frac{0.10}{0.11}\chi_{25}\right)$$
$$= 0.90\chi_0 + 0.10\chi_{25} = b'.$$

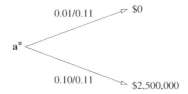

Figure 3.4   Allais Paradox: Lottery a^*.

If we obey the independence axiom, we should also have $b \succ b'$. Therefore, preferences for a over a', and b' over b, violate the independence axiom.

3.6   Extensions

Expected utility maximization has been a fundamental tool in guiding practical decision making under uncertainty. The excellent publications by Fishburn (1970), Fishburn (1989) and Kreps (1988) provide insight, details, and references. The literature on the extensions of the NM theory is also extensive. Good entry points are Fishburn (1982) and Gärdenfors and Sahlin (1988). A general discussion of problems involved in various extensions can be found in Kreps (1988).

Although in our discussion of the NM theory we used a finite set of outcomes $\mathcal{Z}$, similar results can be obtained when $\mathcal{Z}$ is infinite. For example, Fishburn (1970, chapter 10) provides generalization to continuous spaces by requiring additional technical conditions, which effectively allow one to define a utility function that is measurable, along with a new dominance axiom, imposing that if a is preferred to every outcome in a set to which p' assigns probability 1, then a' should not be preferred to a, and vice versa. A remaining limitation of the results discussed by Fishburn (1970) is the fact that the utility function is assumed to be bounded. Kreps (1988) observes that the utility function does not need to be bounded, but only such that plus and minus infinity are not possible expected utilities under the considered class of probability distributions. He also suggests an alternative version for the Archimedean Axiom NM3 that, along with the other axioms, provides a generalization of the von Neumann–Morgenstern representation theorem in which the utility is continuous over a real-valued outcome space.

3.7   Exercises

Problem 3.1   (Kreps 1988) Planners in the war room of the state of Freedonia can express the quality of any war strategy against arch-rival Sylvania by a probability distribution on the three outcomes: Freedonia wins; draws; loses (z_1, z_2, and z_3). Rufus T. Firefly, Prime Minister of Freedonia, expresses his preferences over such probability distributions by the lexicographic preferences

$$a \succ a'$$

whenever

$$p_3 < p'_3 \quad \text{or} \quad [p_3 = p'_3 \quad \text{and} \quad p_2 < p'_2]$$

where $\boldsymbol{p} = (p_1, p_2, p_3)$ is the lottery associated with action a and $\boldsymbol{p}' = (p'_1, p'_2, p'_3)$ the lottery associated with action a'.

Which of the three axioms does this binary relation satisfy (if any) and which does it violate?

Solution

NM1 First, let us check whether the preference relation is complete.

   (i) If $p_3 < p'_3$ then $a \succ a'$ (by the definition of the binary relation).

   (ii) If $p_3 > p'_3$ then $a' \succ a$.

   (iii) If $p_3 = p'_3$ then

      (iii.1) if $p_2 < p'_2$ then $a \succ a'$,

      (iii.2) if $p_2 > p'_2$ then $a' \succ a$,

      (iii.3) if $p_2 = p'_2$ then it must also be that $p_1 = p'_1$ and the two actions are the same.

Therefore, the binary relation is complete. Let us now check whether it is transitive, by considering actions such that $a' \succ a$ and $a'' \succ a'$.

   (i) Suppose $p'_3 < p_3$ and $p''_3 < p'_3$. Then, clearly, $p''_3 < p_3$ and therefore $a'' \succ a$.

   (ii) Suppose $p'_3 < p_3$, $p'_3 = p''_3$, and $p''_2 < p'_2$. We have $p''_3 < p_3$ and therefore $a'' \succ a$.

   (iii) Suppose $p'_3 = p_3, p'_2 < p_2$, and $p''_3 < p'_3$. It follows that $p''_3 < p_3$. Thus, $a'' \succ a$.

   (iv) Suppose $p'_3 = p_3, p'_2 < p_2$, and $p''_3 = p'_3$ and $p''_2 < p'_2$. We now have $p_3 = p''_3$ and $p''_2 < p_2$. Therefore, $a'' \succ a$.

From (i)–(iv) we can conclude that the binary relation is also transitive. So Axiom NM1 is not violated.

NM2 Take $\alpha \in (0, 1]$:

   (i) Suppose $a \succ a'$ due to $p_3 < p'_3$. For any value of $\alpha \in (0, 1]$ we have $\alpha p_3 < \alpha p'_3 \Rightarrow \alpha p_3 + (1 - \alpha)p''_3 < \alpha p'_3 + (1 - \alpha)p''_3$. Thus, $\alpha a + (1 - \alpha)a'' \succ \alpha a' + (1 - \alpha)a''$.

   (ii) Suppose $a \succ a'$ because $p_3 = p'_3$ and $p_2 < p'_2$. For any value of $\alpha \in (0, 1]$ we have:

      (ii.1) $\alpha p_3 = \alpha p'_3 \Rightarrow \alpha p_3 + (1 - \alpha)p''_3 = \alpha p'_3 + (1 - \alpha)p''_3$.

      (ii.2) $\alpha p_2 < \alpha p'_2 \Rightarrow \alpha p_2 + (1 - \alpha)p''_2 < \alpha p'_2 + (1 - \alpha)p''_2$.
      From (ii.1) and (ii.2) we can conclude that $\alpha a + (1 - \alpha)a'' \succ \alpha a' + (1 - \alpha)a''$.

Therefore, from (i) and (ii) we can see that the binary relation satisfies Axiom NM2.

NM3 Consider a, a', and a'' in $\mathcal{A}$ and such that $a \succ a' \succ a''$.

(i) Suppose that $a \succ a' \succ a''$ because $p_3 < p'_3 < p''_3$.
Since $p_3 < p'_3 < p''_3$, $\exists \alpha$, $\beta \in (0, 1)$ such that $\alpha p_3 + (1 - \alpha)p''_3 < p'_3 < \beta p_3 + (1 - \beta)p''_3$ since this is a property of real numbers. So, $\alpha a + (1 - \alpha)a'' \succ a' \succ \beta a + (1 - \beta)a''$.

(ii) Suppose that $a \succ a' \succ a''$ because $p_2 < p'_2 < p''_2$ (and $p_3 = p'_3 = p''_3$).
Since $p_3 = p'_3 = p''_3$, $\forall \alpha$, $\beta \in (0, 1)$: $\alpha p_3 + (1 - \alpha)p''_3 = p'_3 = \beta p_3 + (1 - \beta)p''_3$.
Also, $p_2 < p'_2 < p''_2$ implies that $\exists \alpha$, $\beta \in (0, 1)$ such that $\alpha p_2 + (1 - \alpha)p''_2 < p'_2 < \beta p_2 + (1 - \beta)p''_2$. Thus, $\alpha a + (1 - \alpha)a'' \succ a' \succ \beta a + (1 - \beta)a''$.

(iii) Suppose that $a \succ a' \succ a''$ because of the following conditions:

  (a) $p_3 < p'_3$,

  (b) $p'_3 = p''_3$ and $p'_2 < p''_2$.

  We have $\alpha p_3 + (1 - \alpha)p''_3 = \alpha p_3 + (1 - \alpha)p'_3 < \alpha p'_3 + (1 - \alpha)p'_3 = p'_3$. This implies that $\alpha p_3 + (1 - \alpha)p''_3 < p'_3$. Therefore, we cannot have $a' \succ \alpha a + (1 - \alpha)a''$.

By (iii) we can observe that Axiom NM3 is violated.   □

Problem 3.2   (Berger 1985) An investor has 1000$ to invest in speculative stocks. The investor is considering investing a dollars in stock A and $(1000 - a)$ dollars in stock B. An investment in stock A has a 0.6 chance of doubling in value, and a 0.4 chance of being lost. An investment in stock B has a 0.7 chance of doubling in value, and a 0.3 chance of being lost. The investor's utility function for a change in fortune, z, is $u(z) = \log(0.0007z + 1)$ for $-1000 \leq z \leq 1000$.

  (a) What is $\mathcal{Z}$ (for a fixed a)? (It consists of four elements.)

  (b) What is the optimal value of a in terms of expected utility? (Note: This perhaps indicates why most investors opt for a diversified portfolio of stocks.)

Solution

  (a) $\mathcal{Z} = \{-1000, 2a - 1000, 1000 - 2a, 1000\}$.

  (b) Based on Table 3.2 we can compute the expected utility, that is

$$\mathcal{U} = 0.12 \log(0.3) + 0.18 \log(0.0014a + 0.3)$$
$$+ 0.28 \log(1.7 - 0.0014a) + 0.42 \log(1.7).$$

Now, let us look for the value of a which maximizes the expected utility (that is, the optimum value a^*). We have

$$\frac{d\mathcal{U}}{da} = \frac{0.18}{0.0014a + 0.3} 0.0014 + \frac{0.28}{1.7 - 0.0014a}(-0.0014)$$
$$= \frac{0.000\,252}{0.0014a + 0.3} - \frac{0.000\,392}{1.7 - 0.0014a}.$$

Table 3.2   Rewards, utilities, and probabilities for Problem 3.2, assuming that the investment in stock A is independent of the investment of stock B.

z	-1000	$2a - 1000$	$1000 - 2a$	1000
Utility	$\log(0.3)$	$\log(0.0014a + 0.3)$	$\log(1.7 - 0.0014a)$	$\log(1.7)$
Probability	$(0.4)(0.3)$	$(0.6)(0.3)$	$(0.4)(0.7)$	$(0.6)(0.7)$

Setting the derivative equal to zero and evaluating it at a^* leads to the following equation:

$$\frac{0.000\,252}{0.0014a^* + 0.3} = \frac{0.000\,392}{1.7 - 0.0014a^*}.$$

By solving this equation we obtain $a^* = 344.72$. Also, since the second derivative

$$\frac{d^2\,\mathcal{U}}{d\,a^2} = -\frac{0.000\,252(0.0014)}{(0.0014a + 0.3)^2} - \frac{0.000\,392(0.0014)}{(1.7 - 0.0014a)^2} < 0,$$

we conclude that $a^* = 344.72$ maximizes the expected utility $\mathcal{U}$.

□

Problem 3.3   (French 1988) Consider the actions described in Table 3.1 in which the consequences are monetary payoffs. Convert this problem into one of choosing between lotteries, as defined in the von Neumann–Morgenstern theory. The decision maker holds the following indifferences with reference lotteries:

£100 $\sim$ £120 with probability 1/2; £90 with probability 1/2;
£110 $\sim$ £120 with probability 4/5; £90 with probability 1/5.

Assume that $\pi(\theta_1) = \pi(\theta_3) = 1/4$ and $\pi(\theta_2) = 1/2$. Which is the optimal action according to the expected utility principle?

Problem 3.4   Prove part (a) of Lemma 1.

Problem 3.5   (Berger 1985) Assume that Mr. A and Mr. B have the same utility function for a change, z, in their fortune, given by $u(z) = z^{1/3}$. Suppose now that one of the two men receives, as a gift, a lottery ticket which yields either a reward of r dollars ($r > 0$) or a reward of 0 dollars, with probability 1/2 each. Show that there exists a number $b > 0$ having the following property: regardless of which man receives the lottery ticket, he can sell it to the other man for b dollars and the sale will be advantageous to both men.

Problem 3.6   Prove part (b) of Lemma 1. The proof is by contradiction. Define

$$\alpha^* = \sup\{\alpha \in [0, 1] \ : \ a' \succsim \alpha a + (1 - \alpha)a''\}$$

and consider separately the three cases:

$$\alpha^* a + (1 - \alpha^*)a'' \succ a' \succ a''$$
$$a \succ a' \succ \alpha^* a + (1 - \alpha^*)a''$$
$$a' \sim \alpha^* a + (1 - \alpha^*)a''.$$

The proof is in Kreps (1988). Do not look it up, this was a big enough hint.

Problem 3.7   Prove the von Neumann–Morgenstern theorem in the $\Leftarrow$ direction.

Problem 3.8 (From Kreps 1988)   Kahneman and Tversky (1979) give the following example of a violation of the von Neumann–Morgenstern expected utility model. Ninety-five subjects were asked:

> Suppose you consider the possibility of insuring some property against damage, e.g., fire or theft. After examining the risks and the premium, you find that you have no clear preference between the options of purchasing insurance or leaving the property uninsured.
>
> It is then called to your attention that the insurance company offers a new program called probabilistic insurance. In this program you pay half of the regular premium. In case of damage, there is a 50 percent chance that you pay the other half of the premium and the insurance company covers all the losses; and there is a 50 percent chance that you get back your insurance payment and suffer all the losses. . .
>
> Recall that the premium is such that you find this insurance is barely worth its cost.
>
> Under these circumstances, would you purchase probabilistic insurance?"

And 80 percent of the subjects said that they would not. Ignore the time value of money. (Because the insurance company gets the premium now, or half now and half later, the interest that the premium might earn can be consequential. We want you to ignore such effects. To do this, you could assume that if the insurance company does insure you, the second half of the premium must be increased to account for the interest the company has forgone. While if it does not, when the company returns the first half premium, it must return it with the interest it has earned. But it is easiest simply to ignore these complications altogether.) The question is: does this provide a violation of the von Neumann–Morgenstern model, if we assume (as is typical) that all expected utility maximizers are risk neutral?

4

Utility in action

The von Neumann and Morgenstern (NM) theory is an axiomatization of the expected utility principle, but its logic also provides the basis for measuring an individual decision maker's utilities for outcomes. In this chapter, we discuss how NM utility theory is typically applied to utility elicitation in practical decision problems. In Section 4.1 we discuss a general utility elicitation approach, and then review basic applications in economics and medicine.

In Section 4.2, we apply NM utility theory to situations where the set of rewards consists of alternative sums of money. An important question in this scenario is how attitudes towards future uncertainties change as current wealth changes. The key concept in this regard is risk aversion. We review some intuitive results which describe risk aversion mathematically and characterize properties of utility functions for money. Much of our discussion is based on Keeney et al. (1976) and Kreps (1988).

In Section 4.3, we discuss how the lottery approach can be used to elicit patients' preferences when the outcomes are related to future health. The main issue we will consider is the trade-off between length and quality of life. We also introduce the concept of a quality-adjusted life year (QALY), which is defined as the period of time in good health that a patient says is equivalent to a year in ill health, and commonly used in medical decision-making applications. We discuss its relationship with utility theory, and illustrate both with a simple example. A seminal application of this methodology in medicine is McNeil et al. (1981).

Featured articles:

Pratt, J. (1964). Risk aversion in the small and in the large, *Econometrica* **32**: 122–136.

Torrance, G., Thomas, W. and Sackett, D. (1972). A utility maximization model for evaluation of health care programs, *Health Services Research* **7**: 118–133.

Decision Theory: Principles and Approaches   G. Parmigiani, L. Y. T. Inoue
© 2009 John Wiley & Sons, Ltd

Useful general readings are Kreps (1988) for the economics; Pliskin *et al.* (1980) and Chapman and Sonnenberg (2000) for the medicine.

4.1   The "standard gamble"

When we presented the proof of the NM representation theorem we encountered one step where the utility of each outcome was constructed by identifying the value u that would make receiving that outcome for sure indifferent to a lottery giving the best outcome with probability u and the worse with probability $1 - u$. This step is also the essence of a widely used approach to utility elicitation called "standard gamble."

As a refresher, recall that in the NM theory $\mathcal{Z}$ is the set of rewards. The set of actions (or lotteries, or gambles) is the set probability distributions on $\mathcal{Z}$. It is called $\mathcal{A}$ and its typical elements are things like a, a', a''. In particular, χ_z denotes the degenerate lottery with mass 1 at reward z. We also define $u : \mathcal{Z} \to \Re$ as the decision maker's utility function. Given a utility function u, the expected utility of lottery a is

$$\mathcal{U}(a) = \sum_z u(z)p(z). \tag{4.1}$$

Preferences are described by the binary relation $\succ$, satisfying the NM axioms, so that $a \succ a'$ if and only if

$$\sum_z u(z)p(z) > \sum_z u(z)p'(z).$$

When using the standard gamble approach to utility elicitation, the decision maker lists all outcomes that can occur and ranks them in order of preference. If we avoid the boring case in which all outcomes are valued equally by the decision maker, the weak ordering assumption allows us to identify a worst outcome z_0 and a best outcome z^0. For example, in assessing the utility of health states, "death" is often chosen as the worse outcome and "full health" as the best, although in some problems there are health outcomes that could be ranked worse than death (Torrance 1986). Worst and best outcomes need not be unique. Because all utility functions that are positive linear transformations of the same utility function lead to the same preferences, we can arbitrarily set $u(z_0) = 0$ and $u(z^0) = 1$, which results in a convenient and interpretable utility scale.

The decision maker's utility for each intermediate outcome z can be inferred by eliciting the value of u such that he or she is indifferent between two actions:

$a :$ outcome z for certain;
$a' :$ outcome z_0 with probability $1 - u$, or
          outcome z^0 with probability u.

Another way of writing action a' is $u\chi_{z^0} + (1 - u)\chi_{z_0}$. The existence of a value of u reaching indifference is implied by the Archimedean and independence properties of the decision maker's preferences, as we have seen from Lemma 1. As you can check, the expected utility of both actions is u, and therefore $u(z) = u$.

4.2   Utility of money

4.2.1   Certainty equivalents

After a long digression we are ready to go back to the notion, introduced by Bernoulli, that decisions about lotteries should be made by considering the moral expectation, that is the expectation of the value of money to the decision maker. We are now ready to explore in more detail how utility functions for money typically look.

Say you are about to ship home a valuable rug you just bought in Samarkand for $9000. The probability that it will be lost during transport is 3% according to your intelligence in Uzbekistan. At which price would you be indifferent between buying the insurance or taking the risk? This price defines your *certainty equivalent* of the lottery defined by shipping without insurance.

Formally, assume that the outcome space $\mathcal{Z}$ is an open interval, that is $\mathcal{Z} = (z_0, z^0) \subseteq \mathfrak{R}$. Then, using the same notation as in Chapter 3, a certainty equivalent is any reward $z \in \mathfrak{R}$ that makes you indifferent between that reward for sure, or choosing action a.

Definition 4.1 (Certainty equivalent)   *A certainty equivalent of lottery a is any amount z^* such that*

$$\chi_{z^*} \sim a, \tag{4.2}$$

or equivalently,

$$u(z^*) = \sum_z u(z)p(z). \tag{4.3}$$

A certainty equivalent is also referred to as "cash equivalent" and "selling price" (or "asking price") of a lottery.

4.2.2   Risk aversion

Meanwhile, in Samarkand, you calculate the expected monetary loss from the uninsured shipment, that is $9000 times 0.03 or $270. Would you be willing to pay more to buy insurance? If you do, you qualify as a risk-averse individual.

If you define

$$\bar{z} = \sum_z zp(z) \tag{4.4}$$

as the expected reward under lottery a, then:

Definition 4.2 (Risk aversion)   *A decision maker is* strictly risk averse *if*

$$\chi_{\bar{z}} \succ a. \tag{4.5}$$

Someone holding the reverse preference is called *strictly risk seeking* while someone who is indifferent is called *risk neutral*.

It turns out that the definition above is equivalent to saying that the decision maker is risk averse if $\chi_{\tilde{z}} \succ \chi_{z^*}$; that is, if he or she prefers the expected reward for sure to receiving for sure a certainty equivalent.

Proposition 4.1   *A decision maker is strictly risk averse if and only if his or her utility function is strictly concave, that is*

$$\chi_{\tilde{z}} \succ a \Longleftrightarrow u\left(\sum_z zp(z)\right) > \sum_z u(z)p(z). \tag{4.6}$$

Proof: Suppose first that the decision maker is risk averse and consider a lottery that yields either z_1 or z_2 with probabilities p and $1 - p$, respectively, where $0 < p < 1$. From Definition 4.2, $\chi_{\tilde{z}} \succ a$. The NM theory implies that

$$u(pz_1 + (1 - p)z_2) > pu(z_1) + (1 - p)u(z_2), \qquad 0 < p < 1, \tag{4.7}$$

which is the definition of strict concavity. Conversely, consider a lottery a over $\mathcal{Z}$. Since u is strictly concave, we know that

$$u\left(\sum_z zp(z)\right) > \sum_z u(z)p(z) \tag{4.8}$$

and it follows once again from (4.5) that u is risk averse. $\square$

Analogously, strict risk seeking and risk neutrality can be defined in terms of convexity and linearity of u, respectively, as illustrated in Figure 4.1.

Certainty equivalents exist and are unique under relatively mild conditions:

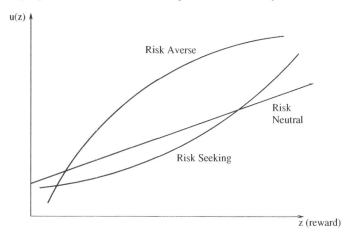

Figure 4.1   Utility functions corresponding to different risk behaviors.

Proposition 4.2   *(a) If u is strictly increasing, then z^* is unique.*

(b) If u is continuous, then there is at least one certainty equivalent.

(c) If u is concave, then there is at least one certainty equivalent.

For the remainder of this chapter, we will assume that u is a concave strictly increasing utility function, so that the certainty equivalent is

$$z^*(a) = u^{-1}\left(\sum_z u(z)p(z)\right). \tag{4.9}$$

Back in Samarkand you have done some thinking and concluded that you would be willing to pay up to \$350 to insure the rug. The extra \$80 amount you are willing to pay, on top of the expected value of \$270, is called the insurance premium. If insurers did not ask for this extra amount they would need a day job. The risk premium is the flip side of the insurance premium:

Definition 4.3 (Risk premium)   *The risk premium associated with a lottery a is the difference between the lottery's expected value and its certainty equivalent:*

$$RP(a) = \bar{z} - z^*(a). \tag{4.10}$$

Figure 4.2 shows this. The negative of the risk premium, or $-RP(a)$, is the *insurance premium*. This works out in our example too. You have bought the rug already, so we are talking about negative sums of money at this point.

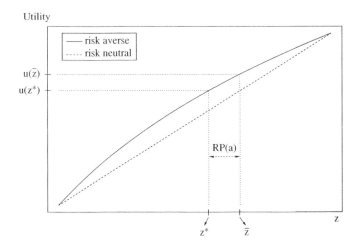

Figure 4.2   Risk premium.

4.2.3   A measure of risk aversion

We discuss next how the decision maker's behavior changes when his or her initial wealth changes proportionally. Let ω be the decision maker's wealth prior to the gamble and let $a + \omega$ be a shorthand notation to represent the final wealth position. Then the decision maker seeks to maximize

$$\mathcal{U}(a + \omega) = \sum_z u(z)p(z + w). \tag{4.11}$$

For example, suppose that the choices are: (i) gamble represented by a, or (ii) sure payment in the amount z. Then, we can represent his or her final wealth as $a + \omega$ or $z + \omega$. The choice will depend on whether $\mathcal{U}(a + \omega)$ is greater, equal, or lower than $u(z + \omega)$.

Definition 4.4 (Decreasing risk aversion)   *A utility function u is said to be decreasingly absolute risk averse if for all $a \in \mathcal{A}, z \in \mathfrak{R}, \omega, \omega' \in \mathfrak{R}$, such that $a + \omega, a + \omega', a + z$, and $\omega' + z$ all lie in $\mathcal{Z}$ and $\omega' > \omega$,*

$$\text{if } \mathcal{U}(a + \omega) > u(z + \omega), \quad \text{then} \quad \mathcal{U}(a + \omega') > u(z + \omega'). \tag{4.12}$$

The idea behind this definition is that, for instance, if you are willing to take the lottery over the sure amount when your wealth is $10\,000$ dollars, you would also take the lottery if you had a larger wealth. The richer you are, the less risk averse you become.

Another way to define decreasing risk aversion is to check whether the risk premium goes down with wealth:

Proposition 4.3   *A utility function u is decreasingly risk averse if and only if for all a in $\mathcal{A}$, the function $\omega \rightarrow RP(a + \omega)$ is nonincreasing in ω.*

A third way to think about decreasing risk aversion is to look at the derivatives of the utility function:

Theorem 4.1   *A utility function u is decreasingly risk averse if and only if the function*

$$\lambda(z) = -\frac{u''(z)}{u'(z)} = -\frac{d}{dz}(\log u'(z)) \tag{4.13}$$

is nonincreasing in z.

While we do not provide a complete proof (see Pratt 1964), the following two results do a large share of the work:

Proposition 4.4   *If u_1 and u_2 are such that $\lambda_1(z) \geq \lambda_2(z)$ for all z in $\mathcal{Z}$, then $u_1(z) = f(u_2(z))$ for some concave, strictly increasing function f from the range of u_2 to $\mathfrak{R}$.*

Proposition 4.5   *A decision maker, with utility function* u_1, *is at least as risk averse as another decision maker, with utility function* u_2 *if and only if, for all* z *in* $\mathcal{Z}$, $\lambda_1(z) \geq \lambda_2(z)$.

The function $\lambda(z)$, defined in Theorem 4.1, is known as the *Arrow–Pratt measure of risk aversion*, or *local risk aversion at* z. Pratt writes:

> we may interpret $\lambda(z)$ as a measure of the *local risk aversion* or *local propensity to insure* at the point z under the utility function u; $-\lambda(z)$ would measure locally liking for risk or propensity to gamble. Notice that we have not introduced any measure of risk aversion in the large. Aversion to ordinary (as opposed to infinitesimal) risks might be considered measured by $RP(a)$, but RP is a much more complicated function than λ. Despite the absence of any simple measure of risk aversion in the large, we shall see that comparisons of aversion to risk can be made simply in the large as well in the small. (Pratt 1964, p. 126 with notational changes)

Example 4.1   Let ω be the decision maker's wealth prior to a gamble a with small range for the rewards and such that the expected value is 0. Define

$$E[a + \omega] = \sum_z (z + w)p(z + w)$$

and

$$E[a^m] = \sum_z z^m p(z).$$

The decision maker's risk premium for $a + \omega$ is

$$RP(a + \omega) = E[a + \omega] - z^*(a + \omega),$$

which implies

$$z^*(a + \omega) = E[a + \omega] - RP(a + \omega)$$

$$= \omega - RP(a + \omega).$$

By the definition of certainty equivalent,

$$E[u(a + \omega)] = u(z^*(a + \omega))$$

$$= u(\omega - RP(a + \omega)).$$

Let us now expand both terms of this equality by using Taylor's formula (in order to simplify the notation let $RP(a + \omega) = RP$). We obtain

$$u(\omega - RP) = u(\omega) - RPu'(\omega) + \frac{RP^2}{2!}u''(\omega) + \ldots \tag{4.14}$$

$$E[u(\omega + a)] = E\left[u(\omega) + au'(\omega) + \frac{1}{2!}a^2u''(\omega) + \frac{1}{3!}a^3u'''(\omega) + \ldots\right]$$

$$= u(\omega) + \frac{1}{2}E[a^2]u''(\omega) + \frac{1}{3!}E[a^3]u'''(\omega) + \ldots. \tag{4.15}$$

By setting equation (4.14) equal to equation (4.15) while neglecting higher-order terms we obtain

$$-RPu'(\omega) \approx \frac{1}{2}E[a^2]u''(\omega). \tag{4.16}$$

Since $E[a^2] = \text{Var}[a]$, and $\lambda(\omega) = -u''(\omega)/u'(\omega)$,

$$\lambda(\omega) \approx 2RP/\text{Var}[a]; \tag{4.17}$$

that is, the decision maker's risk aversion $\lambda(\omega)$ is twice the risk premium per unit of variance for small risks.   ★

Corollary 1   *A utility function u is constantly risk averse if and only if $\lambda(z)$ is constant, in which case there exist constants $a > 0$ and b such that*

$$u(z) = \begin{cases} az + b & \text{if } \lambda(z) = 0, \text{ or} \\ -ae^{-\lambda z} + b & \text{if } \lambda(z) = \lambda > 0. \end{cases} \tag{4.18}$$

If the amount of money involved is small compared to the decision maker's initial wealth, a constant risk aversion is often considered an acceptable rule.

The risk-aversion function captures the information on preferences in the following sense

Theorem 4.2   *Two utility functions u_1 and u_2 have the same preference ranking for any two lotteries if and only if they have the same risk-aversion function.*

Proof: If u_1 and u_2 have the same preference ranking for any two lotteries, they are affine transformations of one another. That is, $u_1(z) = a + bu_2(z)$. Therefore, $u_1'(z) = bu_2'(z)$ and $u_1''(z) = bu_2''(z)$, so

$$\lambda_1(z) = -\frac{u_1''(z)}{u_1'(z)} = -\frac{bu_2''(z)}{bu_2'(z)} = \lambda_2(z). \tag{4.19}$$

Conversely, $\lambda(z) = -(d/dz)(\log u'(z))$ and then

$$\int -\lambda(z)dz = \log u'(z) + c$$

and $\exp(-\int \lambda(z)dz) = e^c u'(z)$, which implies

$$\int \exp(-\int \lambda(z)dz)dz = \int e^c u'(z)dz = e^c u(z) + d. \qquad (4.20)$$

Since e^c and d are constants, $\lambda(z)$ specifies $u(z)$ up to positive linear trans-formations. ☐

With regard to this result, Pratt comments that:

> the local risk aversion λ associated with any utility function u contains all essential information about u while eliminating everything arbitrary about u. However, decisions about ordinary (as opposed to "small") risks are determined by λ only through u ... so it is not convenient entirely to eliminate u from consideration in favor of λ. (Pratt 1964, p. 126 with notational changes)

4.3   Utility functions for medical decisions

4.3.1   Length and quality of life

Decision theory is used in medicine in two scenarios: one is the evaluation of policy decisions that affect groups of individuals, typically carried out by cost-effectiveness, cost–utility or similar analyses. The other is decision making for individuals facing complicated choices regarding their health. Though the two scenarios are very different from a decision-theoretic standpoint, in both cases the foundation is a measurement of utility for future health outcome. The reason why utility plays a key role is that simpler measures of outcome, like duration of life (or in medical jargon survival), fail to capture critical trade-offs. In this regard, McNeil *et al.* (1981) observe:

> The importance of integrating attitudes toward the length of survival and the quality of life is clear. First of all, it is known that persons have different preferences for length of survival: some place greater value on proximate years than on distant years. ... Secondly, the burden of different types of illnesses may vary from patient to patient and from one disorder to another. ... Thirdly, although some people would be willing to trade off some fraction of their lives to avoid morbidities ... if they had normal life expectancy, they might be willing to trade off smaller fractions of their lives if they had a shorter life expectancy. Thus, the importance of estimating the value of different states of health, assuming different potential lengths of survival, is critical. (McNeil *et al.* 1981, p. 987)

Utility elicitation in medical decision making is complex, and the literature is extensive (Naglie *et al.* 1997, Chapman & Sonnenberg 2000). Here we first illustrate the application of the NM theory, then present an alternative approach based on time trade-offs rather than probability trade-offs, and lastly discuss conditions for the two to be equivalent.

4.3.2   Standard gamble for health states

In Section 4.1 we described a general procedure for utility elicitation. We discuss next a mathematically equivalent example of utility elicitation using the standard gamble approach in a medical example.

Example 4.2   A person with severe chronic pain has the option to have surgery that could remove the pain completely, with probability 80%, although there is a 4% risk of death from the surgery. In the remainder of the cases, the surgery has no effect. In this example, the worst outcome is *death* with utility 0, and the best is full recovery with *no pain*, with utility 1. For the intermediate outcome *chronic pain*, the standard gamble is shown in Figure 4.3 . Suppose that the patient's indifference probability α is 0.85. This implies that the utility for chronic pain is 0.85. Thus, the expected utility for surgery is $0.04 \times 0 + 0.16 \times 0.85 + 0.8 \times 1 = 0.936$ which is larger than the expected utility of no surgery, that is 0.85.   ★

4.3.3   The time trade-off methods

In the standard gamble, the trade-off is between an intermediate option for sure and two extreme options with given probabilities. The time trade-off method (Torrance 1971, Torrance *et al.* 1972) plays a similar game with time instead of probability. It is typically used to assess a patient's attitude towards the number of years in ill health he or she is willing to give up in order to live in good health for a shorter number of years. The time in good health equivalent to a year of ill health is called the *quality-adjusted life year* (QALY). To implement this method we first pick an arbitrary period of time t in a particular condition, say chronic pain. Then, we find the amount of time in good health the patient considers equivalent to the arbitrary

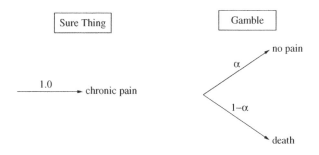

Figure 4.3   Standard gamble for assessing utility for chronic pain.

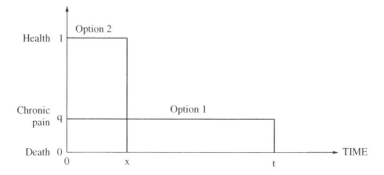

Figure 4.4   Time trade-off for assessing surgical treatment of chronic pain. Figure based on Torrance et al. (1972).

period of time in the condition. For applications in the analysis of clinical trials results, see Glasziou *et al.* (1990).

Example 4.3 (Continuation of the example of Section 4.3.2)   Consider a time horizon of $t = 20$ years, in the ballpark of the patient's life expectancy. This is not a choice to be taken lightly, but for now we will pretend it is easy. The time trade-off method offers two options to the patient, shown in Figure 4.4. Option 1 is $t = 20$ years in chronic pain. Option 2, also deterministic, is $x < t$ years in full health. Time x is varied until the patient is indifferent between the two options. The quality adjustment factor is $q = x/t$. If our patient is willing to give up 5 years if he or she could be certain to live without any pain for 15 years, then $q = 15/20 = 0.75$. The expected quality-adjusted life expectancy without surgery is 12 years, while with surgery it is $0.04 \times 0 + 0.16 \times 15 + 0.8 \times 20 = 18.4$, so surgery is preferred.   ★

4.3.4   Relation between QALYs and utilities

Weinstein *et al.* (1980) identified conditions under which the time trade-off method and the standard gamble method are equivalent. The first condition is that the utility of health states must be such that there is independence between length of life and quality of life. In other words, the trade-offs established on one dimension do not depend on the levels of the other dimension. The second condition is that the utility of health states must be such that trade-offs are proportional, in the following sense: if a person is indifferent between x years of life in good health and x' years of life in chronic pain, then he or she must also be indifferent between spending qx years of life in excellent health and qx' years in chronic pain. The third condition is that the individual must be risk neutral with respect to years of life. This means that the patient's utility for living the next year is the same as for living each subsequent year (that is, his or her marginal utility is constant). The last assumption is not generally considered very realistic. Moreover, that the time trade-off and standard gamble

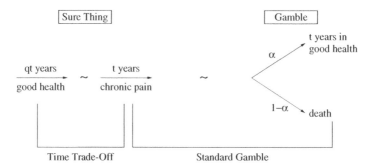

Figure 4.5   Relationship between the standard gamble and time trade-off methods.

could lead to empirically different results even when the decision maker is close to risk neutrality, depending on how different aspects of utilities, such as desirability of states, and time preferences are empirically captured by these methods.

If all these conditions hold, we can establish a correspondence between the quality adjustment factor q elicited using the time trade-off method and the utility α. We wish to establish that, for a given health outcome z, held constant over a period of t years, $\alpha t = qt$ no matter what t is. Figure 4.5 outlines the logic. Using the first condition we can establish trade-offs for length of life and quality of life independently. So, let us first consider the trade-offs for quality of life. Using the standard gamble approach, we find α such that a lottery with probability α for 1 unit of time in z^0 and $(1 - \alpha)$ in z_0 is equivalent to a lottery which yields an intermediate outcome z for 1 unit of time, for sure. Thus, the utility of outcome z for 1 unit of time is α. Using the risk-neutrality condition, the expected utility of t units of time in z is $t \times \alpha$. Next, let us consider the time trade-off method. From the proportional trade-offs condition, if a person is indifferent between 1 unit of time in outcome z and q units of time in z^0, he or she is also indifferent between t units of time in outcome z and qt units of time in z^0. So if these three conditions are true, both methods lead to the same evaluation for living t years in health outcome z.

4.3.5   Utilities for time in ill health

The logic outlined in the previous section can also be used to elicit patient's utilities for a period of time in ill health in a two-step procedure. In the first step the standard gamble strategy is used to elicit the utility of a certain number of years in good health. In the second step the time trade-off is used to elicit patient's preferences between living a shorter time in good health and a longer time in ill health. Alternatively, one could assess patient's utility considering length and quality of life together via lotteries such as those in the NM theory. However, it is usually more difficult to think about length and quality of life at the same time.

Example 4.4   We will illustrate this using an example based on a patient currently suffering for some chronic condition; that is, whose current health state z is stable over time and less desirable than full health. The numbers in the example are from Sox *et al.* (1988). Using the standard gamble approach, we can elicit patient utilities for different lengths of years in perfect health as shown by Table 4.1. This table requires three separate applications of the standard gamble: one for each of the three intermediate lengths of life. Panel (a) of Figure 4.6 shows an interpolation of the resulting mapping between life length and utility, represented by the solid

Table 4.1   Patient's utility function for the length of a healthy life.

Years of perfect health	Utility
0	0.00
3	0.25
7	0.50
12.5	0.75
25	1.00

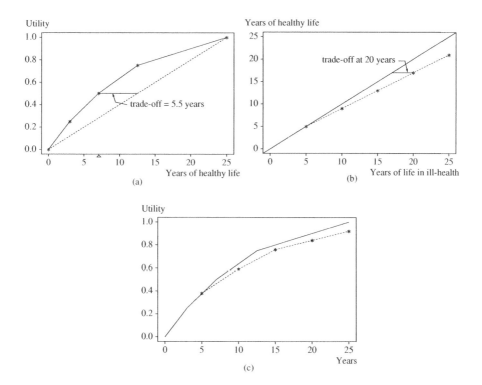

Figure 4.6   Trade-off curves for the elicitation of utilities for time in ill health.

Table 4.2   Equivalent years of life and patient's utility function for various lengths of life in ill health.

Years of disabled life	Equivalent years of perfect health	Utility
5	5	0.38
10	9	0.59
15	13	0.76
20	17	0.84
25	21	0.92

line. The dashed diagonal line represents a person with constant marginal utility. The patient considers a 50/50 gamble between 0 (immediate death) and 25 years (full life expectancy) equivalent to a sure survival of 7 years (represented by a triangle in the figure). Thus, the patient would be willing to "trade off" the difference between the average life expectancy and the certainty equivalent, that is $12.5 - 7 = 5.5$ years to avoid the risk. This is a complicated choice to think about in the abstract, but it is not unrealistic for patients facing surgeries that involve a serious risk of death.

Next, we elicit how long a period of life in full health the patient considered to be equivalent to life in his or her current state. Responses are reported in the first two columns of Table 4.2 and shown in Figure 4.6(b). The solid line represents the relation for a person for whom full health and current health are valued the same. The dashed line represents the relation for a person for whom the current health condition represents an important loss of quality of life. The patient would "trade off" some years in his or her current health for a better health state.

Finally, using both tables (and parts (a) and (b) of Figure 4.6) we can derive the patient's utility for different durations of years of life with the current chronic condition. This is listed in the last column of Table 4.2 and shown in Figure 4.6(c), where the solid line is the utility for being in full health throughout one's life, while the dashed line is the utility for the same duration with the current chronic condition. The derivation is straightforward. Consider the second row of Table 4.2: 10 years in the current health state are equivalent to 9 years in perfect health. Using Table 4.1, we find the utility using a linear interpolation. Let u_x denote the utility for x years of healthy life which gives

$$u_9 = 0.50 + 2 \times \left(\frac{0.75 - 0.50}{5.5} \right) = 0.59.$$

Similar calculations give the remaining utility values shown in Table 4.2 and Figure 4.6(c).   ★

4.3.6   Difficulties in assessing utility

The elicitation of reliable utilities is one of the major obstacles in the application of decision analysis to medical problems. Research on descriptive decision making compares preferences revealed by empirical observations of decisions to utilities ascertained by elicitation methods. The conclusion is generally that the assessed utility function may not reflect the patient's true preferences. The difficulties arise from the fact that in many circumstances, important components of the decision problem may not be reliably quantified. Factors such as attitudes in coping with disease and death, ethical or religious beliefs, etc., can lead to preferences that are more complex than what we can describe within the axiomatization of expected utility theory.

Even if that is not the case, cognitive aspects may challenge utility assessment. Three categories of difficulties are often cited: framing effects, problems with small probability, and changes in utilities over time. Framing effects are artifacts and inconsistencies that arise from how questions are formulated during elicitation, or on the description of the scenarios for the outcomes. For example, "the framing of benefit (or risk) in relative versus absolute terms may have a major influence on patient preferences" (Malenka *et al.* 1993).

Very small or very high probability is a challenge even for decision makers who may have an intuitive understanding of the concept of probability. Yet these are pervasive in medicine: for example, all major surgeries involve a chance of seriously adverse outcomes; all vaccination decisions are based on trading off short-term symptoms for sure against a serious illness with very small probability. In such cases, a *sensitivity analysis* can be used to evaluate how changes in the utilities would affect the ultimate decision making. A related alternative is inverse decision theory (Swartz *et al.* 2006, Davies *et al.* 2007). This is applicable to cases where a small number of actions are available, and is based on partitioning the utility and probability inputs into sets, each collecting all the inputs that lead to the same decision. For example, in Section 4.3.2 we would determine values of u for which surgery is the best strategy. The expected utility of no surgery is u, while for surgery it is $0.04 \times 0 + 0.16 \times u + 0.8 \times 1$. Thus, the optimal strategy is surgery whenever $u < 0.95$. A more general version would partition the probabilities as well.

The patient's attitudes towards length of life may change with age, or life expectancy. Should a patient anticipate these changes and decide now based on the prediction of what his or her utility are expected to become? This sounds rational but prohibitive to implement. The default approach is that the patient's current preferences should be the basis for making decisions.

Lastly, utilities vary widely across individuals, partly because their preferences do and partly as a result of challenges in measurement. For certain health states, such as the consequences of a major stroke, patients distribute across the whole range— with some patients ranking the health state worse than death, and others considering it close to full health (Samsa *et al.* 1988). This makes it difficult to use replication across individuals to increase the precision of estimates.

For an extended review of these issues see also Chapman & Elstein (2000).

4.4   Exercises

Problem 4.1 (Kreps 1988, problem 7)   Suppose a decision maker has constant absolute risk aversion over the range $-\$100$ to $\$1000$. We ask her for her certainty equivalent for gamble with prizes $\$0$ and $\$1000$, each with probability one-half, and she says that her certainty equivalent for this gamble is $\$488$. What, then, should she choose, if faced with the choice of:

> a: a gamble with prizes $-\$100, \$300$, and $\$1000$, each with probability $1/3$;
> a': a gamble with prize $\$530$ with probability $3/4$ and $\$0$ with probability $1/4$; or
> a'': a gamble with a sure thing payment of $\$385$?

Solution
Since the decision maker has constant absolute risk aversion over the range $-\$100$ to $\$1000$, we have

$$u(z) = -ae^{-\lambda z} + b, \text{ for all } z \text{ in } [-100, 1000]. \tag{4.21}$$

We know that the certainty equivalent for a 50/50 gamble with prizes $\$0$ and $\$1000$ is $\$488$. Therefore,

$$u(488) = u(0)\frac{1}{2} + u(1000)\frac{1}{2}. \tag{4.22}$$

Suppose $u(0) = 0, u(1000) = 1$, and consider equations (4.21) and (4.22). We have

$$0 = -ae^{-\lambda 0} + b$$
$$1 = -ae^{-\lambda 1000} + b$$
$$1/2 = -ae^{-\lambda 488} + b.$$

This system implies that $a = b = 10.9207$ and $\lambda = 0.000\,096\,036\,9$. Therefore, we have

$$u(z) = 10.9207(1 - e^{-0.000\,096\,036\,9z}), \text{ for all } z \text{ in } [-100, 1000]. \tag{4.23}$$

From equation (4.23) we obtain

$$u(-100) = -0.105\,384$$
$$u(300) = 0.310\,148$$
$$u(385) = 0.396\,411$$
$$u(530) = 0.541\,949.$$

The expected utilities associated with gambles a, a', and a'' are

$$\mathcal{U}(a) = \frac{1}{3}(u(-100) + u(300) + u(1000)) = 0.401\,588$$

$$\mathcal{U}(a') = \frac{3}{4}u(530) + \frac{1}{4}u(0) = 0.406\,462$$

$$\mathcal{U}(a'') = u(385) = 0.396\,411.$$

Based on these values we conclude that the decision maker should choose gamble a' since it has the maximum expected utility.   □

Problem 4.2   Find the following:

(a) A practical decision problem where it is reasonable to assume that there are only four relevant outcomes $z_1, \ldots, z_4$.

(b) A friend willing to waste 15 minutes.

(c) Your friend's utilities $u(z_1), \ldots, u(z_4)$.

Here's the catch. You can only ask your friend questions about preferences for von Neumann–Morgenstern lotteries. You can assume that your friend believes that Axioms NM1, NM2, and NM3 are reasonable.

Problem 4.3   Prove part (b) of Proposition 4.2.

Problem 4.4   Was Bernoulli risk averse? Decreasingly risk averse?

Problem 4.5 (Lindley 1985)   A doctor has the task of deciding whether or not to carry out a dangerous operation on a person suspected of suffering from a disease. If he has the disease and does operate, the chance of recovery is only 50%; without an operation the similar chance is only 1 in 20. On the other hand if he does not have the disease and the operation is performed there is 1 chance in 5 of his dying as a result of the operation, whereas there is no chance of death without the operation. Advise the doctor (you may assume there are always only two possibilities, death or recovery).

Problem 4.6   You are eliciting someone's utility for money. You know this person has constant risk aversion in the range $0 to $1000. You propose gambles of the form $0 with probability p and $1000 with probability $1 - p$ for the following four values of p: 1/10, 1/3, 2/3, and 9/10. You get the following certainty equivalents: 0.25, 0.60, 0.85, and 0.93. Verify that these are not consistent with constant risk aversion. Assuming that the discrepancies are due to difficulty in the exact elicitation on certainty equivalents, rounding, etc., find a utility function with constant risk aversion that closely approximates the elicited certainty equivalents. Justify briefly the method you use for choosing the approximation.

Problem 4.7 (From French 1988)   An investor has $1000 to invest in two types of shares. If he invests a in share A, he will invest $(1000 - a)$ in share B. An investment in share A has a 0.7 chance of doubling value and a 0.3 chance of being lost altogether. An investment in share B has a 0.6 chance of doubling value and a 0.4 chance of being lost altogether. Outcomes of the two investments are statistically independent. Determine the optimal value of a if the utility function is $u(z) = \log(z + 3000)$.

Problem 4.8   An interesting special case of Example 4.1 happens when $\mathcal{Z} = \{-k, k\}$, $k > 0$. As in that example, assume that ω is the decision maker's initial wealth. The *probability premium* $p(\omega, k)$ of a is defined as the difference $a(k) - a(-k)$ which makes the decision maker indifferent between the status "quo" and a risk z in $\{-k, k\}$. Prove that $\lambda(\omega)$ is twice the probability premium per unit risked for small risks (Pratt 1964).

Problem 4.9   The standard gamble technique has to be slightly modified for chronic states considered worse than death as shown in Figure 4.7. Show that the utility of the chronic state is $u(z) = -\alpha/(1 - \alpha)$. Moreover, show that the quality-adjusted life year is given by $q = x/(x - t)$.

Problem 4.10   A machine can be functioning (F), in repair (R), or dead (D). Under the normal course of operations, the probability of making a transition between any of the three states, in a day, is given by this table:

		TO		
		F	R	D
	F	0.92	0.06	0.02
FROM	R	0.45	0.45	0.10
	D	0.00	0.00	1.0

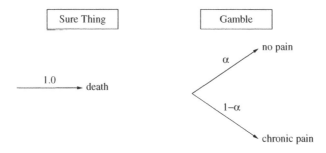

Figure 4.7   Modified standard gamble for chronic states considered worse than death.

These probabilities only depend on where the machine is today, and not on its past history of repairs, age, etc. This is obviously an unreasonable assumption; if you want to relax it, be our guest.

When the machine functions, the daily income is $1000. When it is in repair, the daily cost is $150. There is no income after the machine is dead. Suppose you can put in place a regimen that gives a lower daily income ($900), but also decreases the probability that the machine needs repair or dies. Specifically, the transition table under the new regimen is

		TO		
		F	R	D
	F	0.96	0.03	0.01
FROM	R	0.45	0.45	0.10
	D	0.00	0.00	1.0

New regimen or old? Assume the decision maker is risk neutral and that the machine is functioning today. You can use an analytic approach or a simulation.

Problem 4.11   Describe how you would approach the problem of estimating the cost-effectiveness of regulating emission of pollutants into the environment. Pick a real or hypothetical pollutant/regulation combination, and briefly describe data collection, modeling, and utility elicitation.

5

Ramsey and Savage

In the previous two chapters we have studied the von Neumann–Morgenstern expected utility theory (NM theory) where all uncertainties are represented by objective probabilities. Our ultimate goal, however, is to understand problems where decision makers have to deal with both the utility of outcomes, as in the NM theory, and the probabilities of unknown states of the world, as in de Finetti's coherence theory.

For an example, suppose your roommate is to decide between two bets: bet (i) gives $5 if Duke and not UNC wins this year's NCAA final, and $0 otherwise; bet (ii) gives $10 if a fair die that I toss comes up 3 or greater. The NM theory is not rich enough to help in deciding which lottery to choose. If we agree that the die is fair, lottery (ii) can be thought of as a NM lottery, but what about lottery (i)? Say your roommate prefers lottery (ii) to lottery (i). Is it because of the rewards or because of the probabilities of the two events? Your roommate could be almost certain that Duke will win and yet prefer (ii) because he or she desperately needs the extra $5. Or your roommate could believe that Duke does not have enough of a chance of winning.

Generally, based on agents' preferences, it is difficult to understand their probability without also considering their utility, and vice versa. "The difficulty is like that of separating two different co-operating forces" (Ramsey 1926, p. 172). However, several axiom systems exist that achieve this. The key is to try to hold utility considerations constant while constructing a probability, and vice versa. For example, you can ask your roommate whether he or she would prefer lottery (i) to (iii), where you win $0 if Duke wins this year's NCAA championship, and you win $5 otherwise. If your roommate is indifferent then it has to be that he or she considers Duke winning and not winning to be equally likely. You can then use this as though it were a NM lottery in constructing a utility function for different sums of money.

Decision Theory: Principles and Approaches   G. Parmigiani, L. Y. T. Inoue
© 2009 John Wiley & Sons, Ltd

In a fundamental publication titled *Truth and probability*, written in 1926 and published posthumously in 1931, Frank Plumpton Ramsey developed the first axiom system on preferences that yields an expected utility representation based on unique subjective probabilities and utilities (Ramsey 1926). A full formal development of Ramsey's outline had to wait until the book *The foundations of statistics*, by Leonard J. Savage, "The most brilliant axiomatic theory of utility ever developed" (Fishburn 1970). After summarizing Ramsey's approach, in this chapter we give a very brief overview of the first five chapters of Savage's book. As his book's title implies, Savage set out to lay the foundation to statistics as a whole, not just to decision making under uncertainty. As he points out in the 1971 preface to the second edition of his book:

> The original aim of the second part of the book ... is ... a personal-istic justification ... for the popular body of devices developed by the enthusiastically frequentist schools that then occupied almost the whole statistical scene, and still dominate it, though less completely. The sec-ond part of this book is indeed devoted to personalistic discussion of frequentist devices, but for one after the other it reluctantly admits that justification has not been found. (Savage 1972, p. iv)

We will return to this topic in Chapter 7, when discussing the motivation for the minimax approach.

Featured readings:

Ramsey, F. (1926). *The Foundations of Mathematics*, Routledge & Kegan Paul, London, chapter Truth and Probability, pp. 156–211.

Savage, L. J. (1954). *The foundations of statistics*, John Wiley & Sons, Inc., New York.

Critical reviews of Savage's work include Fishburn (1970), Fishburn (1986), Shafer (1986), Lindley (1980), Drèze (1987), and Kreps (1988). A memorial col-lection of writings also includes biographical materials and transcriptions of several of Savage's very enjoyable lectures (Savage 1981b).

5.1   Ramsey's theory

In this section we review Ramsey's theory, following Muliere & Parmigiani (1993). A more extensive discussion is in Fishburn (1981). Before we present the technical development, it is useful to revisit some of Ramsey's own presentation. In very few pages, Ramsey not only laid out the game plan for theories that would take decades to unfold and are still at the foundation of our field, but also anticipated many of the key difficulties we still grapple with today:

> The old-established way of measuring a person's belief is to propose a bet, and see what are the lowest odds that he will accept. This method I regard as fundamentally sound, but it suffers from being insufficiently general and from being necessarily inexact. It is inexact, partly because of the diminishing return of money, partly because the person may have a special eagerness or reluctance to bet, because he either enjoys or dislikes excitement or for any other reason, e.g. to make a book. The difficulty is like that of separating two different co-operating forces. Besides, the proposal of a bet may inevitably alter his state of opinion; just as we could not always measure electric intensity by actually introducing a charge and seeing what force it was subject to, because the introduction of the charge would change the distribution to be measured.
>
> In order therefore to construct a theory of quantities of belief which shall be both general and more exact, I propose to take as a basis a general psychological theory, which is now universally discarded, but nevertheless comes, I think, fairly close to the truth in the sort of cases with which we are most concerned. I mean the theory that we act in the way we think most likely to realize the objects of our desires, so that a person's actions are completely determined by his desires and opinions. This theory cannot be made adequate to all the facts, but it seems to me a useful approximation to the truth particularly in the case of our self-conscious or professional life, and it is presupposed in a great deal of our thought. It is a simple theory and one which many psychologists would obviously like to preserve by introducing unconscious desires and unconscious opinions in order to bring it more into harmony with the facts. How far such fictions can achieve the required result I do not attempt to judge: I only claim for what follows approximate truth, or truth in relation to this artificial system of psychology, which like Newtonian mechanics can, I think, still be profitably used even though it is known to be false. (Ramsey 1926, pp. 172–173)

Ramsey develops his theory for outcomes that are numerically measurable and additive. Outcomes are not monetary nor necessarily numerical, but each outcome is assumed to carry a value. This is a serious restriction and differs significantly from NM theory and, later, Savage, but it makes sense for Ramsey, whose primary goal was to develop a logic of the probable, rather than a general theory of rational behavior. The set of outcomes is $\mathcal{Z}$. As before we consider a subject that has a weak order on outcome values. Outcomes in the same equivalence class with respect to this order are indicated by $z_1 \sim z_2$, while strict preference is indicated by $\succ$.

The general strategy of Ramsey is: first, find a neutral proposition with subjective probability of one-half, then use this to determine a real-valued utility of outcomes, and finally, use the constructed utility function to measure subjective probability.

Ramsey considers an agent making choices between options of the form "z_1 if θ is true, z_2 if θ is not true." We indicate such an option by

$$a(\theta) = \begin{cases} z_1 & \text{if } \theta \\ z_2 & \text{if not } \theta. \end{cases} \tag{5.1}$$

The outcome z and the option

$$\chi_z = \begin{cases} z & \text{if } \theta \\ z & \text{if not } \theta \end{cases} \tag{5.2}$$

belong to the same equivalence class, a property only implicitly assumed by Ramsey, but very important in this context as it represents the equivalent of reflexivity.

In Ramsey's definition, an ethically neutral proposition is one whose truth or falsity is not "an object of desire to the subject." More precisely, a proposition θ_0 is ethically neutral if two possible worlds differing only by the truth of θ_0 are equally desirable. Next Ramsey defines an ethically neutral proposition with probability $\frac{1}{2}$ based on a simple symmetry argument: $\pi(\theta_0) = \frac{1}{2}$ if for every pair of outcomes (z_1, z_2),

$$a(\theta_0) = \begin{cases} z_1 & \text{if } \theta_0 \\ z_2 & \text{if not } \theta_0 \end{cases} \sim a'(\theta_0) = \begin{cases} z_2 & \text{if } \theta_0 \\ z_1 & \text{if not } \theta_0. \end{cases} \tag{5.3}$$

The existence of one such proposition is postulated as an axiom:

Axiom R1   There is an ethically neutral proposition believed to degree 1/2.

The next axiom states that preferences among outcomes do not depend on which ethically neutral proposition with probability $\frac{1}{2}$ is chosen. There is no loss of generality in taking $\pi(\theta) = \frac{1}{2}$: the same construction could have been performed with a proposition of arbitrary probability, as long as such probability could be measured based solely on preferences.

Axiom R2   The indifference relation (5.3) still holds if we replace θ_0 with any other ethically neutral event θ_0'.

Axiom R2a   If θ is an ethically neutral proposition with probability $\frac{1}{2}$, we have that if

$$\begin{cases} z_1 & \text{if } \theta_0 \\ z_4 & \text{if not } \theta_0 \end{cases} \sim \begin{cases} z_2 & \text{if } \theta_0 \\ z_3 & \text{if not } \theta_0 \end{cases}$$

then $z_1 \succ z_2$ if and only if $z_3 \succ z_4$, and $z_1 \sim z_2$ if and only if $z_3 \sim z_4$.

Axiom R3   The indifference relation between actions is transitive.

Axiom R4   If

$$\begin{cases} z_1 & \text{if } \theta_0 \\ z_4 & \text{if not } \theta_0 \end{cases} \sim \begin{cases} z_2 & \text{if } \theta_0 \\ z_3 & \text{if not } \theta_0 \end{cases}$$

and

$$
\begin{cases} z_3 & \text{if } \theta_0 \\ z_5 & \text{if not } \theta_0 \end{cases} \sim \begin{cases} z_4 & \text{if } \theta_0 \\ z_6 & \text{if not } \theta_0 \end{cases}
$$

then

$$
\begin{cases} z_1 & \text{if } \theta_0 \\ z_5 & \text{if not } \theta_0 \end{cases} \sim \begin{cases} z_2 & \text{if } \theta_0 \\ z_6 & \text{if not } \theta_0. \end{cases}
$$

Axiom R5   For every z_1, z_2, z_3, there exists a unique outcome z such that

$$
\begin{cases} z_1 & \text{if } \theta_0 \\ z_2 & \text{if not } \theta_0 \end{cases} \sim \begin{cases} z & \text{if } \theta_0 \\ z_3 & \text{if not } \theta_0. \end{cases}
$$

Axiom R6   For every pair of outcomes (z_1, z_2), there exists a unique outcome z such that

$$
\begin{cases} z_1 & \text{if } \theta_0 \\ z_2 & \text{if not } \theta_0 \end{cases} \sim \chi_z.
$$

This is a fundamental assumption: as $\chi_z \sim z$, this implies that there is a unique certainty equivalent to a. We indicate this by $z^*(a)$.

Axiom R7   Axiom of continuity: Any progression has a limit (ordinal).

Axiom R8   Axiom of Archimedes.

Ramsey's original paper provides little explanation regarding the last two axioms, which are reported verbatim here. Their role is to make the space of outcomes rich enough to be one-to-one with the real numbers. Sahlin (1990) suggests that continuity should be the analogue of the standard completeness axiom of real numbers. Then it would read like "every bounded set of outcomes has a least upper bound." Here the ordering is given by preferences.

We can imagine formalizing the Archimedean axiom as follows: for every $z_1 \succ z_2 \succ z_3$, there exist z and z' such that

$$
\begin{cases} z & \text{if } \theta_0 \\ z_2 & \text{if not } \theta_0 \end{cases} \prec z_3
$$

and

$$
\begin{cases} z' & \text{if } \theta_0 \\ z_2 & \text{if not } \theta_0 \end{cases} \succ z_1.
$$

Axioms R1–R8 are sufficient to guarantee the existence of a real-valued, one-to-one, utility function on outcomes, designated by u, such that

$$
\begin{cases} z_1 & \text{if } \theta_0 \\ z_4 & \text{if not } \theta_0 \end{cases} \sim \begin{cases} z_2 & \text{if } \theta_0 \\ z_3 & \text{if not } \theta_0 \end{cases} \quad \Leftrightarrow \quad u(z_1) - u(z_2) = u(z_3) - u(z_4). \tag{5.4}
$$

Ramsey did not complete the article before his untimely death at 26, and the posthumously published draft does not include a complete proof. Debreu (1959) and Pfanzagl (1959), among others, designed more formal axioms based on which they obtain results similar to Ramsey's.

In (5.4), an individual's preferences are represented by a continuous and strictly monotone utility function u, determined up to a positive affine transformation. Continuity stems from Axioms R2a and R6. In particular, consistent with the principle of expected utility,

$$\begin{cases} z_1 & \text{if } \theta_0 \\ z_4 & \text{if not } \theta_0 \end{cases} \quad \sim \quad \begin{cases} z_2 & \text{if } \theta_0 \\ z_3 & \text{if not } \theta_0 \end{cases} \quad \Leftrightarrow \quad \frac{u(z_1) + u(z_4)}{2} = \frac{u(z_2) + u(z_3)}{2}. \quad (5.5)$$

Having defined a way of measuring utility, Ramsey can now derive a way of measuring probability for events other than the ethically neutral ones. Paraphrasing closely his original explanation, if the option of z^* for certain is indifferent with that of z_1 if θ is true and z_2 if θ is false, we can define the subject's probability of θ as the ratio of the difference between the utilities of z^* and z_2 to that between the utilities of z_1 and z_2, which we must suppose the same for all the z^*, z_1, and z_2 that satisfy the indifference condition. This amounts roughly to defining the degree of belief in θ by the odds at which the subject would bet on θ, the bet being conducted in terms of differences of value as defined.

Ramsey also proposed a definition for conditional probabilities of θ_1 given that θ_2 occurred, based on the indifference between the following actions:

$$\begin{cases} z_1 & \text{if } \theta_2 \\ z_2 & \text{if not } \theta_2 \end{cases} \quad \text{versus} \quad \begin{cases} z_3 & \text{if } \theta_2 \text{ and } \theta_1 \\ z_4 & \text{if } \theta_2 \text{ and not } \theta_1 \\ z_2 & \text{if not } \theta_2. \end{cases} \quad (5.6)$$

Similarly to what we did with called-off bets, the conditional probability is the ratio of the difference between the utilities of z_1 and z_2 to that between the utilities of z_3 and z_4, which we must again suppose the same for any set of z that satisfies the indifference condition above. Interestingly, Ramsey observes:

This is not the same as the degree to which he would believe θ_1, if he believed θ_2 for certain; for knowledge of θ_2 might for psychological reasons profoundly alter his whole system of beliefs. (Ramsey 1926, p. 180)

This comment is an ancestor of the concern we discussed in Section 2.2, about temporal coherence. We will return to this issue in Section 5.2.3.

Finally, he proceeds to prove that the probabilities so derived satisfy what he calls the fundamental laws, that is the Kolmogorov axioms and the standard definition of conditional probability.

5.2   Savage's theory

5.2.1   Notation and overview

We are now ready to approach Savage's version of this story. Compared to the formulations we encountered so far, Savage takes a much higher road in terms of the mathematical generality, dealing with continuous variables and general outcomes and parameter spaces. Let $\mathcal{Z}$ denote the set of outcomes (or consequences, or rewards) with a generic element z. Savage described an outcome as "a list of answers to all the questions that might be pertinent to the decision situation at hand" (Savage 1981a). As before, Θ is the set of possible states of the world. Its generic element is θ. An *act* or *action* is a function $a : \Theta \to \mathcal{Z}$ from states to outcomes. Thus, $a(\theta)$ is the consequence of taking action a if the state of the world turns out to be θ. The set of all acts is $\mathcal{A}$. Figure 5.1 illustrates the concept. This is a good place to point out that it is not in general obvious how well the consequences can be separated from the states of the world. This difficulty motivates most of the discussion of Chapter 6 so we will postpone it for now.

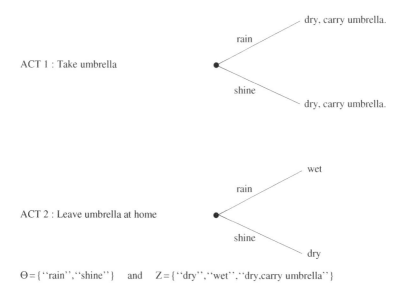

$\Theta = \{\text{"rain","shine"}\}$ and $Z = \{\text{"dry","wet","dry,carry umbrella"}\}$

Figure 5.1   Two acts for modeling the decision of whether or not to take an umbrella to work.

Before we dive into the axioms, we outline the general plan. It has points in common with Ramsey's except it can count on the NM theory and maneuvers to construct the probability first, to be able to leverage it in constructing utilities:

1. Use preferences over acts to define a "more probable than" relationship among states of the world (Axioms 1–5).

2. Use this to construct a probability measure on the states of the world (Axioms 1–6).

3. Use the NM theory to get an expected utility representation.

One last reminder: there are no compound acts (a rather artificial, but very powerful construct) and no physical chance mechanisms. All probabilities are subjective and will be derived from preferences, so we will have some hard work to do before we get to familiar places.

We follow the presentation of Kreps (1988), and give the axioms in his same form, though in slightly different order. When possible, we also provide a notation of the name the axiom has in Savage's book—those are names like P1 and P2 (for postulate). Full details of proofs are given in Savage (1954) and Fishburn (1970).

5.2.2   The sure thing principle

We start, as usual, with a binary preference relation and a set of states that are not so boring to generate complete indifference.

Axiom S1 $\succ$ on $\mathcal{A}$ is a preference relation (that is, the $\succ$ relation is complete and transitive).

Axiom S2 There exist z_1 and z_2 in $\mathcal{Z}$ such that $z_1 \succ z_2$.

These are Axioms P1 and P5 in Savage.

A cornerstone of Savage's theory is the sure thing principle:

> A businessman contemplates buying a certain piece of property. He considers the outcome of the next presidential election relevant to the attractiveness of the purchase. So, to clarify the matter to himself, he asks whether he would buy if he knew that the Republican candidate were going to win, and decides that he would do so. Similarly, he considers whether he would buy if he knew that the Democratic candidate were going to win, and again finds that he would do so. Seeing that he would buy in either event, he decides that he should buy, even though he does not know which event obtains, or will obtain, as we would ordinarily say. It is all too seldom that a decision can be arrived at on the basis of the principle used by this businessman, but, except possible for the assumption of simple ordering, I know of no other extralogical principle governing decisions that finds such ready acceptance.

Having suggested what I will tentatively call the **sure-thing principle**, let me give it relatively formal statement thus: If the person would not prefer a_1 to a_2 either knowing that the event Θ_0 obtained, or knowing that the event Θ_0^c, then he does not prefer a_1 to a_2. (Savage 1954, p. 21, with notational changes)

Implementing this principle requires a few more steps, necessary to define more rigorously what it means to "prefer a_1 to a_2 knowing that the event Θ_0 obtained." Keep in mind that the deck we are playing with is a set of preferences on acts, and those are functions defined on the whole set of states Θ. To start thinking about conditional preferences, consider the acts shown in Figure 5.2. Acts a_1 and a_2 are different from each other on the set Θ_0, and are the same on its complement Θ_0^c. Say we prefer a_1 to a_2. Now look at a_1' versus a_2'. On Θ_0 this is the same comparison we had before. On Θ_0^c, a_1' and a_2' are again equal to each other, though their actual values have changed from the previous comparison. Should we also prefer a_1' to a_2'? Axiom S3 says yes: if two acts are equal on a set (in this case Θ_0^c), the preference between them is not allowed to depend on the common value taken in that set.

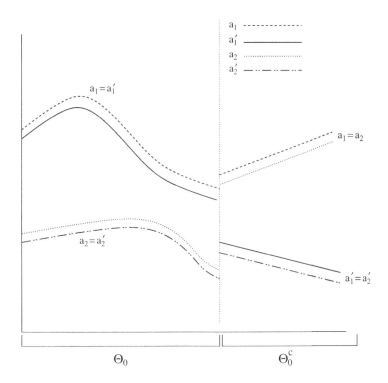

Figure 5.2   The acts that are being compared in Axiom S3. Curves that are very close to one another and parallel are meant to be on top of each other, and are separated by a small vertical shift only so you can tell them apart.

Axiom S3   Suppose that a_1, a_1', a_2, a_2' in $\mathcal{A}$ and $\Theta_0 \subseteq \Theta$ are such that

(i) $a_1(\theta) = a_1'(\theta)$ and $a_2(\theta) = a_2'(\theta)$ for all θ in Θ_0; and

(ii) $a_1(\theta) = a_2(\theta)$ and $a_1'(\theta) = a_2'(\theta)$ for all θ in Θ_0^c.

Then $a_1 \succ a_2$ if and only if $a_1' \succ a_2'$.

This is Axiom P2 in Savage, and resembles the independence axiom NM2, except that here there are no compound acts and the mixing is done by the states of nature.

Axiom S3 makes it possible to define the notion, fundamental to Savage's theory, of conditional preference. To define preference conditional on Θ_0, we make the two acts equal outside of Θ_0 and then compare them. Because of Axiom S3, it does not matter how they are equal outside of Θ_0.

Definition 5.1 (Conditional preference)   *We say that* $a_1 \succ a_2$ *given* Θ_0 *if and only if* $a_1' \succ a_2'$, *where* $a_1' = a_1$ *and* $a_2' = a_2$ *on* Θ_0 *and* $a_2' = a_1'$ *on* Θ_0^c.

Figure 5.3 describes schematically this definition. The setting has the flavor of a called-off bet, and it represents a preference stated before knowing whether Θ_0

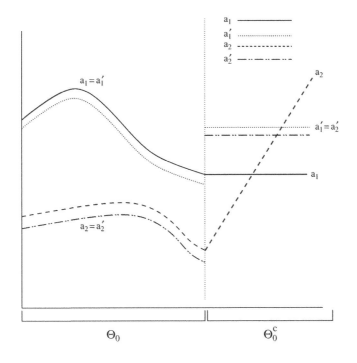

Figure 5.3   Schematic illustration of conditional preference. Acts a_1' and a_2' do not necessarily have to be constant on Θ_0^c. Again, curves that are very close to one another and parallel are meant to be on top of each other, and are separated by a small vertical shift only so you can tell them apart.

occurs or not. Problem 5.1 helps you tie this back to the informal definition of the sure thing principle given in Savage's quote. In particular, within the Savage axiom system, if $a \succ a'$ on Θ_0 and $a \succ a'$ on Θ_0^c, then $a \succ a'$.

5.2.3 Conditional and a posteriori preferences

In Sections 2.1.4 and 5.1 we discussed the correspondence between conditional probabilities, as derived in terms of called-off bets, and beliefs held after the conditioning event is observed. A similar issue arises here when we consider preferences between actions, given that the outcome of an event is known. This is a critical concern if we want to make a strong connection between axiomatic theories and statistical practice. Pratt *et al.* (1964) comment that while conditional preferences before and after the conditioning event is observed can reasonably be equated, the two reflect two different behavioral principles. Therefore, they suggest, an additional axiom is required. Restating their axiom in the context of Savage's theory, it would read like:

Before/after axiom   The a posteriori preference $a_1 \succ a_2$ given knowledge that Θ_0 has occurred holds if and only if the conditional preference $a_1 \succ a_2$ given Θ_0 holds.

Here the conditional preference could be defined according to Definition 5.1. This axiom is not part of Savage's formal development.

5.2.4 Subjective probability

In this section we are going to focus on extracting a unique probability distribution from preferences. We do this in stages, first defining qualitative probabilities, or "more likely than" statements, and then imposing additional restrictions to derive the quantitative version. Dealing with real-valued Θ makes the analysis somewhat complicated. We are showing the tip of a big iceberg here, but the submerged part is secondary to the main story of our book. Real analysis aficionados will enjoy it, though: Kreps (1988) has an entire chapter on it, and so does DeGroot (1970). Interestingly, DeGroot uses a similar technical development, but assumes utilities and probabilities (as opposed to preferences) as primitives in the theory, thus bypassing entirely the need to axiomatize preferences.

The first definition we need is that of a null state. If asked to compare two acts conditional on a null state, the agent will always be indifferent. Null states will turn out to have a subjective probability of zero.

Definition 5.2 (Null state) $\Theta_0 \subseteq \Theta$ *is called* null *if* $a \sim a'$ *given* Θ_0 *for all* a, a' *in* $\mathcal{A}$.

The next two axioms are making sure we can safely hold utility-like considerations constant in teasing probabilities out of preferences. The first, Axiom S4:

> is so couched as not only to assert that knowledge of an event cannot establish a new preference among consequences or reverse an old

one, but also to assert that, if the event is not null, no preference among consequences can be reduced to indifference by knowledge of an event. (Savage 1954, p. 26)

Axiom S4   If:

(i) Θ_0 is not null; and

(ii) $a(\theta) = z_1$ and $a'(\theta) = z_2$, for all $\theta \in \Theta_0$,

then $a \succ a'$ given Θ_0 if and only if $z_1 \succ z_2$.

This is Axiom P3 in Savage. Figure 5.4 describes schematically this condition.

Next, Axiom S5 goes back to the simple binary comparisons we used in the preface of this chapter to understand your roommate probabilities. Say he or she prefers a_1: \$10 if Duke wins the NCAA championship and \$0 otherwise to a_2: \$10 if UNC wins and \$0 otherwise. Should your roommate's preferences remain the same if you change the \$10 to \$8 and the \$0 to \$1? Axiom S5 says yes, and it makes it a lot easier to separate the probabilities from the utilities.

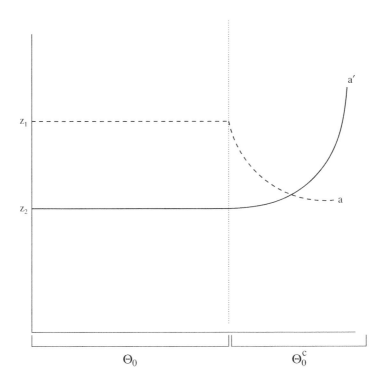

Figure 5.4   Schematic illustration of Axiom S4.

Axiom S5   Suppose that $z_1, z_2, z_1', z_2' \in \mathcal{Z}$, a_1, a_1', a_2, a_2' in $\mathcal{A}$, and $\Theta_0, \Theta_1 \subseteq \Theta$ are such that:

(i) $z_1 \succ z_2$ and $z_1' \succ z_2'$;

(ii) $a_1(\theta) = z_1$ and $a_2(\theta) = z_1'$ on Θ_0 and $a_1(\theta) = z_2$ and $a_2(\theta) = z_2'$ on Θ_0^c;

(iii) $a_1'(\theta) = z_1$ and $a_2'(\theta) = z_1'$ on Θ_1 and $a_1'(\theta) = z_2$ and $a_2'(\theta) = z_2'$ on Θ_1^c;

then $a_1 \succ a_1'$ if and only if $a_2 \succ a_2'$.

Figure 5.5 describes schematically this axiom. This is Axiom P4 in Savage.

We are now equipped to figure out, from a given set of preferences between acts, which of two events an agent considers more likely, using precisely the simple comparison considered in Axiom S5.

Definition 5.3 (More likely than)   *For any two Θ_0, Θ_1 in Θ, we say that Θ_0 is more likely than Θ_1 (denoted by $\Theta_0 \succ \Theta_1$) if for every z_1, z_2 such that $z_1 \succ z_2$ and a, a' defined as*

$$a(\theta) = \begin{cases} z_1 & \text{if } \theta \in \Theta_0 \\ z_2 & \text{if } \theta \in \Theta_0^c \end{cases}, \quad a'(\theta) = \begin{cases} z_1 & \text{if } \theta \in \Theta_1 \\ z_2 & \text{if } \theta \in \Theta_1^c \end{cases}$$

then $a \succ a'$.

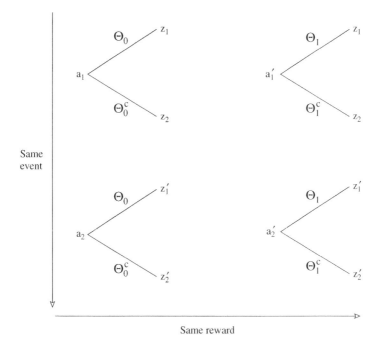

Figure 5.5   The acts in Axiom S5.

This is really nothing but our earlier Duke/UNC example dressed up in style. For your roommate, a Duke win must be more likely than a UNC win.

If we tease out this type of comparison for enough sets, we can define what is called a *qualitative probability*. Formally:

Definition 5.4 (Qualitative probability)   *The binary relation $\dot\succ$ between sets is a qualitative probability whenever:*

1. *$\dot\succ$ is asymmetric and negatively transitive;*

2. *$\Theta_0 \dot\succ \emptyset$ for all Θ_0 (subset of Θ);*

3. *$\Theta \dot\succ \emptyset$; and*

4. *if $\Theta_0 \cap \Theta_2 = \Theta_1 \cap \Theta_2 = \emptyset$, then $\Theta_0 \dot\succ \Theta_1$ if and only if $\Theta_0 \cup \Theta_2 \dot\succ \Theta_1 \cup \Theta_2$.*

As you may expect, we have done enough work to get a qualitative probability out of the preferences:

Theorem 5.1   *If S1–S5 hold then the binary relation $\dot\succ$ on Θ is a qualitative probability.*

Qualitative probabilities are big progress, but Savage is aiming for the good old quantitative kind. It turns out that if you have a quantitative probability, and you say that an event Θ_0 is more likely than Θ_1 if Θ_0 has a larger probability than Θ_1, then you define a legitimate qualitative probability. In general, however, the qualitative probabilities will not give you quantitative ones without additional technical conditions that can actually be a bit strong.

One way to get more quantitative is to be able to split up the set Θ into an arbitrarily large number of equivalent subsets. Then the quantitative probability of each of these must be the inverse of this number. De Finetti (1937) took this route for subjective probabilities, for example. Savage steps in this direction somewhat reluctantly:

> It may fairly be objected that such a postulate would be flagrantly ad hoc.
> On the other hand, such a postulate could be made relatively acceptable
> by observing that it will obtain if, for example, in all the world there is
> a coin that the person is firmly convinced is fair, that is, a coin such that
> any finite sequence of heads and tails is for him no more probable than
> any other sequence of the same length; though such a coin is, to be sure,
> a considerable idealization. (Savage 1954, p. 33)

To follow this thought one more step, we could impose the following sort of condition: given any two sets, you can find a partition fine enough that you can take the less likely of the two sets, and merge it with any element of the partition without altering the fact that it is less likely. More formally:

Definition 5.5 (Finite partition condition)   *If Θ_0, Θ_1 in Θ are such that $\Theta_0 \succ \Theta_1$, then there exists a finite partition $\{E_1, \ldots, E_n\}$ of Θ such that $\Theta_0 \succ \Theta_1 \cup E_k$, for every $k = 1, \ldots, n$.*

This would gets us to the finish line as far as the probability part is concerned, because we could prove the following.

Theorem 5.2   *The relationship $\succ$ is a qualitative probability and satisfies the axiom on the existence of a finite partition if and only if there exists a* unique *probability measure π such that*

(i) $\Theta_0 \succ \Theta_1$ *if and only if $\pi(\Theta_0) > \pi(\Theta_1)$;*

(ii) *for all $\Theta_0 \subseteq \Theta$ and $k \in [0,1]$, there exists $\Theta_1 \subseteq \Theta$, such that $\pi(\Theta_1) = k\pi(\Theta_0)$.*

See Kreps (1988) for details.

Savage approaches this slightly differently and embeds the finite partition condition into his Archimedean axiom, so that it is stated directly in terms of preferences—a much more attractive approach from the point of view of foundations than imposing restrictions on the qualitative probabilities directly. The requirement is that you can split up the set Θ in small enough pieces so that your preference will be unaffected by an arbitrary change of consequences within any one of the pieces.

Axiom S6 (Archimedean axiom)   For all a_1, a_2 in $\mathcal{A}$ such that $a_1 \succ a_2$ and $z \in \mathcal{Z}$, there exists a finite partition of Θ such that for all Θ_0 in the partition:

(i) $a_1'(\theta) = z$ for $\theta \in \Theta_0$ and $a_1'(\theta) = a_1(\theta)$ for $\theta \in \Theta_0^c$; then $a_1' \succ a_2$; *or*

(ii) $a_2'(\theta) = z$ for $\theta \in \Theta_0$ and $a_2'(\theta) = a_2(\theta)$ for $\theta \in \Theta_0^c$; then $a_1 \succ a_2'$.

This is P6 in Savage. Two important consequences of this are that there are no consequences that are infinitely better or worse than others (similarly to NM3), and also that the set of states of nature is rich enough that it can be split up into very tiny pieces, as required by the finite partition condition. A discrete space may not satisfy this axiom for all acts.

It can in fact be shown that in the context of this axiom system the Archimedean condition above guarantees that the finite partition condition is met:

Theorem 5.3   *If S1–S5 and S6 hold then $\succ$ satisfies the axiom on the existence of a finite partition of Θ.*

It then follows from Theorem 5.2 that the above gives a necessary and sufficient condition for a unique probability representation.

5.2.5   Utility and expected utility

Now that we have a unique π on Θ, and the ability to play with very fine partitions of Θ, we can follow the script of the NM theory to derive utilities and a representation of preferences. To be sure, one more axiom is needed, to require that if you prefer a to having any consequence of a' for sure, then you prefer a to a'.

Axiom S7   For all $\Theta_0 \subseteq \Theta$:

  (i) If $a \succ a'(\theta)$ given Θ_0 for all θ in Θ_0, then $a \succ a'$ given Θ_0.

  (ii) If $a'(\theta) \succ a$ given Θ_0 for all θ in Θ_0, then $a' \succ a$ given Θ_0.

This is Axiom P7 in Savage, and does a lot of the work needed to apply the theory to general consequences. Similar conditions, for example, can be used to generalize the NM theory beyond the finite $\mathcal{Z}$ we discussed in Chapter 3.

We finally laid out all the conditions for an expected utility representation.

Theorem 5.4   *Axioms (S1–S7) are sufficient for the following conclusions:*

  (a) $\succ$ as defined above is a qualitative probability, and there exists a unique probability measure π on Θ such that $\Theta_0 \succ \Theta_1$ if and only if $\pi(\Theta_0) > \pi(\Theta_1)$.

  (b) For all $\Theta_0 \subseteq \Theta$ and $k \in [0, 1]$, there exists a subset Θ_1 of Θ_0 such that $\pi(\Theta_1) = k\pi(\Theta_0)$.

  (c) For π given above, there is a bounded utility function $u : \mathcal{Z} \to \Re$ such that $a \succ a'$ if and only if

$$\mathcal{U}_\pi(a) = \int_\Theta u(a(\theta))\pi(\theta)d\theta > \int_\Theta u(a'(\theta))\pi(\theta)d\theta = \mathcal{U}_\pi(a').$$

  Moreover, this u is unique up to a positive affine transformation.

When the space of consequences is denumerable, we can rewrite the expected utility of action a in a form that better highlights the parallel with the NM theory. Defining

$$p(z) = \pi\{\theta : a(\theta) = z\}$$

we have the familiar

$$\mathcal{U}_\pi(a) = \sum_z u(z)p(z)$$

except now p reflects a subjective opinion and not a controlled chance mechanism.

Wakker (1993) discusses generalizations of this theorem to unbounded utility.

5.3   Allais revisited

In Section 3.5 we discussed Allais' claim that sober decision makers can still violate the NM2 axiom of independence. The same example can be couched in terms of Savage's theory. Imagine that the lotteries of Figure 3.3 are based on drawing at random an integer number θ between 1 and 100. Then for any decision maker that trusts the drawing mechanism to be random $\pi(\theta) = 0.01$. The lotteries can then be rewritten as shown in Table 5.1.

This representation is now close to Savage's sure thing principle (Axiom S3). If a number between 12 and 100 is drawn, it will not matter, in either situation, which lottery is chosen. So the comparisons should be based on what happens if a number between 1 and 11 is drawn. And then the comparison is the same in both situations. Savage thought that this conclusion "has a claim to universality, or objectivity" (Savage 1954). Allais obviously would disagree. If you want to know more about this debate, which is still going on now, good references are Gärdenfors and Sahlin (1988) Kahneman et al. (1982), Seidenfeld (1988), Kahneman and Tversky (1979), Tversky (1974), Fishburn (1982), and Kreps (1988).

While this example is generally understood to be a challenge to the sure thing principle, we would like to suggest that there also is a connection with the "before/after axiom" of Section 5.2.3. This axiom requires that the preferences for two acts, after one learns that θ is for sure 11 or less, should be the same as those expressed if the two actions are made equal in all cases where θ is 12 or more— irrespective of what the common outcome is. There are individuals who, like Allais, are reversing the preferences depending on the outcome offered if θ is 12 or more. But we suppose that the same individuals would agree that $a \succ a'$ if and only if $b \succ b'$ once it is known for sure that θ is 11 or less. The two comparisons are now

Table 5.1   Representation of the lotteries of the Allais paradox in terms of Savage's theory. Rewards are in units of $100 000. Probabilities of the three columns are respectively 0.01, 0.10, and 0.89.

	$\theta = 1$	$2 \leq \theta \leq 11$	$12 \leq \theta \leq 100$
a	5	5	5
a'	0	25	5
b	5	5	0
b'	0	25	0

identical for sure! So it seems that agents such as those described by Allais would also be likely to violate the "before/after axiom."

5.4   Ellsberg paradox

The next example is due to Ellsberg. Suppose that we have an urn with 300 balls in it: 100 of these balls are red (R) and the rest are either blue (B) or yellow (Y). As in the Allais example, we will consider two pairs of actions and we will have to choose between lotteries a and a', and then between lotteries b and b'. Lotteries are depicted in Table 5.2.

Suppose a decision maker expresses a preference for a over a'. In action a the probability of winning is 1/3. We may conclude that the decision maker considers blue less likely than yellow, and therefore believes to have a higher chance of winning the prize in a. In fact, this is only slightly more complicated than the sort of comparison that is used in Savage's theory to construct the qualitative probabilities.

What Ellsberg observed is that this is not necessarily the case. Many of the same decision makers also prefer b' to b. In b' the probability of winning the prize is 2/3. If one thought that blue is less likely than yellow, it would follow that the probability of winning the prize in lottery b is more than 2/3, and thus that b is to be preferred to b'.

In fact, the observed preferences violate Axiom S3 in Savage's theory. The actions are such that

$$a(R) = b(R) \quad \text{and} \quad a'(R) = b'(R)$$
$$a(B) = b(B) \quad \text{and} \quad a'(B) = b'(B)$$
$$a(Y) = a'(Y) \quad \text{and} \quad b(Y) = b'(Y).$$

So we are again in the realm of Axiom S3, and it should be that if $a \succ a'$, then $b \succ b'$.

Why is there a reversal of preferences in Ellsberg's experience? The answer seems to be that many decision makers prefer gambles where the odds are known to

Table 5.2   Actions available in the Ellsberg paradox. In Savage's notation, we have $\Theta = \{R, B, Y\}$ and $\mathcal{Z} = \{0, 1000\}$. There are 300 balls in the urn and 100 are red (R), so most decision makers would agree that $\pi(R) = 1/3$ while $\pi(B)$ and $\pi(Y)$ need to be derived from preferences.

	R	B	Y
a	1000	0	0
a'	0	1000	0
b	1000	0	1000
b'	0	1000	1000

gambles where the odds are, in Ellsberg's words, ambiguous. Gärdenfors and Sahlin comment:

> The rationale for these preferences seems to be that there is a difference between the quality of knowledge we have about the states. We *know* that the proportion of red balls is one third, whereas we are *uncertain* about the proportion of blue balls (it can be anything between zero and two thirds). Thus this decision situation falls within the unnamed area between decision making under "risk" and decision making under "uncertainty."
>
> The difference in information about the states is then reflected in the preferences in such a way that the alternative for which the exact probability of winning can be determined is preferred to the alternative where the probability of winning is "ambiguous" (Ellsberg's term). (Gärdenfors and Sahlin 1988, p. 12)

Ellsberg himself affirms that the apparently contradictory behavior presented in the previous example is not random at all:

> *none* of the familiar criteria for predicting or prescribing decision-making under uncertainty corresponds to this pattern of choices. Yet the choices themselves do not appear to be careless or random. They are persistent, reportedly deliberate, and they seem to predominate empirically; many of the people who take them are eminently reasonable, and they insist that they *want* to behave this way, even though they may be generally respectful of the Savage axioms. . . .
>
> Responses from confessed violators indicate that the difference is not to be found in terms of the two factors commonly used to determine a choice situation, the relative desirability of the possible pay-offs and the relative likelihood of the events affecting them, but in a third dimension of the problem of choice: the nature of one's information concerning the relative likelihood of events. What is at issue might be called the *ambiguity* of this information, a quality depending on the amount, type, reliability and "unanimity" of information, and giving rise to one's degree of "confidence" in an estimate of relative likelihoods. (Ellsberg 1961, pp. 257–258)

More readings on the Ellsberg paradox are in Fishburn (1983) and Levi (1985). See also Problem 5.4.

5.5   Exercises

Problem 5.1   Savage's "informal" definition of the sure thing principle, from the quote in Section 5.2.2 is: *If the person would not prefer a_1 to a_2 either knowing that* the event Θ_0 obtained, or knowing that the event Θ_0^c, then he does not prefer a_1 to a_2. Using Definition 5.1 for conditional preference, and using Savage's axioms directly (and not the representation theorem), show that Savage's axioms imply the informal sure thing principle above; that is, show that the following is true: if $a \succ a'$ on Θ_0 and $a \succ a'$ on Θ_0^c, then $a \succ a'$. Optional question: what if you replace Θ_0 and Θ_0^c with Θ_0 and Θ_1, where Θ_0 and Θ_1 are mutually exclusive? Highly optional question for the hopelessly bored: what if Θ_0 and Θ_1 are not mutually exclusive?

Problem 5.2   Prove Theorem 5.1.

Problem 5.3   One thing we find unconvincing about the Allais paradox, at least as originally stated, is that it refers to a situation that is completely hypothetical for the subject making the decisions. The sums are huge, and subjects never really get anything, or at most they get $10 an hour if the guy who runs the study had a decent grant. But maybe you can think of a real-life situation where you expect that people may violate the axioms in the same way as in the Allais paradox.

We are clearly not looking for proofs here—just a scenario, but one that gets higher marks than Allais' for realism. If you can add even the faintest hint that real individuals may violate expected utility, you will have a home run. There is a huge bibliography on this, so one fair way to solve this problem is to become an expert in the field, although that is not quite the spirit of this exercise.

Problem 5.4   (DeGroot 1970) Consider two boxes, each of which contains both red balls and green balls. It is known that one-half of the balls in box 1 are red and the other half are green. In box 2, the proportion θ of red balls is not known with certainty, but this proportion has a probability distribution over the interval $0 \le \theta \le 1$.

(a) Suppose that a person is to select a ball at random from either box 1 or box 2. If that ball is red, the person wins $1; if it is green the person wins nothing. Show that under any utility function which is an increasing function of monetary gain, the person should prefer selecting the ball from box 1 if, and only if, $E_\theta[\theta] < 1/2$.

(b) Suppose that a person can select n balls ($n \ge 2$) at random from either of the boxes, but that all n balls must be selected from the same box; suppose that each selected ball will be put back in the box before the next ball is selected; and suppose that the person will receive $1 for each red ball selected and nothing for each green ball. Also, suppose that the person's utility function u of monetary gain is strictly concave over the interval $[0, n]$, and suppose that $E_\theta[\theta] = 1/2$. Show that the person should prefer to select the balls from box 1.

Hint: Show that, if the balls are selected from box 2, then for any given value of θ, $E_\theta[u]$ is a concave function of π on the interval $0 \le \theta \le 1$. This can be done by showing that

$$\frac{d^2}{d\theta^2} E_\theta[u] = n(n-1) \sum_{i=0}^{n-2} [u(i) - 2u(i+1) + u(i+2)] \binom{n-2}{i} \theta^i (1-\theta)^{n-2-i} < 0.$$

Then apply Jensen's inequality to $g(\theta) = E_\theta[u]$.

(c) Switch red and green and try (b) again.

Note: The function f is concave if $f(\alpha x + (1-\alpha)y) \geq \alpha f(x) + (1-\alpha)f(y)$. Pictorially, concave functions are ∩-shaped. Continuously differentiable concave functions have second derivatives ≤ 0. Jensen's inequality states that if g is a convex (concave) function, and θ is a random variable, then $E[g(\theta)] \geq (\leq) g(E[\theta])$.

Comment: This is a very nice example, and is related to the Ellsberg paradox. The idea there is that in the situation of case (a), according to the expected utility principle, if $E_\theta[\theta] = 1/2$ you should be indifferent between the two boxes. But empirically, most people still prefer box 1. What do you think?

Problem 5.5 Suppose you are to choose among two experiments A and B. You are interested in estimating a parameter based on squared error loss using the data from the experiment. Therefore you want to choose the experiment that minimizes the posterior variance of the parameter. If the posterior variance depends on the data that you will collect, expected utility theory recommends that you take an expectation. Suppose experiment A has an expected posterior variance of 1 and B of 1.03, so you prefer A. But what if you looked at the whole distribution of variances, for different possible data sets, and discovered that the variances you get from A range from 0.5 to 94, while the variances you get from B range from 1 to 1.06? (The distribution for experiment A would have to be quite skewed, but that is not unusual.)

Is A really a good choice? Maybe not (if 94 is not big enough, make it 94 000). But then, what is it that went wrong with our choice criterion? Is there a fault in the expected utility paradigm, or did we simply misspecify the losses and leave out an aspect of the problem that we really cared about? Write a few paragraphs explaining what you would do, why, and whether there may be a connection between this example and any of the so-called paradoxes we discussed in this chapter.

6

State independence

Savage's theory provides an axiomatization that yields unique probabilities and utilities, and provides a foundation for Bayesian decision theory. A key element in this enterprise is the definition of constant outcomes: that is, outcomes that have the same value to the decision maker irrespective of the states of the world. Go back to the Samarkand story of Chapter 4: you are about to ship home a valuable rug you just bought for $9000. At which price would you be indifferent between buying a shipping insurance for the full value of the rug, or taking the risk? Compared to where we were in Chapter 4, we can now answer this question with or without an externally given probability for the rug to be lost. Implicitly, however, all the approaches we have available so far assume that the only relevant state of the world for this decision is whether the rug will be lost. For example, we assume we can consider the value of the sum of money paid for the rug, or the value of the sum necessary to buy insurance, as fixed quantities that do not change with the state of the world. But what if the value of the rug to us depended on how much we can resell it for in New York? Or what if we had to pay for the insurance in the currency of Uzbekistan? Both of these considerations may introduce additional elements of uncertainty that make it hard to know the utility of the relevant outcomes without specifying the states of a "bigger world" that includes the exchange rates and the sale of the rug in addition to the shipment outcome. Our solutions only apply to the "small world" whose only states are whether the rug will be lost or not. They are good to the extent that the small world is a good approximation of bigger worlds. Savage (1954, Sec. 5.5) thought extensively about this issue and included in his book a very insightful section, pertinently called "Small worlds." Elsewhere, he comments:

> Informally, or extraformally, the consequences are conceived of as what the person experiences and what he has preferences for even when there

is no uncertainty. This idea of pure experience, as a good or ill, seems philosophically suspect and is certainly impractical. In applications the role of consequence is played by such things as a cash payment or a day's fishing of which the "real consequences" may be very uncertain but which are nonetheless adapted to the role of sure consequences within the context of the specific application. (Savage 1981a, p. 306)

To elaborate on this point, in this chapter we move away from Savage's theory and consider another formulation of subjective expected utility theory, developed by Anscombe and Aumann (1963). Though this theory is somewhat less general than Savage's, here too we have both personal utilities on outcomes and personal probabilities on unknown states of the world. A real nugget in this theory is the clear answer given to the question of independence of utilities and states of the world. We outline this theory in full, and present an example, due to Schervish *et al.* (1990), that illustrates the difficulties in the definition of small worlds.

Featured articles:

Anscombe, F. J. & Aumann, R. J. (1963). A definition of subjective probability, *Annals of Mathematical Statistics* **34**: 199–205.

Useful background readings are Kreps (1988) and Fishburn (1970).

6.1   Horse lotteries

We start by taking a finite set Θ to be the set of all possible states of the world. This will simplify the mathematics and help us home in to the issue of state independence more cleanly. For simplicity of notation we will label states with integers, so that $\Theta = \{1, 2, \ldots, k\}$. Also, $\mathcal{Z}$ will be the set of prizes or rewards. We will assume that $\mathcal{Z}$ is finite, which will enable us to use results from the NM theory. The twist in this theory comes with the definition of acts. We start with simple acts:

Definition 6.1 (Simple act)   *A simple act is a function $a : \Theta \to \mathcal{Z}$.*

So far so good: these are defined as in Savage. Although we are primarily interested in simple acts, we are going to build the theory in terms of more complicated things called horse lotteries. The reason is that in this way we can exploit the machinery of the NM theory to do most of the work in the representation theorems. Let us suppose that we have a randomization device that lets us define objective lotteries like in the NM theory. Let P be the set of probability functions on $\mathcal{Z}$. Anscombe and Aumann (1963) consider acts to be any function from states of the world to one of these probability distributions. They termed these acts "horse lotteries" to suggest that you may get one of k von

Neumann and Morgenstern lotteries depending on which horse (θ) wins a race. Formally

Definition 6.2 (Act, or horse lottery) *An act is a function* $a : \Theta \to P$.

Then every $a \in \mathcal{A}$ can be written as a list of functions:

$$a = (a(1), a(2), \ldots, a(k))$$

and $a(\theta) \in P$, $\theta = 1, \ldots, k$. We also use the notation $a(\theta, z)$ to denote the probability that lottery $a(\theta)$ assigns to outcome z.

This leads to a very clean system of axioms and proofs but it is a little bit artificial and it requires the notion of an objective randomization. In this regard, Anscombe and Aumann write:

> anyone who wishes to avoid a concept of physical chance distinct from probability may reinterpret our construction as a method of defining more difficult probabilities in terms of easier ones. Such a person may consider that probabilities may be assigned directly to the outcome of spins of a roulette wheel, flips of a coin, and suchlike from considerations of symmetry. The probabilities may be so widely agreed on as to be termed impersonal or objective probabilities. Then with some assumptions concerning independence, our construction can be used to define subjective probabilities for other sorts of outcomes in terms of these objective probabilities. (Anscombe and Aumann 1963, p. 204)

The theory, as in NM theory, will require compound acts. These are defined as earlier, state by state. Formally:

Definition 6.3 (Compound acts) *For a and a' from $\mathcal{A}$ and for* $\alpha \in [0, 1]$, *define the act* $a'' = \alpha a + (1 - \alpha)a'$ *by*

$$a''(\theta) = \alpha a(\theta) + (1 - \alpha)a'(\theta) \qquad \forall \theta \in \Theta.$$

Figure 6.1 illustrates this notation. In the figure, $\Theta = \{1, 2\}$ and $\mathcal{Z} = \{10, 15, 20, 25, 30\}$. So, $a(1) = (0.5, 0.3, 0.2, 0.0, 0.0)$, while $a'(2) = (0.0, 0.0, 0.6, 0.0, 0.4)$. If we define a compound act of the form $a'' = 0.6a + 0.4a'$, we have $a''(1) = (0.3, 0.18, 0.32, 0.2, 0.0)$ and $a''(2) = (0.0, 0.0, 0.84, 0.0, 0.16)$.

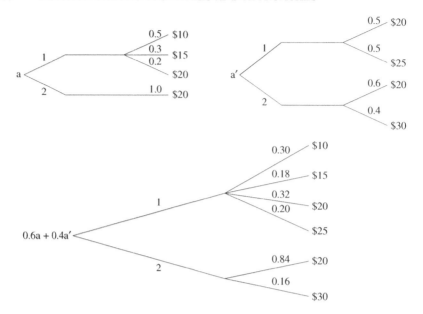

Figure 6.1   Two Anscombe–Aumann actions a and a' and their compound action with $\alpha = 0.6$. Here 1 and 2 are the two possible states of the world.

6.2   State-dependent utilities

Now we are able to introduce a set of axioms with respect to preferences (that is, $\succ$) among elements of $\mathcal{A}$.

Axiom AA1 $\succ$ on $\mathcal{A}$ is a preference relation.

Axiom AA2   If $a \succ a'$ and $\alpha \in (0, 1]$, then

$$\alpha a + (1 - \alpha)a'' \succ \alpha a' + (1 - \alpha)a''$$

for every $a'' \in \mathcal{A}$.

Axiom AA3   If $a \succ a' \succ a''$, then $\exists\, \alpha, \beta \in (0, 1)$ such that

$$\alpha a + (1 - \alpha)a'' \succ a' \succ \beta a + (1 - \beta)a''.$$

These axioms are the same as NM1, NM2, and NM3 except that they apply to the more complicated acts that we are considering here.

Based on results from the NM representation theorem, there is a function $f :\ \mathcal{A} \to \Re$ that represents $\succ$ and satisfies

$$f(\alpha a + (1 - \alpha)a') = \alpha f(a) + (1 - \alpha)f(a').$$

Using this fact and a little more work (Kreps 1988), one can establish the following result.

Theorem 6.1   *Axioms AA1, AA2, and AA3 are necessary and sufficient for the existence of real-valued $u_1, \ldots, u_k$, such that*

$$a \succ a' \iff \sum_{\theta} \sum_{z} u_\theta(z) a(\theta, z) > \sum_{\theta} \sum_{z} u_\theta(z) a'(\theta, z).$$

Also, if $u'_1, \ldots, u'_k$ is another collection of functions satisfying such a condition, then $\exists\, \alpha > 0$ and β_θ, $\theta = 1, \ldots, k$, such that $\alpha u_\theta + \beta_\theta = u'_\theta$.

This theorem is great progress, but we are still far from the standard expected utility representation. The reason is that the functions u depend on both θ and z. These are usually called state-dependent utilities (you have a different utility function in every state of the world) and they are a mix of what we usually call utility (the state-independent ones) and what we call probability. Specifically, if we allow utilities to differ across states, the uniqueness of personal probability no longer holds, and there are multiple probability distributions that satisfy the representation of Theorem 6.1. For example, the representation of Theorem 6.1 can be interpreted to mean that $k u_\theta(z)$ is the (state-dependent) utility and $\pi(\theta) = 1/k$ are the personal probabilities. Alternatively, for any other probability distribution π^* such that $\pi^*(\theta) > 0$ for every state, we can define

$$u_\theta^*(z) = \frac{u_\theta(z)}{\pi^*(\theta)}$$

and still conclude that $a \succ a'$ if and only if

$$\sum_{\theta} \pi^*(\theta) \sum_{z} u_\theta^*(z) a(\theta, z) > \sum_{\theta} \pi^*(\theta) \sum_{z} u_\theta^*(z) a'(\theta, z).$$

We can therefore represent the given preferences with multiple combinations of probabilities and state-dependent utilities.

6.3   State-independent utilities

The additional condition that is needed to disentangle the probability from the utility is state independence. Specifically, we need one more definition and two more axioms. A state θ is called *null* if we no longer care about the outcome of the lottery once θ has occurred. This translates into the condition that we are indifferent between any two acts that differ only in what happens if θ occurs. Formally we have:

Definition 6.4 (Null state)   *The state θ is said to be null if $a \sim a'$ for all pairs a and a' such that $\boldsymbol{a}(\theta') = \boldsymbol{a}'(\theta')$ for all $\theta' \neq \theta$.*

Here are the two new axioms. AA4 is just a structural condition to avoid wasting time with really boring problems where one is indifferent to everything. AA5 is a more serious weapon.

Axiom AA4   There exist a and a' in $\mathcal{A}$ such that $a' \succ a$.

Axiom AA5   Take any $a \in \mathcal{A}$ and any two probability distributions p and q on $\mathcal{Z}$. If

$$(a(1), \ldots, a(\theta - 1), p, a(\theta + 1), \ldots, a(k)) \succ$$
$$(a(1), \ldots, a(\theta - 1), q, a(\theta + 1), \ldots, a(k))$$

for some state θ, then for all non-null θ'

$$(a(1), \ldots, a(\theta' - 1), p, a(\theta' + 1), \ldots, a(k)) \succ$$
$$(a(1), \ldots, a(\theta' - 1), q, a(\theta' + 1), \ldots, a(k)).$$

AA5 is a monotonicity axiom which asks us to consider two comparisons: in the first, the two actions are identical except that in state θ one has lottery p and the other q. In the second, the two actions are again identical, except that now it is in state θ' that one has lottery p and the other q. Suppose that one prefers the action with lottery p in the first comparison. The axiom requires that the preference will hold in the second comparison as well. So the preference for p over q is independent of the state. So Axiom AA5 says that preferences should be state independent. Along with AA1–AA4, this condition will provide a representation theorem, discussed in the next section, with unique probabilities and unique utilities up to an affine transformation.

Before we move to the details of this representation, consider this example, reported by Schervish $et\ al.$ (1990), that shows what could go wrong with AA5. Suppose an agent, who expresses preferences according to Anscombe–Aumann's axioms and has linear utility for money (that is, $u(cz) = cu(z)$), is offered to choose among three simple acts a_1, a_2, a_3 whose payoffs are described in Table 6.1 depending on the states of nature $\theta_1, \theta_2, \theta_3$. Assuming that the agent has state-independent utility, the expected utility of lottery a_i is $u(1)\pi(\theta_i)$, $i = 1, 2, 3$. Furthermore, if all three

Table 6.1   Payoffs for six horse lotteries in the dollar/yen example.

	θ_1	θ_2	θ_3
a_1	\$1	0	0
a_2	0	\$1	0
a_3	0	0	\$1
a_4	¥100	0	0
a_5	0	¥125	0
a_6	0	0	¥150

horse lotteries are equivalent for the agent, then it must be that he or she considers the states to be equally likely, that is $\pi(\theta_i) = 1/3$, $i = 1, 2, 3$.

Now imagine that the agent is also indifferent among lotteries a_4, a_5, a_6 of Table 6.1. Again assuming state-independent linear utility for yen payoffs and assuming that for the agent the lotteries are equivalent, we conclude that $\pi^*(\theta_1)u(100) = \pi^*(\theta_2)u(125) = \pi^*(\theta_3)u(150)$, or $\pi(\theta_1) = 1.25\pi(\theta_2) = 1.5\pi(\theta_3)$ which implies $\pi^*(\theta_1) = 0.4054$, $\pi^*(\theta_2) = 0.3243$, and $\pi^*(\theta_3) = 0.2703$.

So this is in seeming contradiction with the indifference among a_1, a_2, a_3. However, the agent is not necessarily inconsistent, but may simply not be willing to follow axiom AA5. For example, the states could represent exchange rates between dollars and yen ($\theta_1 = \{1$ dollar is worth 100 yen$\}$, $\theta_2 = \{1$ dollar is worth 125 yen$\}$, and $\theta_3 = \{1$ dollar is worth 150 yen$\}$). Then AA5 would be untenable. The reward set is $\mathcal{Z} = \{\$1, ¥100, ¥125, ¥150\}$. In order to make comparisons among the rewards in $\mathcal{Z}$ you must know the state of nature.

You can see that if the agent is (and in this case for a good reason) not ready to endorse AA5, we are left with the question of which is the agent's "personal" probability? The uniqueness of the probability depends on the choice of what counts as a "constant" of utility—the dollar or the yen. Schervish *et al.* (1990), who concocted this example, have an extensive discussion of related issues. They also make an interesting comment on the implications of AA5 for statistical decision theory and all the good stuff in Chapter 7:

> Much of statistical decision theory makes use of utility functions of the form $u(a(\theta))$, where θ is a state of nature and a is a possible decision. The prize awarded when decision a is chosen and the state of nature is θ is not explicitly mentioned. Rather, the utility of this prize is specified without reference to the prize. Although it would appear that $u(a(\theta))$ is a state dependent utility (as well it might be), one has swept comparisons between states "under the rug". For example, if $u(a(\theta)) = -(\theta - a)^2$, one might ask how it was determined that an error of 1 when $\theta = \theta_1$ has the same utility of an error of 1 when $\theta = \theta_2$. (Schervish *et al.* 1990, pp. 846–847 with notational changes)

6.4   Anscombe–Aumann representation theorem

We now are now ready to discuss the Anscombe–Aumann representation theorem and its proof.

Theorem 6.2   *Axioms AA1–AA5 are necessary and sufficient for the existence of a nonconstant function $u : \mathcal{Z} \to \Re$ and a probability distribution π on Θ such that*

$$a \succ a' \quad \Longleftrightarrow \quad \sum_\theta \pi(\theta) \sum_z u(z)a(\theta, z) > \sum_\theta \pi(\theta) \sum_z u(z)a'(\theta, z).$$

Moreover, the probability distribution π is unique, and u is unique up to a positive linear transformation.

This sounds like magic. Where did the probability come from? The proof sheds some light on this. Look out for a revealing point when personal probabilities materialize from the weights in the affine transformations that define the class of solutions to the NM theorem. Our proof follows Kreps (1988).

Proof: The proof in the $\Rightarrow$ direction is the most interesting. It starts by observing that axioms AA1–AA3 imply a representation as given by Theorem 6.1. Axiom AA4 implies that there is at least one nonnull state, say θ_0. For any two probability distributions p and q on $\mathcal{Z}$, any nonnull state θ and an arbitrary a, we have, that

$$\sum_z u_\theta(z) p(z) > \sum_z u_\theta(z) q(z)$$

if and only if

$$(a(1), \ldots, a(\theta - 1), p, a(\theta + 1), \ldots, a(k)) \succ$$
$$(a(1), \ldots, a(\theta - 1), q, a(\theta + 1), \ldots, a(k))$$

if and only if (by application of Axiom AA5)

$$(a(1), \ldots, a(\theta_0 - 1), p, a(\theta_0 + 1), \ldots, a(k)) \succ$$
$$(a(1), \ldots, a(\theta_0 - 1), q, a(\theta_0 + 1), \ldots, a(k))$$

if and only if (by the representation in Theorem 6.1)

$$\sum_z u_{\theta_0}(z) p(z) > \sum_z u_{\theta_0}(z) q(z).$$

For simple lotteries, the result on uniqueness in NM theory guarantees that there are constants $\alpha_\theta > 0$ and β_θ such that

$$\alpha_\theta u_{\theta_0}(.) + \beta_\theta = u_\theta(.).$$

In particular, for null states, $\alpha_\theta = 0$ (because a state θ is null if and only if u_θ is constant). We can define $u(z) = u_{\theta_0}(z)$ (and $\alpha_{\theta_0} = 1$, $\beta_{\theta_0} = 0$) which, along with the representation obtained with Theorem 6.1, implies

$$a \succ a' \iff \sum_\theta \sum_z (\alpha_\theta u(z) + \beta_\theta) a(\theta, z) > \sum_\theta \sum_z (\alpha_\theta u(z) + \beta_\theta) a'(\theta, z)$$

or, equivalently,

$$
\begin{aligned}
a \succ a' \iff \sum_{\theta} \beta_{\theta} + \sum_{\theta} \alpha_{\theta} \left(\sum_{z} u(z) a(\theta, z) \right) \\
> \sum_{\theta} \beta_{\theta} + \sum_{\theta} \alpha_{\theta} \left(\sum_{z} u(z) a'(\theta, z) \right).
\end{aligned}
$$

Subtracting the sum of β_{θ} and dividing the remaining elements on both sides of the inequality by the positive quantity $\sum_{\theta'} \alpha_{\theta'}$ completes the proof if we take $\pi(\theta) = \alpha_{\theta} / \sum_{\theta'} \alpha_{\theta'}$.

The proof of the uniqueness part of the theorem is left as an exercise. The proof that axioms AA1–AA5 are necessary for the representation in Theorem 6.2 is a worked exercise.

Anscombe–Aumann's representation theorem assumes that the set of possible states of the world Θ is finite. An extension of this representation theorem to arbitrary Θ is provided by Fishburn (1970).   □

6.5   Exercises

Problem 6.1 (Schervish 1995, problem 31, p. 212)   Suppose that there are $k \geq 2$ horses in a race and that a gambler believes that π_i is the probability that horse i will win ($\sum_{i=1}^{k} \pi_i = 1$). Suppose that the gambler has decided to wager an amount a to be divided among these k horses. If he or she wagers a_i on horse i and that horse wins, the utility of the gambler is $\log(c_i a_i)$, where $c_1, \ldots, c_k$ are known positive numbers. Find values $a_1, \ldots, a_k$ that maximize the expected utility.

Problem 6.2   Prove the uniqueness of π and u in Theorem 6.2.

Problem 6.3   Prove that if there is a function $u : \mathcal{Z} \to \mathfrak{R}$, and a probability distribution π on Θ such that

$$
\begin{aligned}
a \succ a' \iff \sum_{\theta} \pi(\theta) \sum_{z} u(z) a(\theta, z) \\
> \sum_{\theta} \pi(\theta) \sum_{z} u(z) a'(\theta, z),
\end{aligned}
\tag{6.1}
$$

then $\succ$ satisfies axioms AA1–AA3 and AA5, assuming AA4.

Solution
We do this one axiom at a time.

Axiom AA1 $\succ$ *on $\mathcal{A}$ is a preference relation.*
We know that $\succ$ is a preference relation if $\forall a, a \in \mathcal{A}$, (i) $a \succ a', a \succ a'$, or $a \sim a'$, and (ii) $a \succ a'$ and $a' \succ a''$, then $a \succ a''$. We are going to call $\mathcal{U}, \mathcal{U}'$, and so on the expected utilities associated with actions a, a', and so on. For example,

$$\mathcal{U}' = \sum_{\theta} \pi(\theta) \sum_{z} u(z) a'(\theta, z).$$

(i) Since the $\mathcal{U}$ are real numbers, $\mathcal{U} > \mathcal{U}'$, $\mathcal{U}' > \mathcal{U}$, or $\mathcal{U} = \mathcal{U}'$, and from (6.1) $a \succ a'$, $a' \succ a$, or $a \sim a'$.

(ii) Since $a \succ a' \Longrightarrow \mathcal{U} > \mathcal{U}'$ and $a' \succ a'' \Longrightarrow \mathcal{U}' > \mathcal{U}''$. Consequently, $\mathcal{U} > \mathcal{U}''$, by transitivity on $\mathfrak{R}$.

Axiom AA2   If $a \succ a'$ and $\alpha \in (0, 1]$, then

$$\alpha a + (1 - \alpha) a'' \succ \alpha a' + (1 - \alpha) a'', \text{ for every } a'' \in \mathcal{A}.$$

If $a \succ a'$ then

$$\alpha \sum_{\theta} \pi(\theta) \sum_{z} u(z) a(\theta, z) > \alpha \sum_{\theta} \pi(\theta) \sum_{z} u(z) a'(\theta, z)$$

for all $\alpha \in (0, 1]$. Therefore,

$$\alpha \sum_{\theta} \pi(\theta) \sum_{z} u(z) a(\theta, z) + (1 - \alpha) \sum_{\theta} \pi(\theta) \sum_{z} u(z) a''(\theta, z) >$$

$$\alpha \sum_{\theta} \pi(\theta) \sum_{z} u(z) a'(\theta, z) + (1 - \alpha) \sum_{\theta} \pi(\theta) \sum_{z} u(z) a''(\theta, z)$$

or

$$\sum_{\theta} \pi(\theta) \sum_{z} u(z) \left[\alpha a(\theta, z) + (1 - \alpha) a''(\theta, z) \right] >$$

$$\sum_{\theta} \pi(\theta) \sum_{z} u(z) \left[\alpha a'_{\theta}(z) + (1 - \alpha) a''(\theta, z) \right]$$

and it follows from (6.1) that

$$\alpha a + (1 - \alpha) a'' \succ \alpha a' + (1 - \alpha) a''.$$

Axiom AA3   If $a \succ a' \succ a''$, then $\exists\, \alpha, \beta \in (0, 1)$ such that

$$\alpha a + (1 - \alpha) a'' \succ a' \succ \beta a + (1 - \beta) a''.$$

From (6.1), if $a \succ a' \succ a''$,

$$\mathcal{U} = \sum_{\theta} \pi(\theta) \sum_{z} u(z) a(\theta, z) > \mathcal{U}' = \sum_{\theta} \pi(\theta) \sum_{z} u(z) a'(\theta, z)$$

$$> \mathcal{U}'' = \sum_{\theta} \pi(\theta) \sum_{z} u(z) a''(\theta, z)$$

where $\mathcal{U},\mathcal{U}',\mathcal{U}'' \in \mathfrak{R}$. By continuity on $\mathfrak{R}$, there exist $\alpha, \beta \in (0, 1]$ such that

$$\beta\mathcal{U} + (1 - \beta)\mathcal{U}'' < \mathcal{U}' < \alpha\mathcal{U} + (1 - \alpha)\mathcal{U}''$$

or

$$\sum_\theta \pi(\theta) \sum_z u(z)[\beta a(\theta, z) + (1 - \beta)a''(\theta, z)]$$

$$< \sum_\theta \pi(\theta) \sum_z u(z)a'(\theta, z)$$

$$< \sum_\theta \pi(\theta) \sum_z u(z)\left[\alpha a(\theta, z) + (1 - \alpha)a''(\theta, z)\right]$$

and then $\alpha a + (1 - \alpha)a'' \succ a' \succ \beta a + (1 - \beta)a''$.

Axiom AA5   If $a \in \mathcal{A}$ and $p, q \in P$ are such that

$$(a(1), \ldots, a(i - 1), p, a(i + 1), \ldots, a(k)) \succ$$

$$(a(1), \ldots, a(i - 1), q, a(i + 1), \ldots, a(k))$$

for some i, then for all nonnull j

$$(a(1), \ldots, a(j - 1), p, a(j + 1), \ldots, a(k)) \succ$$

$$(a(1), \ldots, a(j - 1), q, a(j + 1), \ldots, a(k)).$$

Redefine

$$a = (a(1), \ldots, a(i - 1), p, a(i + 1), \ldots, a(k))$$
$$a' = (a(1), \ldots, a(i - 1), q, a(i + 1), \ldots, a(k))$$
$$a'' = (a(1), \ldots, a(j - 1), p, a(j + 1), \ldots, a(k))$$
$$a''' = (a(1), \ldots, a(j - 1), q, a(j + 1), \ldots, a(k))$$

where i and j are nonnull states. Suppose that

$$a \succ a' \text{ but } a''' \succ a''.$$

Then from (6.1) $a \succ a'$ if and only if

$$\sum_\theta \pi(\theta) \sum_z u(z)a(\theta, z) > \sum_\theta \pi(\theta) \sum_z u(z)a'(\theta, z).$$

Since $a(\theta) = a'(\theta), \quad \theta = 1, 2, \ldots, i - 1, i + 1, \ldots, k, a(i) = p$, and $a'(i) = q$ we can see that

$$\sum_z u(z)p(z) > \sum_z u(z)q(z)$$

or $p \succ q$, in NM terminology. Analogously,

$$a''' \succ a'' \implies \sum_\theta \pi(\theta) \sum_z u(z)a'''(\theta, z) > \sum_\theta \pi(\theta) \sum_z u(z)a''(\theta, z)$$

$$\implies \sum_z u(z)q(z) > \sum_z u(z)p(z)$$

$$\underset{\implies}{NM} \ q \succ p$$

which is a contradiction, since p must be preferred to q in any nonnull state. Therefore, $a'' \succ a'''$.   □

Problem 6.4 (Kreps 1988, problem 5, p. 112)   What happens to this theory if the outcome space $\mathcal{Z}$ changes with the state? That is, suppose that in state θ the possible outcomes are given by a set $\mathcal{Z}_\theta$. How much of the development above can you adapt to this setting? If you know that there are at least two prizes that lie in each of the $\mathcal{Z}_\theta$, how much of this chapter's development can you carry over?

Hint: Spend a finite amount of time on this and then write down your thoughts.

Part Two
Statistical Decision Theory

7

Decision functions

This chapter reviews the architecture of statistical decision theory—a formal attempt at providing a rational foundation to the way we learn from data. Our overview is broad, and covers concepts developed over several decades and from different viewpoints. The seed of the ideas we present, as Ferguson (1976) points out, can be traced back to Bernoulli (1738), Laplace (1812), and Gauss (1821). During the late 1880s, an era when the utilitarianism of Bentham and Mill had a prominent role in economics and social sciences, Edgeworth commented:

> the higher branch of probabilities projects into the field of Social Science. Conversely, the Principle of Utility is at the root of even the more objective portions of the Theory of Observations. The founders of the Science, Gauss and Laplace, distinctly teach that, in measuring a physical quantity, the quaesitum is not so much that value which is *most probably* right, as that which may *most advantageously* be assigned—taking into account the frequency and the seriousness of the error incurred (in the long run of metretic operations) by the proposed method of reduction. (Edgeworth 1887, p. 485)

This idea was made formal and general through the conceptual framework known today as statistical decision theory, due essentially to Abraham Wald (Wald 1945, Wald 1949). Historical and biographical details are in Weiss (1992). In his 1949 article on *statistical decision functions*, a prelude to a book of the same title to appear in 1950, Wald proposed a unifying framework for much of the existing statistical theory, based on treating statistical inference as a special case of game theory. A mathematical theory of games that provided the foundation for economic theory had been proposed by von Neumann and Morgenstern in the same 1944 book that contained the axiomatization of utility theory discussed in Chapter 3. Wald framed statistical inference as a two-person, zero-sum game. One of the players is *Nature* and the other player is the *Statistician*. Nature chooses the probability distribution for the experimental evidence that will be observed by the Statistician. The Statistician, on the other hand, observes experimental results and chooses a decision—for example, a hypothesis or a point estimate. In a zero-sum game losses to one player are gains to the other. This leads to statistical decisions based on the *minimax principle* which we will discuss in this chapter. The minimax principle tends to provide conservative statistical decision strategies, and is often justified this way. It is also appealing from a formal standpoint in that it allows us to borrow the machinery of game theory to prove an array of results on optimal decisions, and to devise approaches that do not require a priori distribution on the unknowns. However, the intrinsically pessimistic angle imposed by the zero-sum nature of the game has backlashes which we will begin to illustrate as well.

We find it useful to distinguish two aspects of Wald's contributions: one is the formal architecture of the statistical decision problem, the other is the rationality principle invoked to solve it. The formal architecture lends itself to statistical decision making under the expected utility principle as well. We will define and contrast these two approaches in Sections 7.1 and 7.2. We then illustrate their application by covering common inferential problems: classification and hypothesis testing in Section 7.5, point and interval estimation in Section 7.6.1. Lastly, in Section 7.7, we explore the theoretical relationship between expected utility and minimax rules and show how, even when using a frequentist concept of optimality, expected utility rules are often preferable.

Featured article:

Wald, A. (1949). Statistical decision functions, *Annals of Mathematical Statistics* **20**: 165–205.

There are numerous excellent references on this material, including Ferguson (1967), Berger (1985), Schervish (1995), and Robert (1994). In places, we will make use of concepts and tools from basic parametric Bayesian inference, which can be reviewed, for example, in Berger (1985) or Gelman *et al.* (1995).

7.1   Basic concepts

7.1.1   The loss function

Our decision maker, in this chapter the Statistician, has to choose among a *set of actions*, whose *consequences* depend on some unknown *state of the world*, or *state of nature*. As in previous chapters, the set of actions is called $\mathcal{A}$, and its generic member is called a. The set of states of the world is called Θ, with generic element θ. The basis for choosing among actions is a quantitative assessment of their consequences. Because the consequences also depend on the unknown state of the world, this assessment will be a function of both a and θ. So far, we worked with *utilities* $u(a(\theta))$, attached to the outcomes of the action. Beginning with Wald, statisticians are used to thinking about consequences in terms of the loss associated with each pair $(\theta, a) \in (\Theta \times \mathcal{A})$ and define a *loss* function $L(\theta, a)$.

In Wald's theory, and in most of statistical decision theory, the loss incurred by choosing an action a when the true state of nature is θ is relative to the losses incurred with other actions. In one of the earliest instances in which Wald defined a loss function (then referred to as weight function), he writes:

> The weight function $L(\theta, a)$ is a real valued non-negative function defined for all points θ of Θ and for all elements a of $\mathcal{A}$, which expresses the relative importance of the error committed by accepting a when θ is true. If θ is contained in a, $L(\theta, a)$ is, of course, equal to 0. (Wald 1939, p. 302 with notational changes)

The last comment refers to a decision formulation of confidence intervals, in which a is a subset of the parameter space (see also Section 7.6.2). If θ is contained in a, then the interval covers the parameter. However, the concept is general: if the "right" decision is made for a particular θ, the loss should be zero.

If a utility function is specified, we could restate utilities as losses by considering the negative utility, and by defining the loss function directly on the space $(\theta, a) \in (\Theta \times \mathcal{A})$, as in

$$u(a(\theta)) = -L_u(\theta, a). \tag{7.1}$$

Incidentally, if u is derived in the context of a set of axioms such as Savage's or Anscombe and Aumann's, a key role in this definition is played by state independence. In the loss function, there no longer is any explicit consideration for the outcome $z = a(\theta)$ that determined the loss. However, it is state independence that guarantees that losses occurring at different values of θ can be directly compared. See also Schervish, Seidenfeld *et al.* (1990) and Section 6.3.

However, if one starts from a given utility function, there is no guarantee that, for a given θ, there should be an action with zero loss. This condition requires a further transformation of the utility into what is referred to as a *regret loss function* $L(\theta, a)$. This is calculated from the utility-derived loss function $L_u(\theta, a)$ as

$$L(\theta, a) = L_u(\theta, a) - \inf_{a \in \mathcal{A}} L_u(\theta, a). \tag{7.2}$$

The regret loss function measures the inappropriateness of action a under state θ. Equivalently, following Savage, we can define the regret loss function directly from utilities as the conditional difference in utilities:

$$L(\theta, a) = \sup_{a'(\theta)} u(a'(\theta)) - u(a(\theta)). \tag{7.3}$$

From now on we will assume, unless specifically stated, that the loss is in regret form. Before explaining the reasons for this we need to introduce the *minimax principle* and the *expected utility principle* in the next two sections.

7.1.2   Minimax

The minimax principle of choice in statistical decision theory is based on the analogy with game theory, and assumes that the loss function represents the reward structure for both the Statistician and opponent (Nature). Nature chooses first, and so the best strategy for the Statistician is to assume the worst and chose the action that minimizes the maximum loss. Formally:

Definition 7.1 (Minimax action)   *An action a^M is minimax if*

$$a^M = \operatorname*{argmin}_{\theta} \max\; L(\theta, a). \tag{7.4}$$

When necessary, we will distinguish between the minimax action obtained from the loss function $L_u(\theta, a)$, called *minimax loss action*, and that obtained $L(\theta, a)$, called *minimax regret action* (Chernoff and Moses 1959, Berger 1985).

Taken literally, the minimax principle is a bit of a paranoid view of science, and even those of us who have been struggling for years with the most frustrating scientific problems, such as those of cancer biology, find it a poor metaphor for the scientific enterprise. However, it is probably not the metaphor itself that accounts for the emphasis on minimax. First, minimax does not require any knowledge about the chance that each of the states of the world will turn out to be true. This is appealing to statisticians seeking a rationality-based approach, but not willing to espouse subjectivist axioms. Second, minimax statistical decisions are in many cases reasonable, and tend to err on the conservative side.

Nonetheless, the intrinsic pessimism does create issues, some of which motivate Wald's definition of the loss in regret form. In this regard, Savage notes:

> It is often said that the minimax principle is founded on ultra-pessimism, that it demands that the actor assume the world to be in the worst possible state. This point of view comes about because neither Wald nor other writers have clearly introduced the concept of loss as distinguished from negative income. But Wald does frequently say that in most, if not all, applications $u(a(\theta))$ is never positive and it vanishes for each θ if a is chosen properly, which is the condition that $-u(a(\theta)) = L(\theta, a)$. Application of the minimax rule to $-u(a(\theta))$ generally, instead of to $L(\theta, a)$, is indeed ultra-pessimistic; no serious justification for it has ever been suggested, and it can lead to the absurd conclusion in some cases that no amount of relevant experimentation should deter the actor from behaving as though he were in complete ignorance. (Savage 1951, p. 63 with notational changes)

While we have not yet introduced data-based decision, it is easy to see how this may happen. Consider the negative utility loss in Table 7.1. Nature, mean but not dumb, will always pick θ_3, irrespective of any amount of evidence (short of a revelation of the truth) that experimental data may provide in favor of θ_1 and θ_2. The regret loss function is an improvement on this pessimistic aspect of the minimax principle.

Table 7.1 Negative utility loss function and corresponding regret loss function. The regret loss is obtained by subtracting, column by column, the minimum value in the column.

	Negative utility loss				Regret loss		
	θ_1	θ_2	θ_3		θ_1	θ_2	θ_3
a_1	1	0	6	a_1	0	0	1
a_2	3	4	5	a_2	2	4	0

For example, after the regret transformation, shown on the right of Table 7.1, Nature has a more difficult choice to make between θ_2 and θ_3. We will revisit this discussion in Section 13.4 where we address systematically the task of quantifying the information provided by an experiment towards the solution of a particular decision problem.

Chernoff articulates very clearly some of the most important issues with the regret transformation:

> First, it has never been clearly demonstrated that differences in utility do in fact measure what one may call regret. In other words, it is not clear that the "regret" of going from a state of utility 5 to a state of utility 3 is equivalent in some sense to that of going from a state of utility 11 to one of utility 9. Secondly, one may construct examples where an arbitrarily small advantage in one state of nature outweighs a considerable advantage in another state. Such examples tend to produce the same feelings of uneasiness which led many to object to minimax risk. ... A third objection which the author considers very serious is the following. In some examples the minimax regret criterion may select a strategy a_3 among the available strategies a_1, a_2, a_3 and a_4. On the other hand, if for some reason a_4 is made unavailable, the minimax regret criterion will select a_2 among a_1, a_2 and a_3. The author feels that for a reasonable criterion the presence of an undesirable strategy a_4 should not have an influence on the choice among the remaining strategies. (Chernoff 1954, pp. 425–426)

Savage thought deeply about these issues, as his initial book plan was to develop a rational foundation of statistical inference using the minimax, not the expected utility principle. His initial motivation was that:

> To the best of my knowledge no objectivistic motivation of the minimax rule has ever been published. In particular, Wald in his works always frankly put the rule forward without any motivation, saying simply that it may appeal to some. (Savage 1954, p. 168)

He later abandoned this plan but in the same section of his book—a short and very interesting one—he still reports some of his initial hunches as to why it may have worked:

> there are practical circumstances in which one might well be willing to accept the rule—even one who, like myself, holds a personalistic view of probability. It is hard to state the circumstances precisely, indeed they seem vague almost of necessity. But, roughly, the rule tends to seem acceptable when $\min_a \max_\theta L(\theta, a)$ is quite small compared with the values of $L(\theta, a)$ for some acts a that merit some serious consideration, and some values of θ that do not in common sense seem nearly incredible. . . . It seems to me that any motivation of the minimax principle, objectivistic or personalistic depends on the idea that decision problems with relatively small values of $\min_a \max_\theta L(\theta, a)$ often occur in practice. . . . The cost of a particular observation typically does not depend at all on the uses to which it is to be put, so when large issues are at stake an act incorporating a relatively cheap observation may sometime have a relatively small maximum loss. In particular, the income, so to speak, from an important scientific observation may accrue copiously to all mankind generation after generation. (Savage 1954, pp. 168–169)

7.1.3 Expected utility principle

In contrast, the *expected utility principle* applies to expected losses. It requires, or if you will, incorporates, information about how probable the various values of θ are considered to be, and weighs the losses against their probability of occurring. As before, these probabilities are denoted by $\pi(\theta)$. The action minimizing the resulting expectation is called the Bayes action.

Definition 7.2 (Bayes action)   *An action a^* is Bayes if*

$$a^* = \operatorname{argmin} \int_\Theta L(\theta, a)\pi(\theta)d\theta, \qquad (7.5)$$

where we define

$$\mathcal{L}_\pi(a) = \int_\Theta L(\theta, a)\pi(\theta)d\theta \qquad (7.6)$$

as the *prior expected loss*.

Formally, the difference between (7.5) and (7.4) is simply that the expectation operator replaces the maximization operator. Both operators provide a way of handling the indeterminacy of the state of the world.

The Bayes action is the same whether we use the negative utility or the regret form of the loss. From expression (7.2), the two representations differ by the quantity $\inf_{a \in \mathcal{A}} L_u(\theta, a)$; after taking an expectation with respect to θ this is a constant shift in the prior expected loss, and has no effect on the location of the minimum. None of the issues Chernoff was concerned about in our previous section are relevant here.

It is interesting, on the other hand, to read what Wald's perspective was on Bayes actions:

> First, the objection can be made against it, as Neyman has pointed out, that θ is merely an unknown constant and not a variate, hence it makes no sense to speak of the probability distribution of θ. Second, even if we may assume that θ is a variate, we have in general no possibility of determining the distribution of θ and any assumptions regarding this distribution are of hypothetical character. . . . The reason why we introduce here a hypothetical probability distribution of θ is simply that it proves to be useful in deducing certain theorems and in the calculation of the best system of regions of acceptance. (Wald 1939, p. 302)

In Section 7.7 and later in Chapter 8 we will look further into what "certain theorems" are. For now it will suffice to note that Bayes actions have been studied extensively in decision theory from a frequentist standpoint as well, as they can be used as technical devices to produce decision rules with desirable minimax and frequentist properties.

7.1.4   Illustrations

In this section we illustrate the relationship between Bayes and minimax decision using two simple examples.

In the first example, a colleague is choosing a telephone company for international calls. Thankfully, this particular problem has become almost moot since the days we started working on this book, but you will get the point. Company A is cheaper, but it has the drawback of failing to complete an international call $100\theta\%$ of the time. On the other hand, company B, which is a little bit more expensive, never fails. Actions are A and B, and the unknown is θ. Her loss function is as follows:

$$L(\theta, A) = 2\theta, \ \theta \in [0, 1]$$
$$L(\theta, B) = 1.$$

Here the value of 1 represents the difference between the subscription cost of company B and that of company A. To this your colleague adds a linear function of the number of missed calls, implying an additional loss of 0.02 units of utility for each percentage point of missed calls. So if she chooses company B her loss will be known. If she chooses company A her loss will depend on the proportion of times she will fail to make an international call. If θ was, say, 0.25 her loss would be 0.5, but if θ was 0.55 her loss would be 1.1. The minimax action can be calculated without any further input and is to choose company B, since

$$\sup_{\theta} L(\theta, A) = 2 \ > \ 1 = \sup_{\theta} L(\theta, B).$$

This seems a little conservative: company A would have to miss more than half the calls for this to be the right decision. Based on a survey of consumer reports, your colleague quantifies her prior mean for θ as 0.0476, and her prior standard deviation as 0.1487. To keep things simple, she decides that the beta distribution is a reasonable choice for the prior distribution on θ. By matching the first and second moments of a $Beta(\alpha_0, \beta_0)$ distribution the hyperparameters are $\alpha_0 = 0.05$ and $\beta_0 = 1.00$. Thus, the prior probability density function is

$$\pi(\theta) = 0.05 \, \theta^{-0.95} I_{[0,1]}(\theta)$$

and the prior expected loss is

$$\int_0^1 L(\theta, a)\pi(\theta)d\theta = \begin{cases} \int_0^1 2\theta\pi(\theta)d\theta = 2E_\theta[\theta] & \text{if } a = A \\ \int_0^1 1\pi(\theta)d\theta = 1 & \text{if } a = B. \end{cases}$$

Since $2E_\theta[\theta] = 2\times0.05/(1+0.05) \approx 0.095$ is less than 1, the Bayes action is to apply for company A. Bayes and minimax actions give different results in this example, reflecting different attitudes towards handling the fact that θ is not known. Interestingly, if instead of checking consumer reports, your colleague chose $\pi(\theta) = 1$, a uniform prior that may represent lack of information about θ, the Bayes solution would be indifference between company A and company B.

The second example is from DeGroot (1970) and provides a useful geometric interpretation for Bayes and minimax actions when $\mathcal{A}$ and Θ are both finite. Take $\Theta = \{\theta_1, \theta_2\}$ and $\mathcal{A} = \{a_1, \ldots, a_6\}$ with the loss function specified in Table 7.2. For any action a, the possible losses can be represented by the two-dimensional vector

$$y_a = [L(\theta_1, a), L(\theta_2, a)]'.$$

Figure 7.1 visualizes the vectors $y_1, \ldots, y_6$ corresponding to the losses of each of the six actions. Say the prior is $\pi(\theta_1) = 1/3$. In the space of Figure 7.1, actions that have the same expected loss with respect to this prior all lie on a line with equation

$$\frac{1}{3}L(\theta_1, a) + \frac{2}{3}L(\theta_2, a) = k,$$

where k is the expected loss. Three of these are shown as dashed lines in Figure 7.1. Bayes actions minimize the expected loss; that is, minimize the value of k. Geometrically, to find a Bayes action we look for the minimum value of k such that

Table 7.2   Loss function.

	a_1	a_2	a_3	a_4	a_5	a_6
θ_1	10	8	4	2	0	0
θ_2	0	1	2	5	6	10

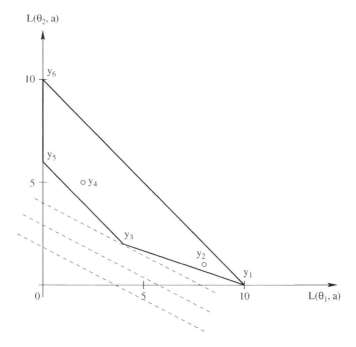

Figure 7.1   Losses and Bayes action. The Bayes action is a_3.

the corresponding line intersects an available point. In our example this happens in correspondence of action a_3, for $k = 8/3$.

Consider now selecting a minimax action, and examine again the space of losses, now reproduced in Figure 7.2. Actions that have the same maximum loss k all lie on the indifference set made of the vertical line for which $L(\theta_1, a) = k$ and $L(\theta_2, a) < k$ and the horizontal line for which $L(\theta_1, a) < k$ and $L(\theta_2, a) = k$. To find the minimax solution we look for the minimum value of k such that the corresponding set intersects an available point. In our example, the solution is again a_3, with a maximum loss of 4. The corresponding set is represented by dotted lines in the figure.

Figures 7.1 and 7.2 also show, in bold, some of the lines connecting the points representing actions. Points on these lines do not correspond to any available option. However, if one was to choose between two actions, say a_3 and a_5, at random, then the expected losses would lie on that line—here the expectation is with respect to the randomization. Rules in which one is allowed to randomly pick actions are called randomized rules; we will discuss them in more detail in Section 7.2. Sometimes using randomized rules it is possible to achieve a lower maximum loss than with any of the available ones, so they are interesting for a minimax agent. For example, in Figure 7.2, to find a minimax randomized action we move the wedge of the indifference set until it contacts the segment joining points a_3 and a_5. This corresponds to a randomized decision selecting action a_3 with probability 3/4 and action a_5 with

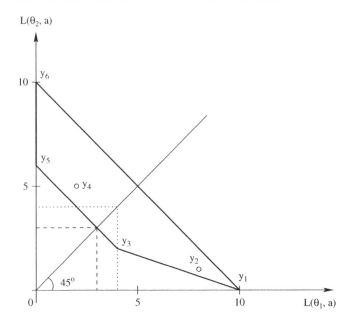

Figure 7.2   Losses and minimax action. The minimax action is a_3, represented by point $\mathbf{y}_3$. The minimax randomized action is to select action a_3 with probability $3/4$ and action a_5 with probability $1/4$.

probability $1/4$. By contrast, suppose the prior is now $\pi(\theta_1) = 1/2$. Then the dashed lines of Figure 7.1 would make contact with the entire segment between points $\mathbf{y}_3$ and $\mathbf{y}_5$ at once. Thus, actions a_3 and a_5, and any randomized decision between a_3 and a_5, would be Bayes actions. However, no gains could be obtained from choosing a randomized action. This is an instance of a much deeper result discussed in Section 7.2.

7.2   Data-based decisions

7.2.1   Risk

From a statistical viewpoint, the interesting questions arise when the outcome of an experiment whose distribution depends on the parameter θ is available. To establish notation, let x denote the experimental outcome with possible values in the set $\mathcal{X}$, and $f(x|\theta)$ is the probability density function (or probability mass, for discrete random variables). This is also called the *likelihood function* when seen as a function of θ. The question is: how should one use the data to make an optimal decision? To explore this question, we define a *decision function* (or *decision rule*) to be any function $\delta(x)$ with domain $\mathcal{X}$ and range in $\mathcal{A}$. A decision function is a recipe for turning data into actions. We denote the class of all decision rules by $\mathcal{D}$. The minimax and Bayes principles provide alternative approaches to evaluating decision rules.

A comment about notation: we use x and θ to denote both the random variables and their realized values. To keep things straight when computing expectations, we use $E_x[g(x, \theta)]$ to denote the expectation of the function g with respect to the marginal distribution of x, $E_{x|\theta}[g(x, \theta)]$ for the expectation of g with respect to $f(x|\theta)$, and $E_\theta[g(x, \theta)]$ for the expectation of g with respect to the prior on θ.

Wald's original theory is based on the expected performance of a decision rule prior to the observation of the experiment, measured by the so-called risk function.

Definition 7.3 (Risk function)   *The risk function of a decision rule δ is*

$$R(\theta, \delta) = \int_{\mathcal{X}} L(\theta, \delta) f(x|\theta) dx. \tag{7.7}$$

The risk function was introduced by Wald to unify existing approaches for the evaluation statistical procedures from a frequentist standpoint. It focuses on the long-term performance of a decision rule in a series of repetitions of the decision problems. Some of the industrial applications that were motivating Wald have this flavor: the Statistician crafts rules that are applied routinely as part of the production process, and are evaluated based on average performance.

7.2.2   Optimality principles

To define the minimax and expected utility principles in terms of decision rules consider the parallel between the risk $R(\theta, \delta)$ and the loss $L(\theta, a)$. Just as L can be used to choose among actions, R can be used to choose among decision functions. Definitions 7.1 and 7.2 can be restated in terms of risk. For minimax:

Definition 7.4 (Minimax decision rule)   *A decision rule δ^M is minimax if*

$$\sup_\theta R(\theta, \delta^M) = \inf_\delta \sup_\theta R(\theta, \delta). \tag{7.8}$$

To define the Bayes rule we first establish a notation for the Bayes risk:

Definition 7.5 (Bayes risk)   *The* Bayes risk *associated with prior distribution π and decision strategy δ is*

$$r(\pi, \delta) = \int_\Theta R(\theta, \delta) \pi(\theta) d\theta. \tag{7.9}$$

The Bayes rules minimize the Bayes risk, that is:

Definition 7.6 (Bayes decision rule)   *A decision rule δ^* is Bayes with respect to π if*

$$r(\pi, \delta^*) = \inf_\delta r(\pi, \delta). \tag{7.10}$$

What is the relationship between the Bayes rule and the expected utility principle? There is a simple and intuitive way to determine a Bayes strategy, which will also clarify this question. For every x, use the Bayes rule to determine the posterior distribution of the states of the world

$$\pi(\theta|x) = \frac{\pi(\theta)f(x|\theta)}{m(x)} \tag{7.11}$$

where

$$m(x) = \int_{\Theta} \pi(\theta)f(x|\theta)d\theta. \tag{7.12}$$

This will summarize what is known about the state of the world given the experimental outcome. Then, just as in Definition 7.2, find the Bayes action by computing the expected loss. The difference now is that the relevant distribution for computing the expectation will not be the prior $\pi(\theta)$ but the posterior $\pi(\theta|x)$ and the function to be minimized will be the posterior expected loss

$$\mathcal{L}_{\pi_x}(a) = \int_{\Theta} L(\theta, a)\pi(\theta|x)d\theta. \tag{7.13}$$

This is a legitimate step as long as we accept that static conditional probability also constrain one's opinion dynamically, once the outcome is observed (see also Chapters 2 and 3). The action that minimizes the posterior expected loss will depend on x. So, in the end, this procedure implicitly defines a decision rule, called the *formal Bayes rule*.

Does the formal Bayes rule satisfy Definition 7.6? Consider the relationship between the posterior expected loss and the Bayes risk:

$$r(\pi, \delta) = \int_{\Theta} \int_{\mathcal{X}} L(\theta, \delta)f(x|\theta)\pi(\theta)dxd\theta \tag{7.14}$$

$$= \int_{\mathcal{X}} \left[\int_{\Theta} L(\theta, \delta)\pi(\theta|x)d\theta \right] m(x)dx, \tag{7.15}$$

assuming we can reverse the order of the integrals. The quantity in square brackets is the posterior expected loss. Our intuitive recipe for finding a Bayes strategy was to minimize the posterior expected loss for every x. But in doing this we inevitably minimize r as well, and this satisfies Definition 7.6. Conversely, if we wish to minimize r with respect to the function δ, we must do so pointwise in x, by minimizing the integral in square brackets.

The conditions for interchangeability of integrals are not to be taken for granted. See Problem 8.7 for an example. Another important point is that the formal Bayes rule may not be unique, and there are examples of nonunique formal Bayes rules whose risk functions differ. More about this in Chapter 8—see for example worked Problem 8.1.

7.2.3   Rationality principles and the Likelihood Principle

The use of formal Bayes rules is backed directly by the axiomatic theory, and has profound implications for statistical inference. First, for a given experimental outcome x, a Bayes rule can be determined without any averaging over the set of all possible alternative experimental outcomes. Only the probabilities of the observed outcome under the various possible states of the world are relevant. Second, all features of the experiment that are not captured in $f(x|\theta)$ do not enter the calculation of the posterior expected loss, and thus are also irrelevant for the statistical decision. The result is actually stronger: $f(x|\theta)$ can be multiplied by an arbitrary nonzero function of x without altering the Bayes rule. Thus, for example, the data can be reduced by sufficiency, as can be seen by plugging the factorization theorem (details in Section 8.4.2) into the integrals above. This applies to both the execution of a decision once data are observed and the calculation of the criterion that is used for choosing between rules.

That all of the information in x about θ is contained in the likelihood function is a corollary of the expected utility paradigm, but is so compelling that it is taken by some as an independent principle of statistical inference, under the name of the Likelihood Principle. This is a controversial principle because the majority of frequentist measures of evidence, including highly prevalent ones such as coverage probabilities, and p-values, violate it. A monograph by Berger and Wolpert (1988) elaborates on this theme, and includes extensive and insightful discussions by several other authors as well. We will return to the Likelihood Principle in relation to optional stopping in Section 15.6.

Example 7.1   While decision rules derived from the expected utility principle satisfy the Likelihood Principle automatically, the same is not true of minimax rules. To illustrate this point in a simple example, return to the loss function in Table 7.1 and consider the regret form. In the absence of data, the minimax action is a_1. Now, suppose you can observe a binary variable x, and you can do so under two alternative experimental designs with sampling distributions f^1 and f^2, shown in Table 7.3. Because there are two possible outcomes and two possible actions, there are four possible decision functions:

$\delta_1(x) = a_1$, that is you choose a_1 regardless of the experimental outcome.

$\delta_2(x) = a_2$, that is you choose a_2 regardless of the experimental outcome.

$$\delta_3(x) = \begin{cases} a_1 & \text{if } x = 1, \\ a_2 & \text{if } x = 0. \end{cases}$$

$$\delta_4(x) = \begin{cases} a_1 & \text{if } x = 0, \\ a_2 & \text{if } x = 1. \end{cases}$$

In the sampling models of Table 7.3 we have $f^2(x = 1|\theta) = 3f^1(x = 1|\theta)$ for all $\theta \in \Theta$; that is, if $x = 1$ is observed, the two likelihood functions are proportional to each other. This implies that when $x = 1$ the expected utility rule will be the same

Table 7.3 Probability functions for two alternative sampling models.

	θ_1	θ_2	θ_3
$f^1(x = 1\|\theta)$	0.20	0.10	0.25
$f^2(x = 1\|\theta)$	0.60	0.30	0.75

under either sampling model. Does the same apply to minimax? Let us consider the risk functions for the four decision rules shown in Table 7.4. Under f^1 the minimax decision rule is $\delta_1^M(x) = \delta_4(x)$. However, under f^2 the minimax decision rule is $\delta_2^M(x) = \delta_1(x) = a_1$. Thus, if we observe $x = 1$ under f^1, the minimax decision is a_2, while under f^2 it is a_1. This is a violation of the Likelihood Principle.

For a concrete example, imagine measuring an ordinal outcome y with categories $0, 1/3, 2/3, 1$, with likelihood as in Table 7.5. Rather than asking about y directly we can use two possible questionnaires giving dichotomous answers x. One, corresponding to f^1, dichotomizes y into 1 versus all else, while the other, corresponding to f^2, dichotomizes y into 0 versus all else. Because categories $1/3, 2/3, 1$ have the same likelihood, f^2 is the better instrument overall. However, if the answer is $x = 1$, then it does not matter which instrument is used, because in both cases we know that the underlying latent variable must be either 1 or a value which is equivalent to it as far as learning about θ is concerned. The fact that in a different experiment the outcome could have been ambiguous about y in one dichotomization and not in the other is not relevant according to the Likelihood Principle. However, the risk function R, which

Table 7.4 Risk functions for the four decision rules, under the two alternative sampling models of Table 7.3.

	Under $f^1(.\|\theta)$			Under $f^2(.\|\theta)$		
	θ_1	θ_2	θ_3	θ_1	θ_2	θ_3
$\delta_1(x)$	0.00	0.00	1.00	0.00	0.00	1.00
$\delta_2(x)$	2.00	4.00	0.00	2.00	4.00	0.00
$\delta_3(x)$	1.60	3.60	0.25	0.80	2.80	0.75
$\delta_4(x)$	0.40	0.40	0.75	1.20	1.20	0.25

Table 7.5 Probability functions for the unobserved outcome y underlying the two sampling models of Table 7.3.

	θ_1	θ_2	θ_3
$f(y = 1\|\theta)$	0.20	0.10	0.25
$f(y = 2/3\|\theta)$	0.20	0.10	0.25
$f(y = 1/3\|\theta)$	0.20	0.10	0.25
$f(y = 0\|\theta)$	0.40	0.70	0.25

depends on the whole sampling distribution, and is concerned about long-run average performance of the rule over repeated experiments, is affected. A couple of famous examples of standard inferential approaches that violate the Likelihood Principle in somewhat embarrassing ways are in Problems 7.12 and 8.5.   ★

7.2.4   Nuisance parameters

The realistic specification of a sampling model often requires parameters other than those of primary interest. These additional parameters are called "nuisance parameters." This is one of the very few reasonably named concepts in statistics, as it causes all kinds of trouble to frequentist and likelihood theories alike. Basu (1975) gives a critical discussion.

From a decision-theoretic standpoint, we can think of nuisance parameters as those which appear in the sampling distribution, but not in the loss function. We formalize this notion from a decision-theoretic viewpoint and establish a general result for dealing with nuisance parameters in statistics. The bottom line is that the expected utility principle justifies averaging the likelihood and the prior over the possible values of the nuisance parameter and taking things from there. In probabilistic terminology, nuisance parameters can be integrated out, and the original decision problem can be replaced by its marginal version. More specifically:

Theorem 7.1   *If θ can be partitioned into (θ^*, η) such that the loss $L(\theta, a)$ depends on θ only through θ^*, then η is a nuisance parameter and the Bayes rule for the problem with likelihood $f(x|\theta)$ and prior $\pi(\theta)$ is the same as the Bayes rule for the problem with likelihood*

$$f^*(x|\theta^*) = \int_H f(x|\theta)\pi(\eta|\theta^*)d\eta$$

and prior

$$\pi(\theta^*) = \int_H \pi(\theta)d\eta$$

where H is the domain of η.

Proof: We assume that all integrals involved are finite. Take any decision rule δ. The Bayes risk is

$$
\begin{aligned}
r(\pi, \delta) &= \int_\Theta \int_\mathcal{X} L(\theta, \delta) f(x|\theta)\pi(\theta)dx d\theta \\
&= \int_\mathcal{X} \int_{\Theta^*} L(\theta^*, \delta) \int_H f(x|\theta)\pi(\theta)d\eta d\theta^* dx \\
&= \int_\mathcal{X} \int_{\Theta^*} L(\theta^*, \delta) \int_H f(x|\theta)\pi(\eta|\theta^*)d\eta\pi(\theta^*)d\theta^* dx \\
&= \int_\mathcal{X} \int_{\Theta^*} L(\theta^*, \delta) f^*(x|\theta^*)\pi(\theta^*)d\theta^* dx,
\end{aligned}
$$

that is the Bayes risk for the problem with likelihood $f^*(x|\theta^*)$ and prior $\pi(\theta^*)$. We used the independence of L on η, and the relation $\pi(\theta) = \pi(\eta, \theta^*) = \pi(\eta|\theta^*)\pi(\theta^*)$. ☐

This theorem puts to rest the issue of nuisance parameters in every conceivable statistical problem, as long as one can specify reasonable priors, and compute integrals. Neither is easy, of course. Priors on high-dimensional nuisance parameters can be very difficult to assess based on expert knowledge and often include surprises in the form of difficult-to-anticipate implications when nuisance parameters are integrated out. Integration in high dimension has made much progress over the last 20 years, thanks mostly to Markov chain Monte Carlo (MCMC) methods (Robert & Casella 1999), but is still hard, in part because we tend to adapt to this progress and specify models that are at the limit of what is computable. Nonetheless, the elegance and generality of the solution are compelling.

Another way to interpret the Bayesian solution to the nuisance parameter problem is to look at posterior expected losses. An argument similar to that used in the proof of Theorem 7.1 would show that one can equivalently compute the posterior expected losses based on the marginal posterior distribution of θ^* given by

$$\pi(\theta^*|x) = \int_H \pi(\theta, \eta|x)d\eta = \int_H \pi(\theta|x, \eta)\pi(\eta|x)d\eta.$$

This highlights the fact that the Bayes rule is potentially affected by any of the features of the posterior distribution $\pi(\eta|x)$ of the nuisance parameter, including all aspects of the uncertainty that remains about them after observing the data. This is in contrast to approaches that eliminate nuisance parameters by "plugging in" best estimates either in the likelihood function or in the decision rule itself. The empirical Bayes approach of Section 9.2.2 is an example.

7.3   The travel insurance example

In this section we introduce a mildly realistic medical example that will hopefully afford the simplest possible illustration of the concepts introduced in this chapter and also give us the excuse to introduce some terminology and graphics from decision analysis. We will return to this example when we consider multistage decision problems in Chapters 12 and 13.

Suppose that you are from the United States and are about to take a trip overseas. You are not sure about the status of your vaccination against a certain mild disease that is common in the country you plan to visit, and need to decide whether to buy medical insurance for the trip. We will assume that you will be exposed to the disease, but you are uncertain about whether your present immunization will work. Based on aggregate data on western tourists, the chance of developing the disease during the trip is about 3% overall. Treatment and hospital abroad would normally cost you, say, 1000 dollars. There is also a definite loss in quality of life in going all the way to an exotic country and being grounded at a local hospital instead of making the most out of your experience, but we are going to ignore this aspect here. On the other hand, if you buy a travel insurance plan, which you can do for 50 dollars, all your expenses will be covered. This is a classical gamble versus sure outcome situation. Table 7.6 summarizes the loss function for this problem.

For later reference we are going to represent this simple case using a decision tree. In a decision tree, a square denotes a *decision node* or *decision point*. The decision maker has to decide among actions, represented by branches stemming out from the decision node. A circle represents a *chance node* or *chance point*. Each branch out of the circle represents, in this case, a state of nature, though circles could also be used for experimental results. On the right side of the decision tree we have the consequences. Figure 7.3 shows the decision tree for our problem.

In a Bayesian mode, you use the expected losses to evaluate the two actions, as follows:

$$\text{No insurance:} \quad \text{Expected loss} = 1000 \times 0.03 + 0 \times 0.97 = 30$$

$$\text{Insurance:} \quad \text{Expected loss} = 50 \times 0.03 + 50 \times 0.97 = 50.$$

Table 7.6   Monetary losses associated with buying and with not buying travel insurance for the trip.

Actions	Events	
	θ_1: ill	θ_2: not ill
Insurance	50	50
No insurance	1000	0

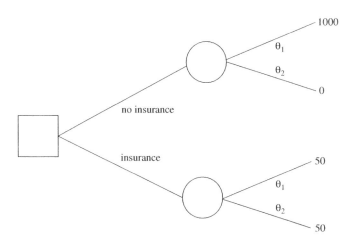

Figure 7.3   Decision tree for the travel insurance example. This is a single-stage tree, because it includes only one decision node along any given path.

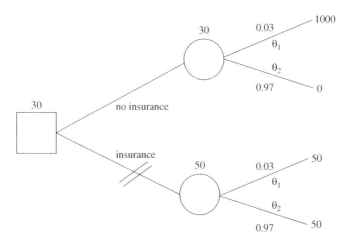

Figure 7.4   Solved decision tree for the medical insurance example. At the top of each chance node we have the expected loss, while at the top of the decision node we have the minimum expected loss. Alongside the branches stemming out from the chance node we have the probabilities of the states of nature. The action that is not optimal is crossed out by a double line.

The Bayes decision is the decision that minimizes the expected loss—in this case not to buy the insurance. However, if the chance of developing the disease was 5% or greater, the best decision would be to buy the insurance. The solution to this decision problem is represented in Figure 7.4.

    You can improve your decision making by gathering data on how likely you are to get the disease. Imagine you have the option of undergoing a medical test that informs you about whether your immunization is likely to work. The test has only two possible verdicts. One indicates that you are prone to the disease and the other indicates that you are not. Sticking with the time-honored medical tradition of calling "positive" the results of tests that suggest the presence of the most devastating illnesses, we will call positive the outcome indicating that you are disease prone. Unfortunately, the test is not perfectly accurate. Let us assume that, after some clinical experiments, it was determined that the probability that the test is positive when you really are going to get the disease is 0.9, while the probability that the test is negative when you are not going to get the disease is 0.77. For a perfect test, these numbers would be both 1. In medical terminology, the first probability represents the *sensitivity* of the test, while the second one represents the *specificity* of the test. Call x the indicator variable for the event, "The test is positive." In the notation of this chapter, the probabilities available so far are

$$\pi(\theta_1) = 0.03$$
$$f(x = 1|\theta_1) = 0.90$$
$$f(x = 0|\theta_2) = 0.77.$$

After the test, your individual chances of illness will be different from the overall 3%. The test could provide valuable information and potentially alter your chosen course of action. The question of this chapter is precisely how to use the results of the test to make a better decision. The test seems reliable enough that we may want to buy the insurance if the test is positive and not otherwise. Is this right?

To answer this question, we will consider decision rules. In our example there are two possible experimental outcomes and two possible actions so there are a total of four possible decisions rules. These are

$\delta_0(x)$: Do not buy the insurance.
$\delta_1(x)$: Buy the insurance if $x = 1$. Otherwise, do not.
$\delta_2(x)$: Buy the insurance if $x = 0$. Otherwise, do not.
$\delta_3(x)$: Buy the insurance.

Decision rules δ_0 and δ_3 choose the same action irrespective of the outcome of the test: they are constant functions. Decision rule δ_1 does what comes naturally: buy the insurance only if the test indicates that you are disease prone. Decision rule δ_2 does exactly the opposite. As you might expect, it will not turn out to be very competitive.

Let us now look at the losses associated with each decision rule ignoring, for now, any costs associated with testing. Of course, the loss for rules δ_1 and δ_2 now depends on the data. We can summarize the situation as shown in Table 7.7.

Two unknowns will affect how good our choice will turn out to be: the test result and whether you will be ill during the trip. As in equation (7.14) we can choose the Bayes rule by averaging out both, beginning with averaging losses by state, and then further averaging the results to obtain overall average losses. The results are shown in Table 7.8. To illustrate how entries are calculated, consider δ_1. The average risk if $\theta = \theta_1$ is

$$1000 f(x = 0|\theta_1) + 50 f(x = 1|\theta_1) = 1000 \times 0.10 + 50 \times 0.90 = 145.0$$

while the average risk if $\theta = \theta_2$ is

$$0 f(x = 0|\theta_2) + 50 f(x = 1|\theta_2) = 0 \times 0.77 + 50 \times 0.23 = 11.5,$$

Table 7.7   Loss table for the decision rules in the travel insurance example.

	θ_1: ill		θ_2: not ill	
	$x = 0$	$x = 1$	$x = 0$	$x = 1$
$\delta_0(x)$	$1000	$1000	$0	$0
$\delta_1(x)$	$1000	$50	$0	$50
$\delta_2(x)$	$50	$1000	$50	$0
$\delta_3(x)$	$50	$50	$50	$50

Table 7.8   Average losses by state and overall for the decision rules in the travel insurance example.

	Average losses by state		Average losses overall
	θ_1: ill	θ_2: not ill	
$\delta_0(x)$	$1000.0	$0.0	$30.0
$\delta_1(x)$	$145.0	$11.5	$15.5
$\delta_2(x)$	$905.0	$38.5	$64.5
$\delta_3(x)$	$50.0	$50.0	$50.0

so that the overall average is

$$145.0 \times \pi(\theta_1) + 11.5 \times \pi(\theta_2) = 15.5 = 145.0 \times 0.03 + 11.5 \times 0.97 = 15.5.$$

Strategy $\delta_1(x)$ is the Bayes strategy as it minimizes the overall expected loss. This calculation is effectively considering the losses in Table 7.7 and computing the expectation of each row with respect to the joint distribution of θ and x. In this sense it is consistent with preferences expressed prior to observing x. You are bound to stick to the optimal rule after you actually observe x only if you also agree with the before/after axiom of Section 5.2.3. Then, an alternative derivation of the Bayes rule could have been worked out directly by computing posterior expected losses given $x = 1$ and $x = 0$, as we know from Section 7.2.2 and equation (7.15).

So far, in solving the decision problem we utilized the Bayes principle. Alternatively, if you follow the minimax principle, your goal is avoiding the largest possible loss. Let us start with the case in which no data are available. In our example, the larges loss is 50 dollars if you buy the insurance and 1000 if you do not. By this principle you should buy the medical insurance. In fact, as we examine Table 7.6, we note that the greatest loss is associated with event θ_1 no matter what the action is. Therefore the maximization step in the minimax calculation will resolve the uncertainty about θ by assuming, pessimistically, that you will become ill, no matter how much evidence you may accumulate to the contrary. To alleviate this drastic pessimism, let us express the losses in "regret" form. The argument is as follows. If you condition on getting ill, the best you can do is a loss of $50, by buying the medical insurance. The alternative action entails a loss of $1000. When you assess the worthiness of this action, you should compare the loss to the best (smallest) loss that you could have obtained. You do indeed lose $1000, but your "regret" is only for the $950 that you could have avoided spending. Applying equation (7.2) to Table 7.6 gives Table 7.9.

When reformulating the decision problem in terms of regret, the expected losses are

No insurance: Expected loss $= 950 \times 0.03 + 0 \times 0.97 = 28.5$

Insurance: Expected loss $= 0 \times 0.03 + 50 \times 0.97 = 48.5.$

Table 7.9   Regret loss table for the actions in the travel insurance example.

		Event	
		θ_1	θ_2
Decision:	insurance	$0	$50
	no insurance	$950	$0

Table 7.10   Risk table for the decision rules in the medical insurance example when using regret losses.

	Risk $R(\theta, \delta)$ by state		Largest risk	Average risk $r(\pi, \delta)$
	θ_1	θ_2		
$\delta_0(x)$	$950	$0	$950	$28.5
$\delta_1(x)$	$95	$11.5	$95	$14.0
$\delta_2(x)$	$855	$38.5	$855	$63.0
$\delta_3(x)$	$0	$50	$50	$48.5

The Bayes action remains the same. The expected losses become smaller, but the expected loss of every action becomes smaller by the same amount. On the contrary, the minimax action may change, though in this example it does not. The minimax solution is still to buy the insurance.

Does the optimal minimax decision change depending on the test results? In Table 7.10 we derive the Bayes and minimax rules using the regret losses. Strategy δ_1 is the Bayes strategy as we had seen before. We also note that δ_2 is dominated by δ_1, that is it has a higher risk than δ_1 irrespective of the true state of the world. Using the minimax approach, the optimal decision is δ_3, that is it is still optimal to buy the insurance irrespective of the test result. This conclusion depends on the losses, sensitivity, and specificity, and different rules could be minimax if these parameters were changed.

This example will reappear in Chapters 12 and 13, when we will consider both the decision of whether to do the test and the decision of what to do with the information.

7.4   Randomized decision rules

In this section we briefly encounter randomized decision rules. From a frequentist viewpoint, the randomized decision rules are important because they guarantee, for example, specified error levels in the development of hypothesis testing procedures and confidence intervals. From a Bayesian viewpoint, we will see that randomized decision rules are not necessary, because they cannot improve the Bayes risk compared to nonrandomized decision rules.

Definition 7.7 (Randomized decision rule)   *A randomized decision rule $\delta^R(x, .)$ is, for each x, a probability distribution on $\mathcal{A}$. In particular, $\delta^R(x, A)$ denotes the probability that an action in A (a subset of $\mathcal{A}$) is chosen. The class of randomized decision rules is denoted by $\mathcal{D}^R$.*

Definition 7.8 (Loss function for a randomized decision rule)   *A randomized decision rule $\delta^R(x, .)$ has loss*

$$L(\theta, \delta^R(x)) = E_{\delta^R(x,.)} L(\theta, a) \tag{7.16}$$

$$= \int_{a \in \mathcal{A}} L(\theta, a) \delta^R(x, a) da.$$

We note that a nonrandomized decision rule is a special case of a randomized decision rule which assigns, for any given x, a specific action with probability one.

In the simple setting of Figure 7.1, no randomized decision rule in $\mathcal{D}^R$ can improve the Bayes risk attained with a nonrandomized Bayes decision rule in $\mathcal{D}$. This turns out to be the case in general:

Theorem 7.2   *For every prior distribution π on Θ, the Bayes risk on the set of randomized estimators is the same as the Bayes risk on the set of nonrandomized estimators, that is*

$$\inf_{\delta \in \mathcal{D}} r(\pi, \delta) = \inf_{\delta^R \in \mathcal{D}^R} r(\pi, \delta^R).$$

For a proof see Robert (1994). This result continues to hold when the Bayes risk is not finite, but does not hold if r is replaced by $R(\theta, \delta)$ unless additional conditions are imposed on the loss function (Berger 1985).

DeGroot comments on the use of randomized decision rules:

> This discussion supports the intuitive notion that the statistician should not base an important decision on the outcomes of the toss of a coin. When two or more pure decisions each yield the Bayes risk, an auxiliary randomization can be used to select one of these Bayes decisions. However, the randomization is irrelevant in this situation since any method of selecting one of the Bayes decisions is acceptable. In any other situation, when the statistician makes use of a randomized procedure, there is a chance that the final decision may not be a Bayes decision.
>
> Nevertheless, randomization has an important use in statistical work. The concepts of selecting a random sample and of assigning different treatments to experimental units at random are basic ones for the performance of effective experimentation. These comments do not really

conflict with those in the preceding paragraph which indicate that the statistician need never use randomized decisions (DeGroot 1970, pp. 129–130)

The next three sections illustrate these concepts in common statistical decision problems: classification, hypothesis testing, point and interval estimation.

7.5   Classification and hypothesis tests

7.5.1   Hypothesis testing

Contrasting their approach to Fisher's significance testing, Neyman and Pearson write:

> no test based upon a theory of probability can by itself provide any valuable evidence of the truth or falsehood of a hypothesis. But we may look at the purpose of tests from another viewpoint. Without hoping to know whether each separate hypothesis is true or false, we may search for rules to govern our behaviour with regard to them, in following which we insure that, in the long run of experience, we shall not often be wrong. (Neyman and Pearson 1933, p. 291)

This insight was one of the foundations of Wald's work, so hypothesis tests are a natural place to start visiting some examples. In statistical decision theory, hypothesis testing is typically modeled as the choice between actions a_0 and a_1, where a_i denotes accepting hypothesis $H_i \colon \theta \in \Theta_i$, with i either 0 or 1. Thus, $\mathcal{A} = \{a_0, a_1\}$ and $\Theta = \Theta_0 \cup \Theta_1$. A discussion of what it really means to accept a hypothesis could take us far astray, but is not a point to be taken lightly. An engaging reading on this subject is the debate between Fisher and Neyman about the meaning of hypothesis tests, which you can track down from Fienberg (1992).

If the hypothesis test is connected to a concrete problem, it may be possible to specify a loss function that quantifies how to penalize the consequences of choosing decision a_0 when H_1 is true, and decision a_1 when H_0 is true. In the simplest formulation it only matters whether the correct hypothesis is accepted, and errors are independent of the variation of the parameter θ within the two hypotheses. This is realistic, for example, when the practical consequences of the decision depend primarily on the direction of an effect. Formally,

$$L(\theta, a_0) = L_0 I_{\{\theta \in \Theta_1\}}$$
$$L(\theta, a_1) = L_1 I_{\{\theta \in \Theta_0\}}.$$

Here L_0 and L_1 are positive numbers and I_A is the indicator of the event A. Table 7.11 restates the same assumption in a more familiar form.

Table 7.11   A simple loss function for the hypothesis testing problem.

	$\theta \in \Theta_0$	$\theta \in \Theta_1$
a_0	0	L_0
a_1	L_1	0

As the decision space is binary, any decision rule δ must split the set of possible experimental results into two sets: one associated with a_0 and the other with a_1. The risk function is

$$R(\theta, \delta) = \int_{\mathcal{X}} L(\theta, \delta) f(x|\theta) dx$$

$$= \begin{cases} L_1 P(\delta(x) = a_1|\theta) = L_1 \alpha_\delta(\theta), & \text{if } \theta \in \Theta_0 \\ L_0 P(\delta(x) = a_0|\theta) = L_0 \beta_\delta(\theta), & \text{if } \theta \in \Theta_1. \end{cases}$$

From the point of view of finding optima, losses can be rescaled arbitrarily, so any solution, minimax or Bayes, will only depend on L_1 and L_0 through their ratio.

If the sets Θ_0 and Θ_1 are singletons (the familiar simple versus simple hypotheses test) we can, similarly to Figures 7.1 and 7.2, represent any decision rule as a point in the space $R(\theta_0, \delta), R(\theta_1, \delta)$, called the risk space. The Bayes and minimax optima can be derived using the same geometric intuition based on lines and wedges. If the data are from a continuous distribution, the lower solid line of Figures 7.1 and 7.2 will often be replaced by a smooth convex curve, and the solutions will be unique. With discrete data there may be a need to randomize to improve minimax solutions.

We will use the terms *null hypothesis* for H_0 and *alternative hypothesis* for H_1 even though the values of L_0 and L_1 are the only real element of differentiation here. In this terminology, α_δ and β_δ can be recognized as the probabilities of type I and II errors. In the classical Neyman–Pearson hypothesis testing framework, rather than specifying L_0 and L_1 one sets the value for $\sup_{\{\theta \in \Theta_0\}} \alpha_\delta(\theta)$ and minimizes $\beta_\delta(\theta), \theta \in \Theta_1$, with respect to δ. Because $\beta_\delta(\theta)$ is one minus the power, a uniformly most powerful decision rule will have to dominate all others in the set Θ_1. For a comprehensive discussion on classical hypothesis testing see Lehmann (1997).

In the simple versus simple case, it is often possible to set L_1/L_0 so that the minimax approach and Neyman–Pearson approach for a fixed α lead to the same rule. The specification of α and that of L_1/L_0 can substitute for each other. In all fields of science, the use of $\alpha = 0.05$ is often taken for granted, without any serious thought being given to the implicit balance of losses. This has far-reaching negative consequences in both science and policy.

Moving to the Bayesian approach, we begin by specifying a prior π on θ. Before making any observation, the expected losses are

$$E[L(\theta, a)] = \int_\Theta L(\theta, a)\pi(\theta)d\theta = \begin{cases} L_0(1 - \pi(\theta \in \Theta_0)), & \text{if } a = a_0, \\ L_1\pi(\theta \in \Theta_0), & \text{if } a = a_1. \end{cases}$$

Thus, the Bayes action is a_0 if

$$\frac{L_0}{L_1} < \frac{\pi(\theta \in \Theta_0)}{1 - \pi(\theta \in \Theta_0)}$$

and otherwise it is a_1. In particular, if $L_0 = L_1$, the Bayes action is a_0 whenever $\pi(\theta \in \Theta_0) > 1/2$. By a similar argument, after data are available, the function

$$E[L(\theta, a)|x] = \begin{cases} L_0(1 - \pi_x(\theta \in \Theta_0)), & \text{if } a = a_0, \\ L_1\pi_x(\theta \in \Theta_0), & \text{if } a = a_1, \end{cases}$$

is the posterior expected loss and the formal Bayes rule is to choose a_0 if

$$\frac{L_0}{L_1} < \frac{\pi_x(\theta \in \Theta_0)}{1 - \pi_x(\theta \in \Theta_0)},$$

and otherwise to choose a_1. The ratio of posterior probabilities on the right hand side can be written as

$$\frac{\pi_x(\theta \in \Theta_0)}{1 - \pi_x(\theta \in \Theta_0)} = \frac{\pi(\theta \in \Theta_0)f(x|\theta \in \Theta_0)}{(1 - \pi(\theta \in \Theta_0))f(x|\theta \in \Theta_1)},$$

where

$$f(x|\theta \in \Theta_i) = \int_{\Theta_i} f(x|\theta)\pi(\theta|\theta \in \Theta_i)d\theta$$

for $i = 0, 1$. The ratio of posterior probabilities can be further expanded as

$$\frac{\pi_x}{1 - \pi_x} = \frac{\pi}{1 - \pi}\frac{f(x|\theta \in \Theta_0)}{f(x|\theta \in \Theta_1)}; \tag{7.17}$$

that is, a product of the prior odds ratio and the so-called Bayes factor $BF = f(x|\theta \in \Theta_0)/f(x|\theta \in \Theta_1)$. The decision depends on the data only through the Bayes factor. In the definition above we follow Jeffreys' definition (Jeffreys 1961) and place the null hypothesis in the numerator, but you may find variation in the literature. Bernardo and Smith (1994) provide further details on the Bayes factors in decision theory.

The Bayes factor has also been proposed by Harold Jeffreys and others as a direct and intuitive measure of evidence to be used in alternative to, say, p-values for quantifying the evidence against a hypothesis. See Kass and Raftery (1995) and Goodman (1999) for recent discussions of this perspective. Jeffreys (1961, Appendix B) proposed the rule of thumb in Table 7.12 to help with the interpretation of the Bayes

Table 7.12   Interpretation of the Bayes factors for comparison between two hypotheses, according to Jeffreys.

Grade	$2\log_{10} BF$	Evidence
0	≥ 0	Null hypothesis supported
1	$-1/2$ to 0	Evidence against H_0, but not worth more than a bare mention
2	-1 to $-1/2$	Evidence against H_0 substantial
3	$-3/2$ to -1	Evidence against H_0 strong
4	-2 to $-3/2$	Evidence against H_0 very strong
5	< -2	Evidence against H_0 decisive

factor outside of a decision context. You can be the judge of whether this is fruitful — in any case, the contrast between this approach and the explicit consideration of consequences is striking.

The Bayes decision rule presented here is broadly applicable beyond a binary partition of the parameter space. For example, it extends easily to nuisance parameters, and can be applied to any pair of hypotheses selected from a discrete set, as we will see in the context of model choice in Section 11.1.2. One case that requires a separate and more complicated discussion is the comparison of a point null with a composite alternative (see Problem 7.3).

7.5.2   Multiple hypothesis testing

A trickier breed of decision problems appears when we wish to jointly test a battery of related hypotheses all at once. For example, in a clinical trial with four treatments, there are six pairwise comparisons we may be interested in. At the opposite extreme, a genome-wide scan for genetic variants associated with a disease may give us the opportunity to test millions of associations between genetic variables and a disease of interest (Hirschhorn and Daly 2005). In this section we illustrate a Bayesian decision-theoretic approach based on Müller et al. (2005). We will not venture here into the frequentist side of multiple testing. A good entry point is Hochberg and Tamhane (1987). Some considerations contrasting the two approaches are in Berry and Hochberg (1999), Genovese and Wasserman (2003), and Müller et al. (2007b).

The setting is this. We are interested in I null hypotheses $\theta_i = 0$, with $i = 1, \ldots, I$, to be compared against the corresponding alternatives $\theta_i = 1$. Available decisions for each i are a rejection of the null hypotheses ($a_i = 1$), or not ($a_i = 0$). In massive comparison problems, rejections are sometimes called discoveries, or selections. I-dimensional vectors θ and a line up all the hypotheses and corresponding decisions. The set of indexes such as $a_i = 1$ is a list of discoveries. In many applications, list making is a good way to think about the essence of the problem. To guide us in the selection of a list we observe data x with distribution $f(x|\theta, \eta)$, where η gathers any remaining model parameters. A key quantity is $\pi_x(\theta_i = 1)$, the marginal posterior probability that the ith null hypothesis is false. The nuisance parameters η can be removed by marginalization at the start, as we saw in Section 7.2.4.

The choice of a specific loss function is complicated by the fact that the experiment involves two competing goals, discovering as many as possible of the components that have $\theta_i = 1$, while at the same time controlling the number of false discoveries. We discuss two alternative utility functions that combine the two goals. These capture, at least as a first approximation, the goals of massive multiple comparisons, are easy to evaluate, lead to simple decision rules, and can be interpreted as generalizations of frequentist error rates. Interestingly, all will lead to terminal decision rules of the same form. Other loss functions for multiple comparisons are discussed in the seminal work of Duncan (1965).

We start with the notation for the summaries that formalize the two competing goals of controlling false negative and false positive decisions. The realized counts of false discoveries and false negatives are

$$FD(a, \theta) = \sum_i a_i (1 - \theta_i)$$

$$FN(a, \theta) = \sum_i (1 - a_i)\theta_i.$$

Writing $D = \sum a_i$ for the number of discoveries, the realized percentages of wrong decisions in each of the two lists, or false discovery rate and false negative rate, are, respectively

$$FDR(a, \theta) = \frac{FD(a, \theta)}{D + \epsilon}$$

$$FNR(a, \theta) = \frac{FD(a, \theta)}{I - D + \epsilon}.$$

$FD(\cdot)$, $FN(\cdot)$, $FDR(\cdot)$, and $FNR(\cdot)$ are all unknown. The additional term ϵ avoids a zero denominator.

We consider two ways of combining the goals of minimizing false discoveries and false negatives. The first two specifications combine false negative and false discovery rates and numbers, leading to the following loss functions:

$$L_N(a, \theta) = k\, FD + FN$$

$$L_R(a, \theta) = k\, FDR + FNR.$$

The loss function L_N is a natural extension of the loss function of Table 7.11 with $k = L_1/L_0$. From this perspective the combination of error rates in L_R seems less attractive, because the losses for a false discovery and a false negative depend on the total number of discoveries or negatives, respectively.

Alternatively, we can model the trade-offs between false negatives and false positives as a multicriteria decision. For the purpose of this discussion you can think of multicriteria decisions as decisions in which the loss function is multidimensional. We have not seen any axiomatic foundation for this approach, which you can learn about in Keeney *et al.* (1976). However, the standard approach to selecting an action in multicriteria decision problems is to minimize one dimension of the expected loss while enforcing a constraint on the remaining dimensions. The Neyman–Pearson approach to maximizing power subject to a fixed type I error, seen in Section 7.5.1, is an example. We will call L_{2N} and L_{2R} the multicriteria counterparts of L_N and L_R.

Conditioning on x and marginalizing with respect to θ, we obtain the posterior expected FD and FN

$$\overline{\mathrm{FD}}_x(a) = \sum a_i(1 - \pi_x(\theta_i = 1))$$

$$\overline{\mathrm{FN}}_x(a) = \sum (1 - a_i)\pi_x(\theta_i = 1)$$

and the corresponding $\overline{\mathrm{FDR}}_x(a) = \overline{\mathrm{FD}}_x(a)/(D + \epsilon)$ and $\overline{\mathrm{FNR}}_x(a) = \overline{\mathrm{FN}}_x(a)(I - D + \epsilon)$. See also Genovese and Wasserman (2002). Using these quantities we can compute posterior expected losses for both loss formulations, and also define the optimal decisions under L_{2N} as minimization of $\overline{\mathrm{FN}}$ subject to $\overline{\mathrm{FD}} \leq \alpha_N$. Similarly, under L_{2R} we minimize $\overline{\mathrm{FNR}}$ subject to $\overline{\mathrm{FDR}} \leq \alpha_R$.

Under all four loss functions the optimal decision about the multiple comparison is to select the dimensions that have a sufficiently high posterior probability $\pi(\theta_i = 1|x)$, using the same threshold for all dimensions:

Theorem 7.3   *Under all four loss functions the optimal decision takes the form*

$$a(t^*) \text{ defined as } a_i = 1 \quad \text{if and only if} \quad \pi(\theta_i = 1|x) \geq t^*.$$

The optimal choices of t^* are

$$
\begin{array}{lll}
t_N^* = k/(k+1) & \text{for} & L_N \\
t_R^*(x) = v_{(I-D^*)} & \text{for} & L_R \\
t_{2N}^*(x) = \min\{s : \overline{\mathrm{FD}}_x(a(s)) \leq \alpha\} & \text{for} & L_{2N} \\
t_{2R}^*(x) = \min\{s : \overline{\mathrm{FDR}}_x(a(s)) \leq \alpha\} & \text{for} & L_{2R}.
\end{array}
\tag{7.18}
$$

In the expression for t_R^*, $v_{(i)}$ is the ith order statistic of the vector $\{\pi(\theta_1 = 1|x), \ldots, \pi(\theta_n = 1|x)\}$, and D^* is the optimal number of discoveries.

The proof is in the appendix of Müller *et al.* (2005).

Under L_R, L_{2N}, and L_{2R} the optimal threshold t^* depends on the observed data. The nature of the terminal decision rule a_i is the same as in Genovese and Wasserman (2002), who discuss a more general rule, allowing the decision to be determined by cutoffs on any univariate summary statistic.

A very early incarnation of this result is in Pitman (1965) who considers, from a frequentist standpoint, the case where one wishes to test I simple hypotheses versus the corresponding simple alternatives. The experiments and the hypotheses have no connection with each other. If the goal is to maximize the average power, subject to a constraint on the average type I error, the optimal solution is a likelihood ratio test for each hypothesis, and the same cutoff point is used in each.

7.5.3   Classification

Binary classification problems (Duda *et al.* 2000) consider assigning cases to one of two classes, based on measuring a series of attributes of the cases. An example is medical diagnosis. In the simplest formulation, the action space has two elements: "diagnosis of disease" (a_1) and "diagnosis of no disease" (a_0). The states of "nature" (the patient, really) are *disease* or *no disease*. The loss function, shown in Table 7.13, is the same we used for hypothesis testing, with L_0 representing the loss of diagnosing a diseased patient as healthy, and L_1 representing the loss of diagnosing a healthy patient as diseased.

The main difference here is that we observe data on the attributes x and correct disease classification y of a sample of individuals, and wish to classify an additional individual, randomly drawn from the same population. The model specifies $f(y_i|x_i, \theta)$ for individual i. If $\tilde{x}$ are the observed features of the new individual to be classified, and $\tilde{y}$ is the unknown disease state, the ingredients for computing the posterior predicted probabilities are

$$\pi(\tilde{y} = 0|y, x, \tilde{x}) = M \int_{\Theta} f(\tilde{y} = 0|\tilde{x}, \theta)\pi(\theta)\, f(y, x|\theta)d\theta$$

$$\pi(\tilde{y} = 1|y, x, \tilde{x}) = M \int_{\Theta} f(\tilde{y} = 1|\tilde{x}, \theta)\pi(\theta)\, f(y, x|\theta)d\theta$$

where M is the marginal probability of (x, y). From steps similar to Section 7.5.1, the Bayes rule is to choose a diagnosis of no disease if

$$\frac{\pi(\tilde{y} = 0|y, x, \tilde{x})}{\pi(\tilde{y} = 1|y, x, \tilde{x})} > \frac{L_0}{L_1}.$$

This classification rule incorporates inference on the population model, uncertainty about population parameters, and relative disutilities of misclassification errors.

Table 7.13   A loss function for the binary classification problem.

	No disease ($\tilde{y} = 0$)	Disease ($\tilde{y} = 1$)
a_0	0	L_0
a_1	L_1	0

7.6   Estimation

7.6.1   Point estimation

Statistical point estimation assigns a single best guess to an unknown parameter. This assignment can be viewed as a decision problem in which $\mathcal{A} = \Theta$. Decision functions map data into point estimates, and they are also called estimators. Even though point estimation is becoming a bit outdated because of the ease of looking at entire distributions of unknowns, it is an interesting simplified setting for examining the implications of various decision principles.

In this setting, the loss function measures the error from declaring that the estimate is a when the correct value is θ. Suppose that $\mathcal{A} = \Theta = \Re$ and that the loss function is quadratic, $L(\theta, a) = (\theta - a)^2$. This loss has been a long-time favorite because it leads to easily interpretable analytic results. In fact its use goes back at least as far as Gauss. With quadratic loss, the risk function can be broken down into two pieces as

$$
\begin{aligned}
R(\theta, \delta) &= \int_{\mathcal{X}} L(\theta, \delta)\, f(x|\theta) dx \\
&= \int_{\mathcal{X}} (\delta(x) - \theta)^2 f(x|\theta) dx \\
&= \int_{\mathcal{X}} \left[(\delta(x) - E[\delta(x)|\theta]) + (E[\delta(x)|\theta] - \theta) \right]^2 f(x|\theta) dx \\
&= \mathrm{Var}[\delta(x)|\theta] + \left[E[\delta(x)|\theta] - \theta \right]^2. \tag{7.19}
\end{aligned}
$$

The first term in the decomposition is the variance of the decision rule and the second term is its bias, squared.

Another interesting fact is that the Bayes rule is simply $\delta^*(x) = E[\theta|x]$, that is the posterior mean. This is because the posterior expected loss is

$$
\begin{aligned}
\mathcal{L}_x(a) &= \int_{\Re} (\theta - a)^2 \pi_x(\theta) d\theta \\
&= \int_{\Re} \left[(\theta - E[\theta|x]) + (a - E[\theta|x]) \right]^2 \pi_x(\theta) d\theta \\
&= \int_{\Re} \left(\theta - E[\theta|x] \right)^2 \pi_x(\theta) d\theta + \left(a - E[\theta|x] \right)^2 \\
&= \mathrm{Var}[\theta|x] + \left(a - E[\theta|x] \right)^2,
\end{aligned}
$$

which is minimized by taking $a^* = E[\theta|x]$. Thus, the posterior expected loss associated with a^* is the posterior variance of θ. If instead $L(\theta, a) = |\theta - a|$, then $\delta^*(x) = $ median of $\pi(\theta|x)$.

A widely accepted estimation paradigm is that of maximum likelihood (Fisher 1925). The decision-theoretic framework is useful in bringing out the implicit value system of the maximum likelihood approach. To illustrate, take a discrete parameter space $\Theta = \{\theta_1, \theta_2, \ldots\}$ and imagine that the estimation problem is such that we gain something only if our estimate is exactly right. The corresponding loss function is

$$L(\theta, a) = I_{\{a \neq \theta\}},$$

where, again, a represents a point estimate of θ. The posterior expected loss is maximized by $a^* = \text{mode}(\theta|x) \equiv \theta^0$ (use your favorite tie-breaking rule if there is more than one mode). If the prior is uniform on Θ (and in other cases as well), the mode of the posterior distribution will coincide with the value of θ that maximizes the likelihood.

Extending this correspondence to the continuous case is more challenging, because the posterior probability of getting the answer exactly right is zero. If we set

$$L(\theta, a) = I_{\{|a - \theta| \geq \epsilon\}},$$

then the Bayes rule is the value of a that has the largest posterior mass in a neighborhood of size ϵ. If the posterior is a density, and the prior is approximately flat, then this a will be close to the value maximizing the likelihood.

In the remainder of this section we present two simple illustrations.

Example 7.2   This example goes back to the "secret number" example of Section 1.1. You have to guess a secret number. You know it is an integer. You can perform an experiment that would yield either the number before it or the number after it, with equal probability. You know there is no ambiguity about the experimental result or about the experimental answer. You can perform the experiment twice. More formally, x_1 and x_2 are independent observations from

$$f(x = \theta - 1|\theta) = f(x = \theta + 1|\theta) = \frac{1}{2}$$

where Θ are the integers. We are interested in estimating θ using the loss function $L(\theta, a) = I_{\{a \neq \theta\}}$. The estimator

$$\delta_0(x_1, x_2) = \frac{x_1 + x_2}{2}$$

is equal to θ if and only if $x_1 \neq x_2$, which happens in one-half of the samples. Therefore $R(\theta, \delta_0) = 1/2$. Also the estimator

$$\delta_1(x_1, x_2) = x_1 + 1$$

is equal to θ if $x_1 < \theta$, which also happens in one-half of the samples. So $R(\theta, \delta_1) = 1/2$. These two estimators are indistinguishable from the point of view of frequentist risk.

What is the Bayes strategy? If $x_1 \neq x_2$ then

$$\pi\left(\theta = \frac{x_1 + x_2}{2} \Big| x_1, x_2\right) = 1$$

and the optimal action is $a^* = (x_1 + x_2)/2$. If $x_1 = x_2$, then

$$\pi(\theta = x_1 + 1|x_1, x_2) = \frac{\pi(x_1 + 1)}{\pi(x_1 + 1) + \pi(x_1 - 1)}$$

and similarly,

$$\pi(\theta = x_1 - 1|x_1, x_2) = \frac{\pi(x_1 - 1)}{\pi(x_1 + 1) + \pi(x_1 - 1)}$$

so that the optimal action is $x_1 + 1$ if $\pi(x_1 + 1) \geq \pi(x_1 - 1)$ and $x_1 - 1$ if $\pi(x_1 + 1) \leq \pi(x_1 - 1)$. If the prior is such that it is approximately Bayes to choose $x_1 + 1$ if $x_1 = x_2$, then the resulting Bayes rule has frequentist risk of $1/4$.   ★

Example 7.3   Suppose that x is drawn from a $N(\theta, \sigma^2)$ distribution with σ^2 known. Let $L(\theta, a) = (\theta - a)^2$ be the loss function. We want to study properties of rules of the form $\delta(x) = cx$, for c a real constant. The risk function is

$$R(\theta, \delta) = \text{Var}[cx|\theta] + [E[cx|\theta] - \theta]^2$$
$$= c^2\sigma^2 + (c - 1)^2\theta^2.$$

First off we can rule out a whole bunch of rules in the family. For example, as illustrated in Figure 7.5, all the δ with $c > 1$ are dominated by the $c = 1$ rule, because $R(\theta, x) < R(\theta, cx)$, $\forall \theta$. So it would be foolish to choose $c > 1$. In Chapter 8 we will introduce technical terminology for this kind of foolishness. The rule $\delta(x) = x$ is a bit special. It is unbiased and, unlike the others, has the same risk no matter what

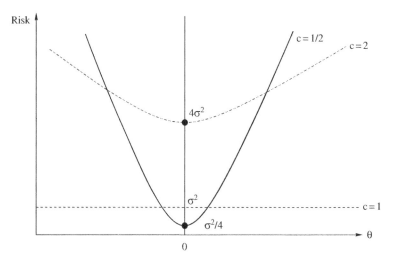

Figure 7.5   Risk functions for $c = 1/2, 1$, and 2. The $c = 2$ rule is dominated by the $c = 1$ rule, while no dominance relation exists between the $c = 1/2$ and $c = 1$ rules.

θ is. This makes it the unique minimax rule in the class, because all others have a maximum risk of infinity. Rules with $0 < c < 1$, however, are not dominated by $\delta(x) = x$, because they have a smaller variance, which may offset the higher bias when θ is indeed small.

One way of understanding why one may want to use a c that is not 1 is to bring in prior information. For example, say $\pi(\theta) \sim N(\mu_0, \tau_0^2)$, where μ_0 and τ_0^2 are known. After x is observed, the posterior expected loss is the posterior variance and the Bayes action is the posterior mean. Thus, using the results from Chapter 16,

$$\delta^*(x) = \frac{\sigma^2 \mu_0 + \tau_0^2 x}{\sigma^2 + \tau_0^2}.$$

When $\mu_0 = 0$ the decision rule belongs to the family we are studying for $c = \tau_0^2/(\sigma^2 + \tau_0^2) < 1$. To see what happens when $\mu_0 \neq 0$ study the class $\delta(x) = c_0 + cx$. ★

7.6.2 Interval inference

A common practice in statistical analysis is to report interval estimates, to communicate succinctly both the likely magnitude of an unknown and the uncertainty remaining after analysis. Interval estimation can also be framed as a decision problem, in fact Wald was doing so as early as 1939. Rice *et al.* (2008) review decision-theoretic approaches to interval estimation and Schervish (1995) provides a through treatment. Rice *et al.* (2008) observe that loss functions for intervals should trade off two competing goals: intervals should be small and close to the true value. One of the illustrations they provide is this. If Θ is the parameter space, we need to choose an interval $a = (a_1, a_2)$ within that space. A loss function capturing the two competing goals of size and closeness is a weighted combination of the half distance between the points and the "miss-distance," that is zero if the parameter is in the interval, and is the distance between the parameter and the nearest extreme if the parameter is outside the interval:

$$L(\theta, a) = L_1 \frac{a_2 - a_1}{2} + L_2 \big[(a_1 - \theta)_+ + (\theta - a_2)_+ \big]. \tag{7.20}$$

Here the subscript $+$ indicates the positive part of the corresponding function: g_+ is g when g is positive and 0 otherwise. If this loss function is used, then the optimal interval is to choose a_1 and a_2 to be the $L_1/(2L_2)$ and $1 - L_1/(2L_2)$ quantiles of the posterior distribution of θ. This provides a decision-theoretic justification for the common practice of computing equal tail posterior probability regions. The tail probability is controlled by the parameters representing the relative importance of size versus closeness. Analogously to the hypothesis test case, the same result can be achieved by specifying L_1/L_2 or the probability α assigned to the two tails in total.

An alternative specification of the loss function leads to a solution based on moments. This loss is a weighted combination of the half distance between the points and the distance of the true parameter from the center of the interval. This is measured as squared error, standardized to the interval's half size:

$$L(\theta, a) = L_1 \frac{a_2 - a_1}{2} + L_2 \left(\theta - \frac{a_2 + a_1}{2}\right) \frac{2}{a_2 - a_1}.$$

The Bayes interval in this case is

$$E[\theta|x] \pm \sqrt{\frac{L_1}{L_2} \text{Var}[\theta|x]}.$$

Finally, Carlin and Louis (2008) consider the loss function

$$L(\theta, a) = I_{\theta \notin a} + c \times \text{volume}(a).$$

Here c controls the trade-off between the volume of a and the posterior coverage probability. Under this loss function, the Bayes rule is the subset of Θ including values with highest posterior density, or HPD region(s).

7.7   Minimax–Bayes connections

When are the Bayes rules minimax? This question has been studied intensely, partly from a game–theoretic perspective in which the prior is nature's randomized decision rule, and the Bayes venue allows a minimax solution to be found. From a statistical perspective, the bottom line is that often one can concoct a prior distribution that is sufficiently pessimistic that the Bayes solution ends up being minimax. Conceptually this is important, because it establishes intersection between the set of all possible Bayes rules, each with its own prior, and minimax rules (which are often not unique), despite the different premises of the two approaches. Technically, matters get tricky very quickly. Our discussion is a quick tour. Much more thorough accounts are given by Schervish (1995) and Robert (1994).

The first result establishes that if a rule δ has a risk R that can be bounded by the average risk r of a Bayes rule, then it is minimax. This also applies to limits of r over sequences of priors.

Theorem 7.4   Let δ_k^* be the Bayes rule with respect to π_k. Suppose that $\lim_{k \to \infty} r(\pi_k, \delta_k^*) = c < \infty$. If there is δ^M such that $R(\theta, \delta^M) \leq c$ for all θ, then δ^M is minimax.

Proof: Suppose by contradiction that δ^M is not minimax. Then, there is δ' and $\epsilon > 0$ such that

$$R(\theta, \delta') \leq \sup_{\theta'} R(\theta', \delta^M) - \epsilon \leq c - \epsilon, \ \forall \theta.$$

Take k^* such that

$$r(\pi_k, \delta_k^*) \geq c - \epsilon/2, \; \forall \, k \geq k^*.$$

Then, for $k \geq k^*$,

$$r(\pi_k, \delta') = \int_\Theta R(\theta, \delta')\pi_k(\theta)d\theta \leq (c - \epsilon) \int_\Theta \pi_k(\theta)d\theta$$

$$= c - \epsilon \leq c - \epsilon/2 \leq r(\pi_k, \delta_k^*),$$

but this contradicts the hypothesis that δ_k^* is a Bayes rule with respect to π_k.   □

Example 7.4   Suppose that $x_1, \ldots, x_p$ are independent and distributed as $N(\theta_i, 1)$. Let $\boldsymbol{x} = (x_1, \ldots, x_p)'$ and $\boldsymbol{\theta} = (\theta_1, \ldots, \theta_p)'$. We are interested in showing that the decision rule $\delta^M(\boldsymbol{x}) = \boldsymbol{x}$ is minimax if one uses an additive quadratic loss function

$$L(\boldsymbol{\theta}, \boldsymbol{a}) = \sum_{i=1}^{p} (\theta_i - a_i)^2.$$

The risk function of δ^M is

$$R(\theta, \delta^M) = \int_\mathcal{X} L(\theta, \delta^M)f(\boldsymbol{x}|\boldsymbol{\theta})d\boldsymbol{x}$$

$$= E_{\boldsymbol{x}|\theta}\left[\sum_{i=1}^{p}(\theta_i - x_i)^2\right]$$

$$= \sum_{i=1}^{p} \mathrm{Var}[x_i|\theta] = p.$$

To build our sequence of priors, we set prior π_k to be such that the θ are independent normals with mean 0 and variance k. With this prior and quadratic loss, the Bayes rule is given by the vector of posterior means, that is

$$\delta_k^*(\boldsymbol{x}) = \frac{k}{k+1}\boldsymbol{x}.$$

Note that the $\theta_i, i = 1, \ldots, p$, are independent a posteriori with variance $k/(k+1)$. Therefore, the Bayes risk of δ_k^* is

$$r(\pi_k, \delta_k^*) = p\frac{k}{k+1}.$$

Taking the limit,

$$\lim_{k \to \infty} r(\pi_k, \delta_k^*) = p.$$

The assumptions of Theorem 7.4 are all met and therefore δ^M is minimax. This example will be developed in great detail in Chapter 9.   ★

We now give a more formal definition of what it means for a prior to be pessimistic for the purpose of our discussion on minimax: it means that that prior implies the greatest possible Bayes risk.

Definition 7.9   *A prior distribution π^M for θ is* least favorable *if*

$$\inf_{\delta} r(\pi^M, \delta) = \sup_{\pi} \inf_{\delta} r(\pi, \delta). \tag{7.21}$$

π^M *is also called the* maximin strategy *for nature.*

Theorem 7.5   *Suppose that δ^* is a Bayes rule with respect to π^M and such that*

$$r(\pi^M, \delta^*) = \int_{\Theta} R(\theta, \delta^*)\pi^M(\theta)d\theta = \sup_{\theta} R(\theta, \delta^*).$$

Then

1. *δ^* is a minimax rule.*

2. *If δ^* is the unique Bayes rule with respect to π, then δ^* is the unique minimax rule.*

3. *π^M is the least favorable prior.*

Proof: To prove *1* and *2* note that

$$\sup_{\theta} R(\theta, \delta^*) = \int_{\Theta} R(\theta, \delta^*)\pi^M(\theta)d\theta$$
$$\leq \int_{\Theta} R(\theta, \delta)\pi^M(\theta)d\theta,$$

where the inequality is for any other decision rule δ, because δ^* is a Bayes rule with respect to π^M. When δ^M is the unique Bayes rule a strict inequality holds. Moreover,

$$\int_{\Theta} R(\theta, \delta)\pi^M(\theta)d\theta \leq \sup_{\theta} R(\theta, \delta).$$

Thus,

$$\sup_{\theta} R(\theta, \delta^*) \leq \sup_{\theta} R(\theta, \delta), \ \forall \, \delta \in \mathcal{D}$$

and δ^M is a minimax rule. If δ^* is the unique Bayes rule, the strict inequality noted above implies that δ^* is the unique minimax rule.

To prove 3, take any other Bayes rule δ with respect to a prior distribution π. Then

$$
\begin{aligned}
r(\pi, \delta) &= \int_\Theta R(\theta, \delta)\pi(\theta)d\theta \\
&\leq \int_\Theta R(\theta, \delta^*)\pi(\theta)d\theta \\
&\leq \sup_\theta R(\theta, \delta^*) \\
&= r(\pi^M, \delta^*),
\end{aligned}
$$

that is π^M is the least favorable prior. $\square$

Example 7.5   Take x to be binomial with unknown θ, assume a quadratic loss function $L(\theta, a) = (\theta - a)^2$, and assume that θ has a priori $Beta(\alpha_0, \beta_0)$ distribution. Under quadratic loss, the Bayes rule, call it δ^*, is the posterior mean

$$
\delta^*(x) = \frac{\alpha_0 + x}{\alpha_0 + \beta_0 + n},
$$

and its risk is

$$
R(\theta, \delta^*) = \frac{1}{(\alpha_0 + \beta_0 + n)^2} \left\{ \theta^2[(\alpha_0 + \beta_0)^2 - n] + \theta[n - 2\alpha_0(\alpha_0 + \beta_0)] + \alpha_0^2 \right\}.
$$

If $\alpha_0 = \beta_0 = \sqrt{n}/2$, we have

$$
\delta^M(x) = \frac{x + \sqrt{n}/2}{n + \sqrt{n}} = \frac{x}{n}\frac{\sqrt{n}}{1 + \sqrt{n}} + \frac{1}{2(1 + \sqrt{n})}. \tag{7.22}
$$

This rule has constant risk

$$
R(\theta, \delta^M) = \frac{1}{4 + 8\sqrt{n} + 4n} = \frac{1}{4(1 + \sqrt{n})^2}. \tag{7.23}
$$

Since the risk is constant, $R(\theta, \delta^M) = r(\pi^M, \delta^M)$ for all θ, and π^M is a $Beta(\sqrt{n}/2, \sqrt{n}/2)$ distribution. By applying Theorem 7.5 we conclude that δ^M is minimax and π^M is least favorable.

The maximum likelihood estimator δ, under quadratic loss, has risk $\theta(1 - \theta)/n$ which has a unique maximum at $\theta = 1/2$. Figure 7.6 compares the risk of the maximum likelihood estimator δ to that of the minimax estimator δ^M for four choices of n. For small values of n, the minimax estimator is better for most values of the parameter. However, for larger values of n the improvement achieved by the minimax estimator is negligible and limited to a narrow range. $\star$

Example 7.6   We have seen how to determine minimax rules for normal and binomial data. Lehmann (1983) shows how to use these results to derive minimax

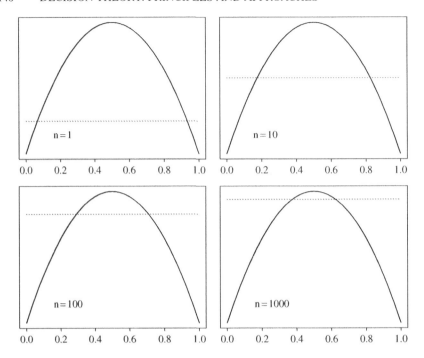

Figure 7.6   Risk functions for n = 1, 10, 100, 1000 as a function of θ. In each panel, the solid line shows the risk function for the maximum likelihood estimator δ, while the dotted line shows the flat risk function for the minimax estimator δ^M.

estimators for the means of arbitrary distributions, under a much more general setting than the parametric assumptions considered so far. The general idea is captured in the following lemma.

Lemma 2   *Let x be a random quantity with unknown distribution F, and let g(F) be a functional defined over a set $\mathcal{F}_1$ of distributions F. Suppose that δ^M is a minimax estimator of g(F) when F is restricted to some subset $\mathcal{F}_0$ of $\mathcal{F}_1$. Then if*

$$\sup_{F \in \mathcal{F}_0} R(F, \delta) = \sup_{F \in \mathcal{F}_1} R(F, \delta), \tag{7.24}$$

then δ^M is minimax also when F is permitted to vary over $\mathcal{F}_1$.

See Lehmann (1983) for details. We will look at two examples where the functional g is the mean of F. Let $x_1, \ldots, x_n$ be independent and identically distributed observations from F, and such that $E[x_i|\theta] = \theta < \infty$, for $i = 1, \ldots, n$. We consider the problem of estimating θ under quadratic loss $L(\theta, a) = (\theta - a)^2$.

For a real-valued observation, we impose the restriction of bounded variance of F, that is

$$\text{Var}[x_i|\theta] = \sigma^2 < \infty. \tag{7.25}$$

Distributions with bounded variance will constitute our class $\mathcal{F}_1$. Bounded variance will guarantee that the maximum risk of reasonable estimators of θ is finite. We choose $\mathcal{F}_0$ to be the family of normal distributions for which restriction (7.25) also holds. Similarly to Example 7.4, assume that the prior distribution of θ is $N(\mu_0, k)$. Under quadratic loss the Bayes estimator is the posterior mean and the Bayes risk is the posterior variance. This gives

$$\delta_k^*(x) = \frac{n\bar{x}/\sigma^2 + \mu_0/k}{n/\sigma^2 + 1/k}$$

$$r(\pi_k, \delta_k^*) = \frac{1}{n/\sigma^2 + 1/k}.$$

Now focus on $\delta^M = \bar{x}$. Since $\lim_{k\to\infty} r(\pi_k, \delta_k^*) = \sigma^2/n$ and $R(\theta, \delta^M) = \sigma^2/n$, by applying Theorem 7.4 we conclude that $\bar{x}$ is minimax for $\mathcal{F}_0$. Using the lemma, δ^M is minimax for $\mathcal{F}_1$.

Now take $\mathcal{F}_1$ to be the class of distributions F such that $F(1) - F(0) = 1$. These distributions have bounded support, a condition which allow us to work with finite risk without bounding the variance. Let $\mathcal{F}_0$ be the Bernoulli family of distributions. These are such that $f(x_i = 1|\theta) = \theta$ and $f(x_i = 0|\theta) = 1 - \theta$, where $0 < \theta < 1$. Let us consider the estimator

$$\delta^M(x) = \bar{x}\frac{\sqrt{n}}{1 + \sqrt{n}} + \frac{1}{2(1 + \sqrt{n})}.$$

As we saw in Example 7.5, δ^M is the minimax estimator of θ as F varies in $\mathcal{F}_0$. To prove that δ^M is minimax with respect to $\mathcal{F}_1$, by virtue of the lemma, it is enough to show that the risk $R(\theta, \delta^M)$ takes on its maximum over $\mathcal{F}_0$. Observe that the risk is

$$R(F, \delta^M) = E\left[\frac{\sqrt{n}}{1 + \sqrt{n}}\bar{x} + \frac{1}{2(1 + \sqrt{n})} - \theta \Big| \theta\right]^2.$$

By adding and subtracting $\sqrt{n}/(1 + \sqrt{n})\theta$ we obtain

$$R(F, \delta^M) = \frac{1}{(1 + \sqrt{n})^2}\left[\text{Var}[x|\theta] + \left(\frac{1}{2} - \theta\right)^2\right].$$

Observe that $0 \leq x \leq 1$. Then $x^2 \leq x$ and $\text{Var}[x|\theta] = E[x^2|\theta] - \theta^2 \leq E[x|\theta] - \theta^2$. Therefore

$$R(F, \delta^M) \leq \frac{1}{4(1 + \sqrt{n})^2}.$$

As seen in Example 7.5, the quantity $1/4(1+\sqrt{n})^2$ is the risk of δ^M over $\mathcal{F}_0$. Since the risk function of δ^M takes its maximum over $\mathcal{F}_0$, we can conclude that δ^M is minimax for the broader family $\mathcal{F}_1$.   ★

We close our discussion with a general sufficient condition under which minimax rules and least favorable distributions exist.

Theorem 7.6 (The minimax theorem)   *Suppose that the loss function L is bounded below and Θ is finite. Then*

$$\sup_{\pi} \inf_{\delta} r(\pi, \delta) = \inf_{\delta} \sup_{\theta} R(\theta, \delta)$$

and there exists a least favorable distribution π^M. If R is closed from below, then there is a minimax rule that is a Bayes rule with respect to π^M.

See Schervish (1995) for a proof. Weiss (1992) elaborates on the significance of this result within Wald's theory. The following example illustrates that the theorem does not necessarily hold if Θ is not finite.

Example 7.7 (Ferguson 1967)   Suppose that $\theta = \mathcal{A} = \{1, 2, \ldots\}$, in a decision problem where no data are available and the loss function is

$$L(\theta, a) = \begin{cases} 1 & \text{if } a < \theta, \\ 0 & \text{if } a = \theta, \\ -1 & \text{if } a > \theta. \end{cases}$$

If we think of this as a game between the Statistician and Nature, both have to pick an integer, and whoever chooses the largest integer wins the game. Priors can formally be thought of as randomized strategies for Nature, so $\pi(\theta_i)$ is the probability of Nature choosing integer i. We have $r(\pi, a) = E_\theta[L(\theta, a)] = P(\theta > a) - P(\theta < a)$. Therefore, $\sup_\pi \inf_a r(\pi, a) = -1$, which differs from $\inf_a \sup_\theta R(\theta, a) = 1$. Thus, we do not have an optimal minimax strategy for this game.   ★

7.8   Exercises

Problem 7.1   Lindley reports this interesting gastronomical conundrum with regard to minimax:

> You enter a restaurant and you study the menu: let us suppose for simplicity it only contains two items, duck and goose. After reflection you decide on goose. However, when the waiter appears he informs you that chicken is also available. After further reflection and the use of the minimax method you may decide on duck. It is hard to see how the availability of a third possibility should make you change your selection from the second to the first of the original alternatives, but such a change can

occur if the minimax method is used. The point to be emphasized here is that two decision problems have been posed and solved and yet the two results do not hang together, they do not cohere. (Lindley 1968b, pp. 317–318)

Construct functions u, and L_u and L, that provide a numerical illustration of Lindley's example. This paradox is induced by the transformation to regret, so you only need to go as far as illustrating the paradox for L.

Problem 7.2   You are collecting n normal observations with mean θ and standard deviation 1 because you are interested in testing $\theta = \theta_0 = -1$ versus $\theta = \theta_1 = 1$. Using the notation of Section 7.5.1:

1. Find the set of values of π, L_0, and L_1 that give you the same decision rule as the most powerful α level test, with $\alpha = 0.04$.

2. Can you find values of π, L_0, and L_1 that give you the same decision rule as the most powerful α level test, with $\alpha = 0.04$, for *both* a sample of size n and a sample of size $2n$?

3. (Optional) If your answer to the second question is no, that means that using the same α irrespective of sample size violates some of the axioms of expected utility theory. Can you say which?

Problem 7.3   Testing problems are frequently cast in terms of comparisons between a point null hypothesis, that is a hypothesis made up of a single point in the parameter space, against a composite alternative, including multiple points. Formally, $\Theta_0 = \theta_0$ and $\Theta_1 \neq \theta_0$. A standard Bayesian approach to this type of problem is to specify a mixture prior for θ that assigns a point mass π_0 to the event $\Theta = \theta_0$ and a continuous distribution $\pi_1(\theta|\theta \in \Theta_1)$. This is sometimes called "point and slab" prior and goes back to Jeffreys (1961). Given data $x \sim f(x|\theta)$, find the Bayes rule in this setting. Read more about this problem in Berger & Delampady (1987).

Problem 7.4   You are collecting normal data with mean θ and standard deviation 1 because you are interested in testing $\theta < 0$ versus $\theta \geq 0$. Your loss is such that mistakes are penalized differently within the alternative compared to the null, that is

$$L(\theta, a_0) = L_0|\theta|1_{\{\theta \geq 0\}},$$
$$L(\theta, a_1) = L_1|\theta|1_{\{\theta < 0\}}.$$

1. Assume a uniform prior on θ and use Bayes' theorem to derive the posterior distribution.

2. Find the posterior expected losses of a_0 and a_1.

3. Find the Bayes decision rule for this problem.

Problem 7.5   Find the least favorable priors for the two sampling distributions in Example 7.1. Does it bother you that the prior depends on the sampling model?

Problem 7.6   In order to choose an action a based on the loss $L(\theta, a)$, you elicit the opinion of two experts about the probabilities of the various outcomes of θ (you can assume if you wish that θ is a discrete random variable). The two experts give you distributions π_1 and π_2. Suppose that the action a^* is Bayes for both distributions π_1 and π_2. Is it true that a^* must be Bayes for all weighted averages of π_1 and π_2; that is, for all the distributions of the form $\alpha\pi_1 + (1 - \alpha)\pi_2$, with $0 < \alpha < 1$?

Problem 7.7   Prove that the Bayes rule under absolute error loss is the posterior median. More precisely, suppose that $E[|\theta|] < \infty$. Show that a number a^* satisfies

$$E[|\theta - a^*|] = \inf_{-\infty < a < \infty} E[|\theta - a|]$$

if and only if a^* is a median of the distribution of θ.

Problem 7.8   Let θ be a random variable with distribution $\pi(\theta)$ that is symmetric with respect to θ_0. Formally, $\pi(\theta + \theta_0) = \pi(\theta_0 - \theta)$ for all $\theta \in R$. Suppose that L is a nonnegative twice differentiable convex loss function on the real line that is symmetric around the value of 0. Also suppose that, for all values of a,

$$\mathcal{L}(a) = \int_{-\infty}^{\infty} L(\theta - a)\pi(\theta)d\theta < \infty.$$

Prove that:

1. $\mathcal{L}$ is convex;

2. $\mathcal{L}$ is symmetric with respect to θ_0;

3. $\mathcal{L}$ is minimized at $a = \theta_0$.

Problem 7.9   Derive an analogous equation to (7.17), using the same assumptions of Section 7.5.1, except now the sampling distribution is $f(x|\theta, \varphi)$, and the prior is $\pi(\theta, \varphi)$.

Problem 7.10   Consider a point estimation problem in which you observe $x_1, \ldots, x_n$ as i.i.d. random variables from the Poisson distribution

$$f(x|\theta) = \frac{1}{x!}\theta^x e^{-\theta}.$$

Assume a squared error of estimation loss $L(\theta, a) = (a - \theta)^2$, and assume a prior distribution on θ given by the gamma density

$$\pi(\theta) = \frac{1}{\Gamma(\alpha_0)\beta_0^{\alpha_0}}\theta^{\alpha_0 - 1}e^{-\theta/\beta_0}.$$

1. Show that the Bayes decision rule with respect to the prior above is of the form

$$\delta^*(x_1, \ldots, x_n) = a + b\bar{x}$$

where $a > 0$, $b \in (0, 1)$, and $\bar{x} = \sum_i x_{i/n}$. You may use the fact that the distribution of $\sum_i x_i$ is Poisson with parameter $n\theta$ without proof.

2. Compute and graph the risk functions of $\delta^*(x_1, \ldots, x_n)$ and that of the MLE $\delta(x_1, \ldots, x_n) = \bar{x}$.

3. Compute the Bayes risk of $\delta^*(x_1, \ldots, x_n)$ and show that it is (a) decreasing in n and (b) it goes to 0 as n gets large.

4. Suppose an investigator wants to collect a sample that is large enough that the Bayes risk after the experiment is half of the Bayes risk before the experiment. Find that sample size.

Problem 7.11 Take x to be binomial with unknown θ and $n = 2$, and consider testing $\theta_0 = 1/3$ versus $\theta_1 = 1/2$. Draw the points corresponding to every possible decision rule in the risk space with coordinates $R(\theta_0, \delta)$ and $R(\theta_1, \delta)$. Identify minimax randomized and nonrandomized rules and the expected utility rule for prior $\pi(\theta_0) = 1/4$. Identify the set of rules that are not dominated by any other rule (those are called admissible in Chapter 8).

Problem 7.12

An engineer draws a random sample of electron tubes and measures the plate voltage under certain conditions with a very accurate voltmeter, accurate enough so that measurement error is negligible compared with the variability of the tubes. A statistician examines the measurements, which look normally distributed and vary from 75 to 99 volts with a mean of 87 and a standard deviation of 4. He makes the ordinary normal analysis, giving a confidence interval for the true mean. Later he visits the engineer's laboratory, and notices that the voltmeter used reads only as far as 100, so the population appears to be "censored." This necessitates a new analysis, if the statistician is orthodox. However, the engineer says he has another meter, equally accurate and reading to 1000 volts, which he would have used if any voltage had been over 100. This is a relief to the orthodox statistician, because it means the population was effectively uncensored after all. But the next day the engineer telephones and says: "I just discovered my high-range voltmeter was not working the day I did the experiment you analyzed for me." The statistician ascertains that the engineer would not have held up the experiment until the meter was fixed, and informs him that a new analysis will be required. The engineer is astounded. He says: "But the experiment turned out just the same as if the high-range meter had been working. I obtained the precise voltages of my sample anyway, so I learned exactly what I would have learned if the high-range meter had been available. Next you'll be asking me about my oscilloscope." (From Pratt's comments to Birnbaum 1962, pp. 314–315)

State a probabilistic model for the situation described by Pratt, and specify a prior distribution and a loss function for the point estimation of the mean voltage. Writing code if necessary, compute the risk function R of the Bayes rule and that of your favorite frequentist rule in two scenarios: when the high-range voltmeter is available and when it is not. Does examining the risk function help you select a decision rule once the data are observed?

8

Admissibility

In this chapter we will explore further the concept of admissibility. Suppose, following Wald, we agree to look at long-run average loss R as the criterion of interest for choosing decision rules. R depends on θ, but a basic requirement is that one should not prefer a rule that does worse than another no matter what the true θ is. This is a very weak requirement, and a key rationality principle for frequentist decision theory. There is a similarity between admissibility and the strict coherence condition presented in Section 2.1.2. In de Finetti's terminology, a decision maker trading a decision rule for another that has higher risk everywhere could be described as a sure loser—except for the fact that the risk difference could be zero in some "lucky" cases.

Admissible rules are those that cannot be dominated; that is, beaten at the risk game no matter what the truth is. Admissibility is a far more basic rationality principle than minimax in the sense that it does not require adhering to the "ultra-pessimistic" perspective on θ. The group of people willing to take admissibility seriously is in fact much larger than those equating minimax to frequentist rationality. To many, therefore, characterization of sets of admissible decision rules has been a key component of the contribution of decision theory to statistics.

It turns out that just by requiring admissibility one is drawn again towards an expected utility perspective. We will discover that to generate an admissible rule, all one needs is to find a Bayes rule. The essence of the story is that a decision rule cannot beat another everywhere and come behind on average. This needs to be made more rigorous, but, for example, it is enough to require that the Bayes rule be unique for it to work out. A related condition is to rule out "know-it-all" priors that are not ready to change in the light of data. This general connection between admissibility and Bayes means there is no way to rule out Bayes rules from the standpoint of admissibility, even though admissibility is based on long-run performance over repeated experiments.

Even more interesting from the standpoint of foundations is that, in certain fairly big classes of decision problems, all admissible rules are Bayes rules or limiting cases of the Bayes rule. Not only can the chief frequentist rationality requirement not rule any Bayesian out, but also one has to be Bayesian (or close) to satisfy it in the first place! Another way to think about this is that no matter what admissible procedure one may concoct, somewhere behind the scenes, there is a prior (or a limit of priors) for which that is the expected utility solution. A Bayesian perspective brings that into the open and contributes to a more forthright scientific discourse.

Featured articles:

Neyman, J. & Pearson, E. S. (1933). On the problem of the most efficient test of statistical hypotheses, *Philosophical Transaction of the Royal Society (Series A)* **231**: 286–337.

This feature choice is a bit of a stretch in the sense that we do not dwell on the Neyman–Pearson theory at all. However, we discover at the end of the chapter that the famed Neyman–Pearson lemma can be reinterpreted as a complete class theorem in the light of all the theory we developed so far. The Neyman–Pearson theory was the spark for frequentist decision theory, and this is a good place to appreciate its impact on the broader theory.

There is a vast literature on the topics covered in this chapter, which we only briefly survey. More extensive accounts and references can be found in Berger (1985), Robert (1994), and Schervish (1995).

8.1   Admissibility and completeness

If we take the risk function as the basis for comparison of two decision rules δ and δ', the following definition is natural:

Definition 8.1 (R-better)   *A decision rule δ is called R-*better *than another decision rule δ' if*

$$R(\theta, \delta) \leq R(\theta, \delta') \qquad \forall \, \theta \in \Theta \qquad (8.1)$$

and $R(\theta, \delta) < R(\theta, \delta')$ for some θ. We also say that δ dominates δ'.

Figure 8.1 shows an example. If two rules have the same risk function they are called R-equivalent.

If we are building a toolbox of rules for a specific decision problem, and δ is R-better than δ', we do not need to include δ' in our toolbox. Instead we should focus on rules that cannot be "beaten" at the R-better game. We make this formal with some additional definitions.

Definition 8.2 (Admissibility)   *A decision rule δ is admissible if there is no R-better rule.*

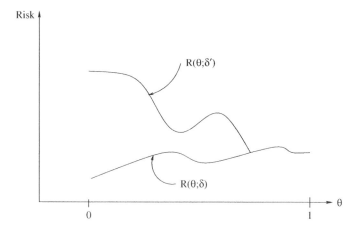

Figure 8.1   Risk functions for decision rules δ and δ'. Here δ dominates δ': the two risk functions are the same for θ close to 1 but the risk of δ is lower near 0 and never greater.

Referring back to Figure 7.2 from Chapter 7, a_5 is R-better than a_6, which is therefore not admissible. On the other hand, a_4 is admissible as long as only nonrandomized actions are allowed, even though it could never be chosen as the optimum under either principle. When randomized rules are allowed, it becomes possible to decrease losses from y_4 in both dimensions, and a_4 becomes inadmissible. It is true that Figure 7.2 talks about losses and not risk, but loss is a special case of risk when there are no data available.

In general, admissibility will eliminate rules but will not determine a unique winner. For example, in Figure 7.2, there are five admissible rules among nonrandomized ones, and infinitely many admissible randomized rules. Nonetheless admissibility has two major strengths that place it in a strategic position in decision theory: it can be used to rule out decision rules, or entire approaches, as inconsistent with rational behavior; and it can be used to characterize broad families of rules that are sure to include all possible admissible rules. This is implemented via the notion of completeness, a property of classes of decision rules that ensures that we are not leaving out anything that could turn out to be useful.

Definition 8.3 (Completeness)   *We give three variants:*

1. *A class $\mathcal{D}$ of decision rules is complete if, for any decision rule $\delta \notin \mathcal{D}$, there is a decision rule $\delta' \in \mathcal{D}$ which is R-better than δ.*

2. *A class $\mathcal{D}$ of decision rules is essentially complete if, for any decision rule $\delta \notin \mathcal{D}$, there is a decision rule $\delta' \in \mathcal{D}$ such that $R(\theta, \delta') \leq R(\theta, \delta)$ for all θ; that is, δ' is R-better or R-equivalent to δ.*

3. *A class $\mathcal{D}$ of decision rules is said to be minimal (essentially) complete if $\mathcal{D}$ is complete and if no proper subset of $\mathcal{D}$ is (essentially) complete.*

From this definition it follows that any decision rule δ that is outside of a complete class is inadmissible. If the class is essentially complete, then a decision rule δ that is outside of the class may be admissible, but one can find a decision δ' in the class with the same risk function. Thus, it seems reasonable to restrict our attention to a complete or essentially complete class.

Complete classes contain all admissible decision rules, but they may contain inadmissible rules as well. A stronger connection exists with minimal complete classes: these classes are such that, if even one rule is taken out, they are no longer complete.

Theorem 8.1   *If a minimal complete class exists, then it coincides with the class of all admissible decision rules.*

We left the proof for you (Problem 8.3).

Complete class results are practical in the sense that if one is interested in admissible strategies, one can safely restrict attention to complete or essentially complete classes and study those further. The implications for foundations are also far reaching. Complete class results are useful for characterizing statistical approaches at a higher level of abstraction than single decision rules. For example, studying the class of all tests based on a threshold of the likelihood ratio, one can investigate the whole Neyman–Pearson theory from the point of view of rationality. For another example, we can explore whether it is necessary to use randomized decision rules:

Theorem 8.2   *If the loss function $L(\theta, a)$ is convex in a, then the class of non-randomized decision rules $\mathcal{D}$ is complete.*

We left this proof for you as well (Problem 8.8).

8.2 Admissibility and minimax

Unfortunately for minimax aficionados, this section is very short. In general, to prove admissibility of minimax rules it is best to hope they are close enough to Bayes or limiting Bayes and use the theory of the next section. This will not always be the case: a famously inadmissible minimax rule is the subject of Chapter 9.

On the flip side, a very nice result is available for proving minimaxity once you already know a rule is admissible.

Theorem 8.3   *If δ^M has constant risk and it is admissible, then δ^M is the unique minimax rule.*

Problem 8.11 asks you to prove this and is a useful one to quickly work out before reading the rest of the chapter.

For an example, say x is a $Bin(n, \theta)$ and

$$L(\theta, a) = \frac{(\theta - a)^2}{\theta(1 - \theta)}.$$

Consider the rule $\delta^*(x) = x/n$. The risk function is

$$R(\theta, \delta^*) = E_{x|\theta}\left[\frac{(\theta - x/n)^2}{\theta(1 - \theta)}\right] = \frac{1}{n},$$

that is δ^* has constant risk. With a little work you can also show that δ^* is the unique Bayes rule with respect to a uniform prior on $(0, 1)$. Lastly, using the machinery of the next section we will be able to prove that δ^* is admissible. Then, using Theorem 8.3, we can conclude that it is minimax.

8.3   Admissibility and Bayes

8.3.1   Proper Bayes rules

Let us now consider more precisely the conditions for the Bayes rules to be admissible. We start with the easiest case, when Θ is discrete. In this case, a sufficient condition is that the prior should not completely rule out any of the possible values of θ.

Theorem 8.4   *If Θ is discrete, and the prior π gives positive probability to each element of Θ, then the Bayes rule with respect to π is admissible.*

Proof: Let δ^* be the Bayes rule with respect to π, and suppose, for a contradiction, that δ^* is not admissible; that is, there is another rule, say δ, that is R-better than δ^*. Then $R(\theta, \delta) \leq R(\theta, \delta^*)$ with strict inequality $R(\theta_0, \delta) < R(\theta_0, \delta^*)$ for some θ_0 in Θ with positive mass $\pi(\theta_0) > 0$. Then,

$$r(\pi, \delta) = \sum_{\theta \in \Theta} R(\theta, \delta)\pi(\theta) < \sum_{\theta \in \Theta} R(\theta, \delta^*)\pi(\theta) = r(\pi, \delta^*).$$

This contradicts the fact that δ^* is Bayes. Therefore, δ^* must be admissible.   □

Incidentally, from a subjectivist standpoint, it is a little bit tricky to accept the fact that there may even be points in a discrete Θ that do not have a positive mass: why should they be in Θ in the first place if they do not? This question opens a well-populated can of worms about whether we can define Θ independently of the prior, but we will close it quickly and pretend, for example, that we have a reasonably objective way to define Θ or that we can take Θ to be the union of the supports over a group of reasonable decision makers.

An analogous result to Theorem 8.4 holds for continuous θ, but you have to start being careful with your real analysis. An easy case is when π gives positive mass to all open sets and the risk function is continuous. See Problem 8.10. The plot of the proof is very similar to that used in the discrete parameter case, with the difference that, to reach a contradiction, we need to use the continuity condition to create a lump of dominating cases with a strictly positive prior probability.

Another popular condition for guaranteeing admissibility of Bayes rules is uniqueness.

Theorem 8.5   *Any unique Bayes estimator is admissible.*

The proof of this theorem is left for you as Problem 8.4. To see the need for uniqueness, look at Problem 8.9.

Uniqueness is by no means a necessary condition: the important point is that when there is more than one Bayes rule, all Bayes rules share the same risk function. This happens for example if two Bayes estimators differ only on a set S such that $P_\theta(S) = 0$, for all θ.

Theorem 8.6   *Suppose that every Bayes rule with respect to a prior distribution, π, has the same risk function. Then all these rules are admissible.*

Proof: Let δ^* denote a Bayes rule with respect to π and $R(\theta, \delta^*)$ denote the risk of any such Bayes rule. Suppose that there exists δ_0 such that $R(\theta, \delta_0) \leq R(\theta, \delta)$ for all θ and with strict inequality for some θ. This implies that

$$\int_\Theta R(\theta, \delta_0)\pi(\theta)d\theta \leq \int_\Theta R(\theta, \delta^*)\pi(\theta)d\theta.$$

Because δ^* is a Bayes rule with respect to π, the inequality must be an equality and δ_0 is also a Bayes rule with risk $R(\theta, \delta^*)$, which leads to a contradiction.   □

8.3.2   Generalized Bayes rules

Priors do not sit well with frequentists, but being able to almost infallibly generate admissible rules does. A compromise that sometimes works is to be lenient on the requirement that the prior should integrate to one. We have seen an example in Section 7.7. A prior that integrates to something other than one, including perhaps infinity, is called an improper prior. Sometimes you can plug an improper prior into Bayes' rule and get a proper posterior-like distribution, which you can use to figure out the action that minimizes the posterior expected loss. What you get out of this procedure is called a *generalized Bayes rule*:

Definition 8.4   *If π is an improper prior, and δ^* is an action which minimizes the posterior expected loss $E_{\theta|x}[L(\theta, \delta(x))]$, for each x with positive predictive density $m(x)$, then δ^* is a* generalized Bayes rule.

For example, if you are estimating a normal mean θ, with known variance, and you assume a prior $\pi(\theta) = 1$, you can work out the generalized Bayes rule to be $\delta^*(x) = \bar{x}$. This is also the maximum likelihood estimator, and the minimum variance unbiased estimator. The idea of choosing priors that are vague enough that the Bayes solution agrees with common statistical practice is a convenient and practical compromise in some cases, but a really tricky business in others (Bernardo and Smith 1994, Robert 1994). Getting into details would be too much of a digression for us now. What is more relevant for our discussion is that a lot of generalized Bayes rules are admissible.

Theorem 8.7   *If Θ is a subset of $\mathfrak{R}^k$ such that every neighborhood of every point in Θ intersects the interior of Θ, $R(\theta, \delta)$ is continuous in θ for all δ, the Lebesgue measure on Θ is absolutely continuous with respect to π, δ^* is a generalized Bayes rule with respect to π, and $L(\theta, \delta^*(x))f(x|\theta)$ is $\nu \times \pi$ integrable, then δ^* is admissible.*

See Schervish (1995) for the proof. A key piece is the integrability of L times f, which gives a finite Bayes risk r. The other major condition is continuity. Nonetheless, this is a very general result.

Example 8.1   (Schervish 1995, p. 158) Suppose x is a $Bin(n, \theta)$ and $\theta \in \Theta = \mathcal{A} = [0, 1]$. The loss is $L(\theta, a) = (\theta - a)^2$ and the prior is $\pi(\theta) = 1/[\theta(1 - \theta)]$, which is improper. We will show that $\delta(x) = x/n$ is admissible using Theorem 8.7. There is an easier way which you can learn about in Problem 8.6.

The posterior distribution of θ, given x, is $Beta(x, n - x)$. Therefore, for $x = 1, 2, \ldots, n - 1$, the (generalized) Bayes rule is $\delta^*(x) = x/n$, that is the posterior mean. Now, if $x = 0$,

$$E_{\theta|x}[L(\theta, \delta(x))] = \int_0^1 (\theta - \delta(x))^2 (1 - \theta)^{n-1}\theta^{-1}d\theta,$$

which is finite only if $\delta^*(x) = 0$. Similarly, if $x = n$,

$$E_{\theta|x}[L(\theta, \delta(x))] = \int_0^1 (\theta - \delta(x))^2 \theta^{n-1}(1 - \theta)^{-1}d\theta,$$

which is finite only if $\delta^*(x) = 1$. Therefore, $\delta^*(x) = x/n$ is the generalized Bayes rule with respect to π. Furthermore,

$$L(\theta, \delta^*(x))f(x|\theta) = \left(\theta - \frac{x}{n}\right)^2 \binom{n}{x}\theta^x(1 - \theta)^{n-x}$$

has prior expectation $1/n$. Therefore, δ^* is admissible. If you are interested in a general characterization of admissible estimators for this problem, read Johnson (1971).   ★

The following theorem from Blyth (1951) is one of the earliest characterizations of admissibility. It provides a sufficient admissibility condition by relating admissibility of an estimator to the existence of a sequence of priors using which we obtain Bayes rules whose risk approximates the risk of the estimator in question. This requires continuity conditions that are a bit strong: in particular the decision rules with continuous risk functions must form a complete class. We will return to this topic in Section 8.4.1.

Theorem 8.8   *Consider a decision problem in which Θ has positive Lebesgue measure, and in which the decision rules with continuous risk functions form a complete* *class. Then an estimator δ_0 (with a continuous risk function) is admissible if there exists a sequence $\{\pi_n\}_{n=1}^{\infty}$ of (generalized) priors such that:*

1. *$r(\pi_n, \delta_0)$ and $r(\pi_n, \delta_n^*)$ are finite for all n, where δ_n^* is the Bayes rule with respect to π_n;*

2. *for any nondegenerate convex set $C \subset \Theta$, there exists a positive constant K and an integer N such that, for $n \geq N$,*

$$\int_C \pi_n(\theta)d\theta \geq K;$$

3. $\lim_{n \to \infty}[r(\pi_n, \delta_0) - r(\pi_n, \delta_n^*)] = 0.$

Proof: Suppose for a contradiction that δ_0 is not admissible. Then, there must exist a decision rule δ' that dominates δ_0, and therefore a θ_0 such that $R(\theta_0, \delta') < R(\theta_0, \delta_0)$. Because the rules with continuous risk functions form a complete class, both δ_0 and δ' have a continuous risk function. Thus, there exists $\epsilon_1 > 0, \epsilon_2 > 0$ such that

$$R(\theta, \delta_0) - R(\theta, \delta') > \epsilon_1,$$

for θ in an open set $C = \{\theta \in \Theta : |\theta - \theta_0| < \epsilon_2\}$.

From condition *1*, $r(\pi_n, \delta_n^*) \leq r(\pi_n, \delta')$, because δ_n^* is a Bayes rule with respect to π_n. Thus, for all $n \geq N$,

$$\begin{aligned}
r(\pi_n, \delta_0) - r(\pi_n, \delta_n^*) &\geq r(\pi_n, \delta_0) - r(\pi_n, \delta') \\
&= E_\theta[R(\theta, \delta_0) - R(\theta, \delta')] \\
&\geq \int_C [R(\theta, \delta_0) - R(\theta, \delta')]\pi_n(\theta)d\theta \\
&\geq \epsilon_1 \int_C \pi_n(\theta)d\theta \\
&\geq \epsilon_1 K,
\end{aligned}$$

from condition 2. That $r(\pi_n, \delta_0) - r(\pi_n, \delta_n) \geq \epsilon_1 K$ is in contradiction with condition 3. Thus we conclude that δ_0 must be admissible.   □

Example 8.2   To illustrate the application of this theorem, we will consider a simplified version of Blyth's own example on the admissibility of the sample average for estimating a normal mean parameter. See also Berger (1985, p. 548). Suppose that $x \sim N(\theta, 1)$ and that it is desired to estimate θ under loss $L(\theta, a) = (\theta - a)^2$. We will show that $\delta_0(x) = x$ is admissible. A convenient choice for π_n is the unnormalized normal density

$$\pi_n(\theta) = (2\pi)^{-1/2} \exp(-0.5\theta^2/n).$$

The posterior distribution of θ, given x, is $N(xn/(n + 1), n/(n + 1))$. Therefore, with respect to π_n, the generalized Bayes decision rule is $\delta_n^*(x) = xn/(n + 1)$, that is the posterior mean.

For condition *1*, observe that

$$R(\theta, \delta_0) = \int_{\mathcal{X}} L(\theta, \delta_0(x)) f(x|\theta) dx$$

$$= E[(\theta - x)^2 | \theta] = \text{Var}[x|\theta] = 1.$$

Therefore,

$$r(\pi_n, \delta_0) = \int_{\Theta} R(\theta, \delta_0) \pi_n(\theta) d\theta = \int_{\Theta} \pi_n(\theta) d\theta = \sqrt{n}.$$

Here $\sqrt{n}$ is the normalizing constant of the prior. Similarly,

$$R(\theta, \delta_n^*) = E\left[\left(\theta - \frac{xn}{n+1}\right)^2\right]$$

$$= \frac{\theta^2}{(n+1)^2} + \frac{n^2}{(n+1)^2}.$$

Thus, we obtain that

$$r(\pi_n, \delta_n^*) = \int_{\Theta} R(\theta, \delta_n^*) \pi_n(\theta) d\theta = \frac{\sqrt{n}n}{1+n}.$$

For condition *2*,

$$\int_C \pi_n(\theta) d\theta \geq \int_C \pi_1(\theta) d\theta = K > 0$$

as long as C is a nondegenerate convex subset of Θ. Note that when $C = \Re$, $K = 1$. The sequence of proper priors $N(0, 1/n)$ would not satisfy this condition.

Finally, for condition *3* we have

$$\lim_{n \to \infty} [r(\pi_n, \delta_0) - r(\pi_n, \delta_n^*)] = \lim_{n \to \infty} \left[\sqrt{n}\left(1 - \frac{n}{n+1}\right)\right]$$

$$= \lim_{n \to \infty} \left[\frac{\sqrt{n}}{n+1}\right] = 0.$$

Therefore, $\delta_0(x) = x$ is admissible. ★

It can be shown that $\delta_0(x) = x$ is admissible when x is bivariate, but this requires a more complex sequence or priors. We will see in the next chapter that the natural generalization of this result to three or more dimensions does not work. In other words, the sample average of samples from a multivariate normal distribution is not admissible.

A necessary and sufficient admissibility condition based on limits of improper priors was established later by Stein (1955).

Theorem 8.9   *Assume that $f(x|\theta)$ is continuous in θ and strictly positive over Θ. Moreover, assume that the loss function $L(\theta, a)$ is strictly convex, continuous, and such that for a compact subset $E \subset \Theta$,*

$$\lim_{||\delta|| \to +\infty} \inf_{\theta \in E} L(\theta, \delta) = +\infty.$$

An estimator δ_0 is admissible if and only if there exists a sequence $\{F_n\}$ of increasing compact sets such that $\Theta = \cup_n F_n$, a sequence $\{\pi_n\}$ of finite measures with support F_n, and a sequence $\{\delta_n^*\}$ of Bayes estimators associated with π_n such that:

1. *there exists a compact set $E_0 \subset \Theta$ such that $\inf_n \pi_n(E_0) \geq 1$;*

2. *if $E \subset \Theta$ is compact, $\sup_n \pi_n(E) < +\infty$;*

3. *$\lim_n r(\pi_n, \delta_0) - r(\pi_n, \delta_n^*) = 0$;*

4. *$\lim_n R(\theta, \delta_n^*) = R(\theta, \delta_0)$.*

Stein's result is stated in terms of continuous losses, while earlier ones we looked at were based on continuous risk. The next lemma looks at conditions for the continuity of loss functions to imply the continuity of the risk.

Lemma 3   *Suppose that Θ is a subset of $\Re^m$ and that the loss function $L(\theta, a)$ is bounded and continuous as a function of θ, for all $a \in \mathcal{A}$. If $f(x|\theta)$ is continuous in θ, for every x, the risk function of every decision rule δ is continuous.*

The proof of this lemma is in Robert (1994, p. 239).

8.4   Complete classes

8.4.1   Completeness and Bayes

The results in this section relate the concepts of admissibility and completeness to Bayes decision rules and investigate conditions for Bayes rules to span the set of all admissible decision rules and thus form a minimal complete class.

Theorem 8.10   *Suppose that Θ is finite, the loss function is bounded below, and the risk set is closed from below. Then the set of all Bayes rules is a complete class, and the set of admissible Bayes rules is a minimal complete class.*

For a proof see Schervish (1995, pp. 179–180). The conditions of this result mimic the typical game-theoretic setting (Berger 1985, chapter 5) and are well illustrated by the risk set discussion of Section 7.5.1. The Bayes rules are also rules whose risk functions are on the lower boundary of the risk set.

Generalizations of this result to parameter sets that are general subsets of $\Re^m$ require more work. A key role is played again by continuity of the risk function.

Theorem 8.11   *Suppose that $\mathcal{A}$ and Θ are closed and bounded subsets of the Euclidean space. Assume that the loss function $L(\theta, a)$ is a continuous function of a for each $\theta \in \Theta$, and that all decision rules have continuous risk functions. Then the Bayes rules form a complete class.*

The proof is in Berger (1985, p. 546).

Continuity of the risk may seem like a restrictive assumption, but it is commonly met and in fact decision rules with continuous risk functions often form a complete class. An example is given in the next theorem. An important class of statistical problems is those with the *monotone likelihood ratio* property, which requires that for every $\theta_1 < \theta_2$, the likelihood ratio

$$\frac{f(x|\theta_2)}{f(x|\theta_1)}$$

is a nondecreasing function of x on the set for which at least one of the densities is nonzero. For problems with monotone likelihood ratio, and continuous loss, we only need to worry about decision rules with finite and continuous risk:

Theorem 8.12   *Consider a statistical decision problem where $\mathcal{X}$, Θ, and $\mathcal{A}$ are subsets of $\Re$ with $\mathcal{A}$ being closed. Assume that $f(x|\theta)$ is continuous in θ and it has the monotone likelihood ratio property. If:*

1. *$L(\theta, a)$ is a continuous function of θ for every $a \in \mathcal{A}$;*

2. *L is nonincreasing in a for $a \leq \theta$ and nondecreasing for $a \geq \theta$; and*

3. *there exist two functions K_1 and K_2 bounded on the compact subsets of $\Theta \times \Theta$ such that*

$$L(\theta_1, a) \leq K_1(\theta_1, \theta_2)L(\theta_2, a) + K_2(\theta_1, \theta_2), \quad \forall \, a \in \mathcal{A},$$

then decision rules with finite and continuous risk functions form a complete class.

For proofs of this theorem see Ferguson (1967) and Brown (1976).

8.4.2   Sufficiency and the Rao–Blackwell inequality

Sufficiency is one of the most important concepts in statistics. A sufficient statistic for θ is a function of the data that summarizes them in such a way that no information concerning θ is lost. Should decision rules only depend on the data via sufficient statistics? We have hinted at the fact that Bayes rules automatically do that, in Section 7.2.3. We examine this issue in more detail in this section. This is a good time to lay out a definition of sufficient statistic.

Definition 8.5 (Sufficient statistic)   *Let $x \sim f(x|\theta)$. A function t of x is said to be a* sufficient statistic *for θ if the conditional distribution of x, given $t(x) = t$, is independent of θ.*

The Neyman factorization theorem shows that t is sufficient whenever the likelihood function can be decomposed as

$$f(x|\theta) = h(x)g(t(x)|\theta). \tag{8.2}$$

See, for example, Schervish (1995), for a discussion of sufficiency and this factorization.

Bahadur (1955) proved a very general result relating sufficiency and rationality. Specifically, he showed under very general conditions that the class of decision functions which depend on the observed sample only through a function t is essentially complete if and only if t is a sufficient statistic. Restating it more concisely:

Theorem 8.13   *The class of all decision rules based on a sufficient statistic is essentially complete.*

The proof in Bahadur (1955) is hard measure-theoretic work, as it makes use of the very abstract definition of sufficiency proposed by Halmos and Savage (1949).

An older, less general result provides a constructive way of showing how to find a rule that is R-better or R-equivalent than a rule that depends on the data via statistics that are not sufficient.

Theorem 8.14 (Rao–Blackwell theorem)   *Suppose that $\mathcal{A}$ is a convex subset of $\Re^m$ and that $L(\theta, a)$ is a convex function of a for all $\theta \in \Theta$. Suppose also that t is a sufficient statistic for θ, and that δ_0 is a nonrandomized decision rule such that $E_{x|\theta}[|\delta_0(x)|] < \infty$. The decision rule defined as*

$$\delta_1(t) = E_{x|t}[\delta_0(x)] \tag{8.3}$$

is R-equivalent to or R-better than δ_0.

Proof: Jensen's inequality states that if a function g is convex, then $g(E[x]) \leq E[g(x)]$. Therefore,

$$L(\theta, \delta_1(t)) = L(\theta, E_{x|t}[\delta_0(x)]) \leq E_{x|t}[L(\theta, \delta_0(x))] \tag{8.4}$$

and

$$\begin{aligned}
R(\theta, \delta_1) &= E_{t|\theta}[L(\theta, \delta_1(t))] \\
&\leq E_{t|\theta}[E_{x|t}[L(\theta, \delta_0(x))]] \\
&= E_{x|\theta}[L(\theta, \delta_0(x))] \\
&= R(\theta, \delta_0)
\end{aligned}$$

completing the proof. $\square$

This result does not say that the decision rule δ_1 is any good, but only that it is no worse than the δ_0 we started with.

Example 8.3   (Schervish 1995, p. 153). Suppose that $x_1, \ldots, x_n$ are independent and identically distributed as $N(\theta, 1)$ and that we wish to estimate the tail area to the left of $c - \theta$ with squared error loss

$$L(\theta, a) = (a - \Phi(c - \theta))^2.$$

Here c is some fixed real number, and $a \in \mathcal{A} \equiv [0, 1]$. A possible decision rule is the empirical tail frequency

$$\delta_0(x) = \sum_{i=1}^{n} I_{(-\infty, c]}(x_i)/n.$$

It can be shown that $t = \bar{x}$ is a sufficient statistic for θ, for example using equation (8.2). Since $\mathcal{A}$ is convex and the loss function is a convex function of a,

$$E_{x|t}[\delta_0(x)] = \frac{1}{n} \sum_{i=1}^{n} E_{x_i|t}[I_{(-\infty, c]}(x_i)]$$

$$= \Phi\left(\frac{c - t}{\sqrt{\frac{n-1}{n}}} \right),$$

because $x_1|t$ is $N(t, [n-1]/n)$.

Because of the Rao–Blackwell theorem, the rule $\delta_1(t) = E_{x|t}[\delta_0(x)]$ is R-better than δ_0. In this case, the empirical frequency does not make any use of the functional form of the likelihood, which is known. Using this functional form we can bring all the data to bear in estimating the tail probability and come up with an estimate with lower risk everywhere. Clearly this is predicated on knowing for sure that the data are normally distributed. Using the entire set of data to estimate the left tail would not be as effective if we did not know the parametric form of the sampling distribution.   ★

8.4.3   The Neyman–Pearson lemma

We now revisit the Neyman–Pearson lemma (Neyman and Pearson 1933) from the point of view of complete class theorems. See Berger (1985) or Schervish (1995) for proofs of the theorem.

Theorem 8.15 (Neyman–Pearson lemma)   *Consider a simple versus simple hypothesis testing problem with null hypothesis $H_0 : \theta = \theta_0$ and alternative hypothesis $H_1 : \theta = \theta_1$. The action space is $\mathcal{A} = \{a_0, a_1\}$ where a_i denotes accepting hypothesis H_i $(i = 0, 1)$. Assume that the loss function is the $0 - K_L$ loss, that is a correct decision costs zero, while incorrectly deciding a_i costs K_L.*

Tests of the form

$$\delta(x) = \begin{cases} 1 & \text{if } f(x|\theta_1) > Kf(x|\theta_0) \\ \gamma(x) & \text{if } f(x|\theta_1) = Kf(x|\theta_0) \\ 0 & \text{if } f(x|\theta_1) < Kf(x|\theta_0) \end{cases} \tag{8.5}$$

where $0 \le \gamma(x) \le 1$ if $0 < K < \infty$ and $\gamma(x) = 0$ if $K = 0$, together with the test

$$\delta(x) = \begin{cases} 1 & \text{if } f(x|\theta_0) = 0 \\ 0 & \text{if } f(x|\theta_0) > 0 \end{cases} \tag{8.6}$$

(corresponding to $K = \infty$ above), form a minimal complete class of decision rules. The subclass of such tests with $\gamma(x) \equiv \gamma$ (a constant) is an essentially complete class.

For any $0 \le \alpha \le \alpha^*$, there exists a test δ of the form (8.5) or (8.6) with $\alpha_0(\delta) = \alpha$ and any such test is a most powerful test of size α (that is, among all tests δ with $\alpha_0(\delta) \le \alpha$ such a test minimizes $\alpha_1(\delta)$).

With this result we have come in a complete circle: the Neyman–Pearson theory was the seed that started Wald's statistical decision theory; minimal completeness is the ultimate rationality endorsement for a statistical approach within that theory—all and only the rules generated by the approach are worth considering. The Neyman–Pearson tests are a minimal complete class. Also, for each of these tests we can find a prior for which that test is the formal Bayes rule. What is left to argue about?

Before you leave the theater with the impression that a boring happy ending is on its way, it is time to start looking at some of the results that have been, and are still, generating controversy. We start with a simple one in the next section, and then devote the entire next chapter to a more complicated one.

8.5 Using the same α level across studies with different sample sizes is inadmissible

An example of how the principle of admissibility can be a guide in evaluating the rigor of common statistical procedures arises in hypothesis testing. It is common practice to use a type I error probability α, say 0.05, across a variety of applications and studies. For example, most users of statistical methodologies are prepared to use $\alpha = 0.05$ irrespective of the sample size of the study. In this section we work out a simple example that shows that using the same α in two studies with different sample sizes—all other study characteristics being the same—results in an inadmissible decision rule.

Suppose we are interested in studying two different drugs, A and B. Efficacy is measured by parameters θ_A and θ_B. To seek approval, we wish to test both H_{0A}: $\theta_A = 0$ versus H_{1A}: $\theta_A = \theta_1$ and H_{0B}: $\theta_B = 0$ versus H_{1B}: $\theta_B = \theta_1$ based on observations sampled from populations $f(x|\theta_A)$ and $f(x|\theta_B)$ that differ only in the value of the parameters. Suppose also that the samples available from the two populations are of sizes $n_A < n_B$.

The action space is the set of four combinations of accepting or rejecting the two hypotheses. Suppose the loss structure for this decision problem is the sum of the two drug specific loss tables

	$\theta_A = 0$	$\theta_A = \theta_1$		$\theta_B = 0$	$\theta_B = \theta_1$
a_{0A}	0	L_0	a_{0B}	0	L_0
a_{1A}	L_1	0	a_{1B}	L_1	0

where L_0 and L_1 are the same for both drugs. This formulation is similar to the multiple hypothesis testing setting of Section 7.5.2. Adding the losses is natural if the consequences of the decisions apply to the same company, and also make sense from the point of view of a regulatory agency that controls the drug approval process.

The risk function for any decision function is defined over four possible combinations of values of θ_A and θ_B. Consider the space of decision functions $\delta = (\delta_A(x_A), \delta_B(x_B))$. These rules use the two studies separately in choosing an action. In the notation of Section 7.5.1, α_δ and β_δ are the probabilities of type I and II errors associated with decision rule δ. So the generic decision rule will have risk function

θ_A	θ_B	$R(\theta, \delta)$
0	0	$L_1\alpha_{\delta_A} + L_1\alpha_{\delta_B}$
0	1	$L_1\alpha_{\delta_A} + L_0\beta_{\delta_B}$
1	0	$L_0\beta_{\delta_A} + L_1\alpha_{\delta_B}$
1	1	$L_0\beta_{\delta_A} + L_0\beta_{\delta_B}$

Consider now the decision rule that specifies a fixed value of α, say 0.05, and selects the rejection region to minimize β subject to the constraint that α is no more than 0.05. Define $\beta(\alpha, n)$ to be the resulting type II error. These rules are admissible in the single-study setting of Section 7.5.1. Here, however, it turns out that using the same α in both $\delta_A(x_A)$ and $\delta_B(x_B)$ is often inadmissible. The reason is that the two studies have the same loss structure. As we have seen, by specifying α one implicitly specifies a trade-off between type I and type II error—which in decision-theoretic terms is represented by L_1/L_0. But this implicit relationship depends on the sample size. In this formulation we have an additional opportunity to trade-off errors with one drug for errors in the other to beat the fixed α strategy no matter what the true parameters are.

Taking the rule with fixed α as the starting point, one is often able to decrease α in the larger study, and increase it in the smaller study, to generate a dominating decision rule. A simple example to study is one where the decrease and the increase cancel each other out, in which case the overall type I error (the risk when both drugs are null) remains α. Specifically, call δ the fixed $\alpha = \alpha_0$ rule, and δ' the rule obtained by choosing $\alpha_0 + \epsilon$ in study A and $\alpha_0 - \epsilon$ in study B, and then choosing rejection regions by minimizing type II error within each study. The risk functions are

θ_A	θ_B	$R(\theta, \delta)$	$R(\theta, \delta')$
0	0	$2L_1\alpha_0$	$2L_1\alpha_0$
0	1	$L_1\alpha_0 + L_0\beta(\alpha_0, n_A)$	$L_1(\alpha_0 + \epsilon) + L_0\beta(\alpha_0 + \epsilon, n_A)$
1	0	$L_1\alpha_0 + L_0\beta(\alpha_0, n_B)$	$L_1(\alpha_0 - \epsilon) + L_0\beta(\alpha_0 - \epsilon, n_B)$
1	1	$L_0(\beta(\alpha_0, n_A) + \beta(\alpha_0, n_B))$	$L_0(\beta(\alpha_0 + \epsilon, n_A) + \beta(\alpha_0 - \epsilon, n_B))$

If the risk of δ' is strictly less than that of δ in the $(0, 1)$ and $(1, 0)$ states, then it turns out that it will also be less in the $(1, 1)$ state. So, rearranging, sufficient conditions for δ' to dominate δ are

$$\frac{1}{\epsilon}\left(\beta(\alpha_0, n_A) - \beta(\alpha_0 + \epsilon, n_A)\right) > \frac{L_1}{L_0}$$

$$\frac{1}{\epsilon}\left(\beta(\alpha_0, n_B) - \beta(\alpha_0 - \epsilon, n_B)\right) > -\frac{L_1}{L_0}.$$

These conditions are often met. For example, say the two populations are $N(\theta_A, \sigma^2)$ and $N(\theta_B, \sigma^2)$, where the variance σ^2, to keep things simple, is known. Then, within each study

$$\beta(\alpha, n) = \Phi\left(\Phi^{-1}(1 - \alpha) - \frac{\sqrt{n}}{\sigma}\theta_1\right).$$

When $\theta_1/\sigma = 1$, $n_A = 5$, and $n_B = 20$, choosing $\epsilon = 0.001$ gives

$$3.213 > \frac{L_1}{L_0}$$

$$-0.073 > -\frac{L_1}{L_0}$$

which is satisfied over a useful range of loss specifications.

Berry and Viele (2008) consider a related admissibility issue framed in the context of a random sample size, and show that choosing α ahead of the study, and then choosing the rejection region by minimizing type II error given n after the study is completed, leads to rules that are inadmissible. In this case the risk function must be computed by treating n as an unknown experimental outcome. The mathematics has similarities with the example shown here, except there is a single hypothesis, so only the $(0, 0)$ and $(1, 1)$ cases are relevant. In the single study case, the expected utility solution specifies, for a fixed prior and loss ratio, how the type I error probability should vary with n. This is discussed, for example, by Seidenfeld *et al.* (1990b), who show that for an expected utility maximizer, $d\alpha_n/dn$ must be constant with n. Berry and Viele (2008) show how varying α in this way can dominate the fixed α rule. See also Problem 8.5.

In Chapter 9, we will discuss borrowing strength across a battery of related problems, and we will encounter estimators that are admissible in single studies but not when the ensemble is considered and losses are added. The example we just discussed is similar in that a key is the additivity of losses, which allows us to trade off errors in one problem with errors in another. An important difference, however, is that here we use only data from study A to decide about drug A, and likewise for B, while in Chapter 9 will will also use the ensemble of the data in determining each estimate.

8.6   Exercises

Problem 8.1   (Schervish 1995) Let $\Theta = (0, \infty)$, $\mathcal{A} = [0, \infty)$, and loss function $L(\theta, a) = (\theta - a)^2$. Suppose also that $x \sim U(0, \theta)$, given θ, and that the prior for θ is the $U(0, c)$ distribution for $c > 0$. Find two formal Bayes rules one of which dominates the other. Try your hand at it before you look at the solution below. What is the connection between this example and the Likelihood Principle? If $x < c$, should it matter what the rule would have said in the case $x > c$? What is the connection with the discussion we had in the introduction about judging an answer by what it says versus judging it by how it was derived?

Solution
     After observing $x < c$, the posterior distribution is

$$\pi(\theta|x) = \frac{1}{\theta \log(c/x)} I_{(x,c)}(\theta)$$

while if $x \geq c$ the posterior distribution is defined arbitrarily. So the Bayes rules are of the form

$$\delta^*(x) = \begin{cases} \dfrac{(c-x)}{\log(c/x)} & \text{if } x < c \\ \text{arbitrary} & \text{if } x \geq c \end{cases}$$

which is the posterior mean, if $x < c$. Let $\delta_0^*(x)$ denote the Bayes rule which has $\delta^*(x) = c$ for $x \geq c$. Similarly, define the Bayes rule $\delta_1^*(x)$ which has $\delta^*(x) = x$ for $x \geq c$. The difference in risk functions is

$$R(\theta, \delta_0^*) - R(\theta, \delta_1^*) = \int_c^\theta [(\theta - c)^2 - (\theta - x)^2] \frac{1}{\theta} dx > 0.$$

Problem 8.2   Suppose T is a sufficient statistic for θ and let δ_0^R be any randomized decision rule in $\mathcal{D}$. Prove that, subject to measurability conditions, there exists a randomized decision rule δ_1^R, depending on T only, which is R-equivalent to δ_0^R.

Problem 8.3   Prove that if a minimal complete class $\mathcal{D}$ exists, then it is exactly the class of all admissible estimators.

Problem 8.4   Prove Theorem 8.5.

Problem 8.5   (Based on Cox 1958) Consider testing H_0: $\theta = 0$ based on a single observation x. The probabilistic mechanism generating x is this. Flip a coin: if head draw an observation from a $N(\theta, \sigma = 1)$; if tail from a $N(\theta, \sigma = 10)$. The outcome of the coin is known. For a concrete example, imagine randomly drawing one of two measuring devices, one more precise than the other.

Decision rule δ_c (a conditional test) is to fix $\alpha = 0.05$ and then reject H_0 if $x > 1.64$ when $\sigma = 1$ and $x > 16.4$ when $\sigma = 10$. An unconditional test is allowed to select the two rejection regions based on properties of both devices. Show that the uniformly most powerful test of level $\alpha = 0.05$ dominates δ_c.

There are at least two possible interpretations for this result. The first says that the decision-theoretic approach, as implemented by considering frequentist risk, is problematic in that it can lead to violating the Likelihood Principle. The second argues that the conditional test is not a rational one in the first place, because it effectively implies using a different loss function depending on the measuring device used. Which do you favor? Can they both be right?

Problem 8.6   Let x be the number of successes in n independent trials with probability of success $\theta \in (0, 1)$. Show that a rule is admissible for the squared error loss

$$L(\theta, a) = (a - \theta)^2$$

if and only if it is admissible for the "standardized" squared error loss

$$L_s(\theta, a) = \frac{(a - \theta)^2}{\theta(1 - \theta)}.$$

Is this property special to the binomial case? To the squared error loss? State and prove a general version of this result. The more general it is, the better, of course.

Problem 8.7   Take $x \sim N(\theta, 1)$, $\theta \sim N(0, 1)$, and

$$L(\theta, a) = (\theta - a)^2 e^{3\theta^2/4}.$$

Show that the formal Bayes rule is $\delta^*(x) = 2x$ and that it is inadmissible. The reason things get weird here is that the Bayes risk r is infinite. It will be easier to think of a rule that dominates $2x$ if you work on Problem 8.6 first.

Problem 8.8   Prove Theorem 8.2.

Problem 8.9   Let x be the number of successes in $n = 2$ independent trials with probability of success $\theta \in [0, 1]$. Consider estimating θ with squared error loss. Say the prior is a point mass at 0. Show that the rule

$$\delta^*(0) = 0, \quad \delta^*(1) = 1, \quad \delta^*(2) = 0$$

is a Bayes rule. Moreover, show that it is not the unique Bayes rule, and it is not admissible.

Problem 8.10   If π gives positive mass to all open sets and the risk function is continuous, then the Bayes rule with respect to this prior is admissible.

Problem 8.11   Prove that if δ^M has constant risk and it is admissible, then δ^M is the unique minimax rule.

9

Shrinkage

In this chapter we present a famous result due to Stein (1955) and further elaborated by James & Stein (1961). It concerns estimating several parameters as part of the same decision problem. Let $x = (x_1, \ldots, x_p)'$ be distributed according to a multivariate p-dimensional $N(\theta, I)$ distribution, and assume the multivariate quadratic loss function

$$L(\theta, a) = \sum_{i=1}^{p} (a_i - \theta_i)^2.$$

The estimator $\delta(x) = x$ is the maximum likelihood estimator of θ, it is unbiased, and it has the smallest risk among all unbiased estimators. This would make it the almost perfect candidate from a frequentist angle. It is also the formal Bayes rule if one wishes to specify an improper flat prior on θ. However, Stein showed that such an estimator is not admissible. Oops.

Robert describes the aftermath as follows:

> One of the major impacts of the Stein paradox is to signify the end of a "Golden Age" for classical statistics, since it shows that the quest for *the* best estimator, i.e., the unique minimax admissible estimator, is hopeless, unless one restricts the class of estimators to be considered or incorporates some prior information. ... its main consequence has been to reinforce the Bayesian–frequentist interface, by inducing frequentists to call for Bayesian techniques and Bayesians to robustify their estimators in terms of frequentist performances and prior uncertainty. (Robert 1994, p. 67)

Decision Theory: Principles and Approaches   G. Parmigiani, L. Y. T. Inoue
© 2009 John Wiley & Sons, Ltd

Efron adds:

> The implications for objective Bayesians and fiducialists have been especially disturbing. . . . If a satisfactory theory of objective Bayesian inference exists, Stein's estimator shows that it must be a great deal more subtle than previously expected. (Efron 1978, p. 244)

The Stein effect occurs even though the dimensions are unrelated, and it can be explained primarily through the joint loss $L(\boldsymbol{\theta}, \boldsymbol{a}) = \sum_{i=1}^{p}(\theta_i - a_i)^2$, which allows the dominating estimator to "borrow strength" across dimensions, and bring individual estimates closer together, trading off errors in one dimension with those in another. This is usually referred to as shrinkage.

Our goal here is to build some intuition for the main results, and convey a sense for which approaches weather the storm in reasonable shape. In the end these include Bayes, but also empirical Bayes and some varieties of minimax. We first present the James–Stein theorem, straight up, no chaser, in the simplest form we know. We then look at some of the most popular attempts at intuitive explanations, and finally turn to more general formulations, both Bayes and minimax.

Featured article:

Stein, C. (1955). Inadmissibility of the usual estimator for the mean of a multivariate normal distribution, *Proceedings of the Third Berkeley Symposium on Mathematical Statistics and Probability* **1**: 197–206.

The literature on shrinkage, its justifications and ramifications, is vast. For a more detailed discussion we refer to Berger (1985), Brandwein and Strawderman (1990), Robert (1994), and Schervish (1995) and the references therein.

9.1   The Stein effect

Theorem 9.1   *Suppose that a p-dimensional vector $\boldsymbol{x} = (x_1, x_2, \ldots, x_p)'$ follows a multivariate normal distribution with mean vector $\boldsymbol{\theta} = (\theta_1, \ldots, \theta_p)'$ and the identity covariance matrix $\boldsymbol{I}$. Let $\mathcal{A} = \Theta = \Re^p$, and let the loss be $L(\boldsymbol{\theta}, \boldsymbol{a}) = \sum_{i=1}^{p}(\theta_i - a_i)^2$. Then, for $p > 2$, the decision rule*

$$\delta^{JS}(\boldsymbol{x}) = \boldsymbol{x}\left[1 - \frac{p-2}{\sum_{i=1}^{p} x_i^2}\right] \tag{9.1}$$

dominates $\delta(\boldsymbol{x}) = \boldsymbol{x}$.

Proof: The risk function of δ is

$$R(\boldsymbol{\theta}, \delta) = \int_{\Re^p} L(\boldsymbol{\theta}, \delta(\boldsymbol{x})) f(\boldsymbol{x}|\boldsymbol{\theta}) d\boldsymbol{x}$$

$$= \sum_{i=1}^{p} \int_{\Re} (\theta_i - x_i)^2 f(x_i|\theta_i) dx_i$$

$$= \sum_{i=1}^{p} \text{Var}[x_i|\theta_i] = p.$$

To compute the risk function of δ^{JS} we need two expressions:

$$E_{\boldsymbol{x}|\boldsymbol{\theta}} \left[\frac{\sum_{i=1}^{p} x_i \theta_i}{\sum_{i=1}^{p} x_i^2} \right] = E_{y|\boldsymbol{\theta}} \left[\frac{2y}{p - 2 + 2y} \right] \tag{9.2}$$

$$E_{\boldsymbol{x}|\boldsymbol{\theta}} \left[\frac{1}{\sum_{i=1}^{p} x_i^2} \right] = E_{y|\boldsymbol{\theta}} \left[\frac{1}{p - 2 + 2y} \right] \tag{9.3}$$

where y follows a Poisson distribution with mean $\sum_{i=1}^{p} \theta_i^2/2$. We will prove these results later. First, let us see how these can be used to get us to the finish line:

$$R(\boldsymbol{\theta}, \delta^{JS}) = \int_{\Re^p} L(\boldsymbol{\theta}, \delta^{JS}(\boldsymbol{x})) f(\boldsymbol{x}|\boldsymbol{\theta}) d\boldsymbol{x}$$

$$= E_{\boldsymbol{x}|\boldsymbol{\theta}} \left\{ \sum \left[x_i - \theta_i - \frac{p-2}{\sum_{i=1}^{p} x_i^2} x_i \right]^2 \right\}$$

$$= E_{\boldsymbol{x}|\boldsymbol{\theta}} \left\{ \sum_{i=1}^{p} \left[(x_i - \theta_i)^2 - 2(x_i - \theta_i) \frac{p-2}{\sum_{i=1}^{p} x_i^2} x_i + \left(\frac{p-2}{\sum_{i=1}^{p} x_i^2} x_i \right)^2 \right] \right\}$$

$$= p - 2(p-2)E_{\boldsymbol{x}|\boldsymbol{\theta}} \left\{ \frac{\sum_{i=1}^{p} (x_i - \theta_i) x_i}{\sum_{i=1}^{p} x_i^2} \right\} + (p-2)^2 E_{\boldsymbol{x}|\boldsymbol{\theta}} \left\{ \frac{1}{\sum_{i=1}^{p} x_i^2} \right\}$$

$$= p - 2(p-2) \left[1 - E_{y|\boldsymbol{\theta}} \left\{ \frac{2y}{p - 2 + 2y} \right\} \right] + (p-2)^2 E_{y|\boldsymbol{\theta}} \left\{ \frac{1}{p - 2 + 2y} \right\}$$

$$= p - 2(p-2) \left[E_{y|\boldsymbol{\theta}} \left\{ \frac{p - 2 + 2y}{p - 2 + 2y} \right\} - E_{y|\boldsymbol{\theta}} \left\{ \frac{2y}{p - 2 + 2y} \right\} \right]$$

$$+ (p-2)^2 E_{y|\boldsymbol{\theta}} \left\{ \frac{1}{p - 2 + 2y} \right\}$$

$$= p - E_{y|\boldsymbol{\theta}} \left\{ \frac{(p-2)^2}{p - 2 + 2y} \right\} < p = R(\boldsymbol{\theta}, \delta)$$

and the theorem is proved.

Now on to the two expectations in equations (9.2) and (9.3). See also James and Stein (1961), Baranchik (1973), Arnold (1981), or Schervish (1995, pp. 163–167).

We will first prove equation (9.3). A random variable U with a noncentral chi-square distribution with k degrees of freedom and noncentrality parameter λ is a mixture with conditional distributions $U|Y = y$ having a central chi-square distribution with $k + 2y$ degrees of freedom and where the mixing variable Y has a Poisson distribution with mean λ. In our problem, $U = \sum_{i=1}^{p} x_i^2$ has a noncentral chi-square distribution with p degrees of freedom and noncentrality parameter equal to $\sum_{i=1}^{p} \theta_i^2/2$. Using the mixture representation of the noncentral chi-square, with $Y \sim \mathrm{Poisson}(\sum_{i=1}^{p} \theta_i^2/2)$ and $U|Y = y \sim \chi_{p+2y}^2$, we get

$$E_{x|\theta}\left[\frac{1}{\sum_{i=1}^{p} x_i^2}\right] = E_{u|\theta}\left[\frac{1}{u}\right] = E_{y|\theta}\left[E_{u|y,\theta}\left[\frac{1}{u}\right]\right] = E_{y|\theta}\left[\frac{1}{p + 2y - 2}\right].$$

The last equality is based on the expression for the mean of a central inverse chi-square distribution.

Next, to prove equation (9.2) we will use the following result:

$$E_{x|\theta}\left[\frac{x_i - \theta_i}{\sum_{i=1}^{n} x_i^2}\right] = \frac{\partial}{\partial \theta_i} E_{x|\theta}\left[\frac{1}{\sum_{i=1}^{n} x_i^2}\right]. \tag{9.4}$$

The proof is left as an exercise (Problem 9.4). Now,

$$\frac{\partial}{\partial \theta_i} E_{x|\theta}\left[\frac{1}{\sum_{i=1}^{n} x_i^2}\right] = \frac{\partial}{\partial \theta_i} E_{y|\theta}\left[\frac{1}{p - 2 + 2y}\right]$$

$$= \frac{\partial}{\partial \theta_i} \sum_{y=0}^{\infty} \frac{1}{y!\,(p - 2 + 2y)} e^{-\sum_{i=1}^{p} \theta_i^2/2}\left(\sum_{i=1}^{p} \theta_i^2/2\right)^{y}$$

$$= \frac{\theta_i}{\sum_{i=1}^{p} \theta_i^2} E_{y|\theta}\left[\frac{2y - \sum_{i=1}^{p} \theta_i^2}{p - 2 + 2y}\right].$$

Note that

$$E_{x|\theta}\left[\frac{\theta_i x_i}{\sum_{i=1}^{p} x_i^2}\right] = \theta_i^2 E_{x|\theta}\left[\frac{1}{\sum_{i=1}^{p} x_i^2}\right] + E_{x|\theta}\left[\frac{\theta_i(x_i - \theta_i)}{\sum_{i=1}^{p} x_i^2}\right]$$

$$= \theta_i^2 E_{y|\theta}\left[\frac{1}{p - 2 + 2y}\right] + \frac{\theta_i^2}{\sum_{i=1}^{p} \theta_i^2} E_{y|\theta}\left[\frac{2y - \sum_{i=1}^{p} \theta_i^2}{p - 2 + 2y}\right].$$

Thus,

$$E_{x|\theta}\left[\frac{\sum_{i=1}^{p} \theta_i x_i}{\sum_{i=1}^{p} x_i^2}\right] = \sum_{i=1}^{p} \theta_i^2 E_{y|\theta}\left[\frac{1}{\sum_{i=1}^{p} \theta_i^2}\left(\frac{\sum_{i=1}^{p} \theta_i^2}{p - 2 + 2y} + \frac{2y - \sum_{i=1}^{p} \theta_i^2}{p - 2 + 2y}\right)\right]$$

$$= E_{y|\theta}\left[\frac{2y}{p - 2 + 2y}\right].$$

□

The assumption of unit variance is convenient but not required—any known p-dimensional vector of variances would lead to the same result as we can standardize the observations. For the same reason, as long as the variance(s) are known, x itself could be a vector of sample means.

The estimator δ^{JS} in Theorem 9.1 is the *James–Stein estimator*. It is a shrinkage estimator in that, in general, the individual components of δ are closer to 0 than the corresponding components of x, so the p-dimensional vector of observations is said to be shrunk towards 0.

How does the James–Stein estimator itself behave under the admissibility criterion? It turns out that it too can be dominated. For example, the so-called *truncated James–Stein estimator*

$$\delta_+^{JS} = x\left[1 - \frac{p-2}{\sum_{i=1}^p x_i^2}\right]_+ \tag{9.5}$$

is a variant that never switches the sign of an observation, and does dominate δ^{JS}. Here the subscript $+$ denotes the positive part of the expression in the square brackets. Put more simply, when the stuff in these brackets is negative we replace it with 0. This estimator is itself inadmissible. One way to see this is to use a complete class result due to Sacks (1963), who showed that any admissible estimate can be represented as a formal Bayes rule of the form

$$\delta^*(x) = \frac{\int \theta e^{-\frac{1}{2}\sum_{i=1}^p (x_i - \theta_i)^2} \Pi(d\theta)}{\int e^{-\frac{1}{2}\sum_{i=1}^p (x_i - \theta_i)^2} \Pi(d\theta)}.$$

These estimators are analytic in x while the positive part estimator is continuous but not differentiable. Baranchik (1970) also provided an estimator which dominates the James–Stein estimator. The James–Stein estimator is important for having revealed a weakness of standard approaches and for having pointed to shrinkage as an important direction for improvement. However, direct applications have not been numerous, perhaps because it is not difficult to find more intuitive, or better performing, implementations.

9.2   Geometric and empirical Bayes heuristics

9.2.1   Is x too big for θ?

An argument that is often heard as "intuitive" support for the need to shrink is captured by Figure 9.1. Stein would mention it in his lectures, and many have reported it since. The story goes more or less like this. In the setting of Theorem 9.1, because $E_{x|\theta}[(x-\theta)'\theta] = 0$, we expect orthogonality between $x - \theta$ and θ. Moreover, because $E_{x|\theta}[x'x] = p + \theta'\theta$, it may appear that x is too big as an estimator of θ and thus that it may help to shorten it. For example, the projection of θ on x might be closer. However, the projection $(1-a)x$ depends on θ through a. Therefore, we need to estimate a. Assuming that (i) the angle between θ and $x - \theta$ is right, (ii) $x'x$ is close to its

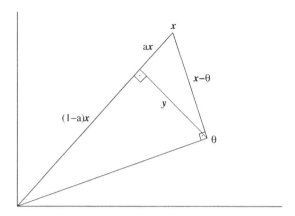

Figure 9.1   Geometric heuristic for the need to shrink. Vectors here live in p-dimensional spaces. Adapted with minor changes from Brandwein and Strawderman (1990).

expected value, $\boldsymbol{\theta}'\boldsymbol{\theta} + p$, and (iii) $(\boldsymbol{x} - \boldsymbol{\theta})'(\boldsymbol{x} - \boldsymbol{\theta})$ is close to its expected value p, then using Pythagoras's theorem in the right subtriangles in Figure 9.1 we obtain

$$\boldsymbol{y}'\boldsymbol{y} = (\boldsymbol{x} - \boldsymbol{\theta})'(\boldsymbol{x} - \boldsymbol{\theta}) - \hat{a}^2\boldsymbol{x}'\boldsymbol{x}$$
$$= p - \hat{a}^2\boldsymbol{x}'\boldsymbol{x}$$

and

$$\boldsymbol{y}'\boldsymbol{y} = \boldsymbol{\theta}'\boldsymbol{\theta} - (1 - \hat{a})^2\boldsymbol{x}'\boldsymbol{x}$$
$$= \boldsymbol{x}'\boldsymbol{x} - p - (1 - \hat{a})^2\boldsymbol{x}'\boldsymbol{x}.$$

By equating the above expressions, $\hat{a} = p/\boldsymbol{x}'\boldsymbol{x}$ is an estimate of a. Thus, an estimator for $\boldsymbol{\theta}$ would be

$$(1 - \hat{a})\boldsymbol{x} = \left(1 - \frac{p}{\boldsymbol{x}'\boldsymbol{x}}\right)\boldsymbol{x} \tag{9.6}$$

similar to the James–Stein estimator introduced in Section 9.1.

While this heuristic does not have any pretense of rigor, it is suggestive. However, a doubt remains about how much insight can be gleaned from the emphasis placed on the origin as the shrinkage point. For example, Efron (1978) points out that if one specifies an arbitrary origin $\boldsymbol{x}_0$ and defines the estimator

$$\delta_{\boldsymbol{x}_0}^{JS} = \boldsymbol{x}_0 + (\boldsymbol{x} - \boldsymbol{x}_0)\left[1 - \frac{p - 2}{(\boldsymbol{x} - \boldsymbol{x}_0)'(\boldsymbol{x} - \boldsymbol{x}_0)}\right] \tag{9.7}$$

to shrink towards the $\boldsymbol{x}_0$ instead of $\boldsymbol{0}$, this estimator also dominates $\boldsymbol{x}$. Berger (1985) has details and references for various implementations of this idea. For many choices of $\boldsymbol{x}_0$ the heuristic we described breaks down, but the Stein effect still works.

One way in which the insight of Figure 9.1 is sometimes summarized is by saying that "$\boldsymbol{\theta}$ is closer to $\mathbf{0}$ than $\boldsymbol{x}$." Is this reasonable? Maybe. But then shrinkage works for every $\boldsymbol{x}_0$. It certainly seems harder to claim that "$\boldsymbol{\theta}$ is closer to $\boldsymbol{x}_0$ than $\boldsymbol{x}$" irrespective of what $\boldsymbol{x}_0$ is. Similarly, some skepticism is probably useful when considering heuristics based on the fact that the probability that $\boldsymbol{x}'\boldsymbol{x} > \boldsymbol{\theta}'\boldsymbol{\theta}$ can be quite large even for relatively small p and large $\boldsymbol{\theta}$. Of course this is true, but it is less clear whether this is why shrinkage works. For more food for thought, observe that, overall, shrinkage is more pronounced when the data $\boldsymbol{x}$ are closer to the origin. Perhaps a more useful perspective is that of "borrowing strength" across dimensions, and shrinking when dimensions look similar. This is made formal in the Bayes and empirical Bayes approaches considered next.

9.2.2   Empirical Bayes shrinkage

The idea behind empirical Bayes approaches is to estimate the prior empirically. When one has a single estimation problem this is generally quite hard, but when a battery of problems are considered together, as is the case in this chapter, this is possible, and often quite useful. The earliest example is due to Robbins (1956) while Efron and Morris highlighted the relevance of this approach for shrinkage and for the so-called hierarchical models (Efron and Morris 1973b, Efron and Morris 1973a).

A relatively general formulation could go as follows. The p-dimensional observation vector $\boldsymbol{x}$ has distribution $f(\boldsymbol{x}|\boldsymbol{\theta})$, while parameters $\boldsymbol{\theta}$ have distribution $\pi(\boldsymbol{\theta}|\tau)$ with τ unknown and of dimension generally much lower than p. A classic example is one where x_i represents a noisy measurement of θ_i—say, the observed and real weight of a tree. The distribution $\pi(\boldsymbol{\theta}|\tau)$ describes the variation of the unobserved true measurements across the population. This model is an example of a multilevel model, with one level representing the noise in the measurements and the other the population variation. Depending on the study design, the $\boldsymbol{\theta}$ may be called random effects. Multilevel models are now a mainstay of applied statistics and the primary venue for shrinkage in practical applications (Congdon 2001, Ferreira and Lee 2007).

If interest is in $\boldsymbol{\theta}$, an empirical Bayes estimator can be obtained by deriving the marginal likelihood

$$m(\boldsymbol{x}|\tau) = E_{\boldsymbol{\theta}|\tau}\Big[f(\boldsymbol{x}|\boldsymbol{\theta})\pi(\boldsymbol{\theta}|\tau)\Big]$$

and using it to identify a reasonable estimator of τ, say $\hat{\tau}(\boldsymbol{x})$. This estimator is then plugged back into π, so that a prior is now available for use, via the Bayes rule, to obtain the Bayes estimator for $\boldsymbol{\theta}$. Under squared error loss this is

$$\delta^{EB}(\boldsymbol{x}) = E\big[\boldsymbol{\theta}\big|\boldsymbol{x}, \hat{\tau}(\boldsymbol{x})\big].$$

An empirical Bayes analysis does not correspond to a coherent Bayesian updating, since the data are used twice, but it has nice properties that contributed to its wide use.

While we use the letter π for the distribution of θ given τ, this distribution is a somewhat intermediate creation between a prior and a likelihood: its interpretation can vary significantly with the context, and it can be empirically estimable in some problems. A Bayesian analysis of this model would also assign a prior to τ, and proceed with coherent updating and expected loss minimization as in Theorem 7.1. If p is large and the likelihood is more concentrated than the prior, that is the noise is small compared to the variation across the population, the Bayes and empirical Bayes estimators of θ will be close.

Returning to our p-dimensional vector x from a multivariate normal distribution with mean vector θ and covariance matrix I, we can derive easily an empirical Bayes estimator as follows. Assume that a priori $\theta \sim N(0, \tau_0^2 I)$. The Bayes estimator of θ, under a quadratic loss function, is the posterior mean, that is

$$\delta_{\tau_0}^*(x) = \frac{1}{1 + \tau_0^{-2}} x, \tag{9.8}$$

assuming τ_0^2 is fixed. The empirical Bayes estimator of θ is found as follows. First, we find the unconditional distribution of x which is, in this case, normal with mean 0 and variance $(1 + \tau_0^2)I$. Second, we find a reasonable estimator for τ_0. Actually, in this case, it makes sense to aim directly at the shrinkage factor $(1 + \tau_0^2)^{-1}$. Because

$$\sum x_i^2 \sim (1 + \tau_0^2)\chi_p^2$$

and

$$E\left[(p - 2)/\sum x_i^2\right] = (1 + \tau_0^2)^{-1}$$

then $(p - 2)/\sum x_i^2$ is an unbiased estimator of $1/(1 + \tau_0^2)$. Plugging this directly into equation (9.8) gives the empirical Bayes estimator

$$\delta^{EB}(x) = \left(1 - \frac{p - 2}{\sum_{i=1}^p x_i^2}\right) x. \tag{9.9}$$

This is the James–Stein estimator! We can reinterpret it as a weighted average between a prior mean of 0 and an observed measurement x, with weights that are learned completely from the data. The prior mean does not have to be zero: in fact versions of this where one shrinks towards the empirical average of x can also be shown to dominate the MLE, and to have an empirical Bayes justification.

Instead of using an unbiased estimator of the shrinkage factor, one could, alternatively, find an estimator for τ^2 using the maximum likelihood approach, which leads to

$$\hat{\tau}^2 = \begin{cases} \sum_{i=1}^p x_i^2/p - 1 & \text{if} \quad \sum_{i=1}^p x_i^2 > p \\ 0 & \text{otherwise} \end{cases} \tag{9.10}$$

and the corresponding empirical Bayes estimator is

$$\delta_+^{EB}(x) = \hat\tau^2 x/(1 + \hat\tau^2) \tag{9.11}$$

$$= x\left[1 - \frac{p}{\sum_{i=1}^p x_i^2}\right]_+ \tag{9.12}$$

similar to the estimator of equation (9.5).

9.3   General shrinkage functions

9.3.1   Unbiased estimation of the risk of $x + g(x)$

Stein developed a beautiful way of providing insight into the subtle issues raised in this chapter, by setting very general conditions for estimators of the form $\delta_g(x) = x + g(x)$ to dominate $\delta(x) = x$. We begin with two preparatory lemmas. All the results in this section are under the assumptions of Theorem 9.1.

Lemma 1   *Let y be distributed as a $N(\theta, 1)$. Let h be the indefinite integral of a measurable function h' such that*

$$\lim_{y \to \pm\infty} h(y)\exp\left(-\frac{1}{2}(y - \theta)^2\right) = 0. \tag{9.13}$$

Then

$$E_{y|\theta}[h(y)(y - \theta)] = E_{y|\theta}[h'(y)]. \tag{9.14}$$

Proof: Starting from the left hand side,

$$\frac{1}{\sqrt{2\pi}}\int_{-\infty}^{+\infty} h(y)(y - \theta)\exp\left(-\frac{(y - \theta)^2}{2}\right)dy$$

$$= \frac{1}{\sqrt{2\pi}}\int_{-\infty}^{+\infty} h(y)\frac{d}{dy}\left[\exp\left(-\frac{(y - \theta)^2}{2}\right)\right]dy$$

$$= -h(y)\frac{1}{\sqrt{2\pi}}\exp\left(-\frac{(y - \theta)^2}{2}\right)\Big|_{-\infty}^{+\infty} + \frac{1}{\sqrt{2\pi}}\int_{-\infty}^{+\infty} h'(y)\exp\left(-\frac{(y - \theta)^2}{2}\right)dy$$

$$= E_{y|\theta}[h'(y)]. \tag{9.15}$$

Assumption (9.13) is used in the integration by parts.   □

Weaker versions of assumption (9.13) are used in Stein (1981). Incidentally, the converse of this lemma is also true, so if equation (9.14) holds for all reasonable h, then y must be normal.

The next lemma cranks this result up to p dimensions, and looks at the covariance between the error $x - \theta$ and the shrinkage function $g(x)$. It requires a technical differentiability condition and some more notation: a function $h : \mathfrak{R}^p \to \mathfrak{R}$ is almost differentiable if there exists a function $\nabla h : \mathfrak{R}^p \to \mathfrak{R}^p$ such that

$$h(x + z) - h(x) = \int_0^1 z' \nabla h(x + tz) dt,$$

where z is in $\mathfrak{R}^p$ and ∇ can also be thought of as the vector differential operator

$$\nabla = \left(\frac{\partial h}{\partial x_1}, \ldots, \frac{\partial h}{\partial x_p} \right).$$

A function $g : \mathfrak{R}^p \to \mathfrak{R}^p$ is almost differentiable if every coordinate is.

Lemma 2   *If h is almost differentiable and $E_{x|\theta}[(\nabla h(x))'(\nabla h(x))] < \infty$, then*

$$E_{x|\theta}[h(x)'(x - \theta)] = E_{x|\theta}[\nabla h(x)]. \tag{9.16}$$

Using these two lemmas, the next theorem provides a usable closed form for the risk of estimators of the form $\delta_g(x) = x + g(x)$. Look at Stein (1981) for proofs.

Theorem 9.2 (Unbiased estimation of risk)   *If $g : \mathfrak{R}^p \to \mathfrak{R}^p$ is almost differentiable and such that*

$$E_{x|\theta}\left[\sum_{i=1}^p |\nabla_i g_i(x)| \right] < \infty$$

then

$$E_{x|\theta}\left[(x + g(x) - \theta)'(x + g(x) - \theta) \right] = p + E_{x|\theta}\left[(g(x))'(g(x)) \right.$$
$$\left. + 2 \sum_i \nabla_i g_i(x) \right]. \tag{9.17}$$

The risk is decomposable into the risk of the estimator x plus a component that depends on the shrinkage function g. A corollary of this result, and the key of our exercise, is that if we can specify g so that

$$(g(x))'(g(x)) + 2 \sum_i \nabla_i g_i(x) \leq 0 \tag{9.18}$$

for all values of x (with at least a strict inequality somewhere), then $\delta_g(x)$ dominates $\delta(x) = x$. This argument encapsulates the bias–variance trade-off in shrinkage estimators: shrinkage works when the negative covariance between errors $x - \theta$ and corrections $g(x)$ more than offsets the bias induced by $g(x)$. Because of the additivity of the losses across dimensions, we can trade off bias in one dimension for variance in another. This aspect more than any other sets the Stein estimation setting apart from problems with a single parameter of interest.

Before we move to discussing ways to choose g, note that the right hand side of equation (9.17) is the risk of δ_g, so the quantity

$$p + (g(x))'(g(x)) + 2 \sum_i \nabla_i g_i(x)$$

is an unbiased estimator of the unknown $R(\delta_g, \theta)$. The technique of finding an unbiased estimator of the risk directly can be useful in general. Stein, for example, suggests that it could be used for choosing estimators that, after the data are observed, have small estimated risk. More discussion appears in Stein (1981).

9.3.2 Bayes and minimax shrinkage

There is a solid connection between estimators like δ_g and Bayes rules. As usual, call $m(x)$ the marginal density of x, that is $m(x) = \int f(x|\theta)\pi(\theta)d\theta$. Then it turns out that, using the gradient representation,

$$E[\theta|x] = x + \nabla \log m(x).$$

This is because

$$\nabla \log m(x) = \frac{-\int (x-\theta)\exp(-(x-\theta)'(x-\theta)/2)\pi(\theta)\,d\theta}{\int \exp(-(x-\theta)'(x-\theta)/2)\pi(\theta)\,d\theta}$$

$$= -x + \frac{\int \theta \exp(-(x-\theta)'(x-\theta)/2)\pi(\theta)\,d\theta}{\int \exp(-(x-\theta)'(x-\theta)\pi(\theta)\,d\theta}$$

$$= -x + E[\theta|x].$$

Thus setting $g(x) = \nabla \log m(x)$, for some prior π, is a very promising choice of g, as are, more generally, functions m from $\Re^p$ to the positive real line. When we restrict attention to these, the condition for dominance given in inequality (9.18) becomes, after a little calculus,

$$(g(x))'(g(x)) + 2 \sum_i \nabla_i g_i(x) = \frac{4\nabla^2 \sqrt{m(x)}}{\sqrt{m(x)}} < 0.$$

Thus, to produce an estimator that dominates x it is enough to find a function m with $\nabla^2 \sqrt{m(x)} \leq 0$ or $\nabla^2 m(x) \leq 0$—the latter are called *superharmonic* functions.

If we do not require that $m(x)$ is a proper density, we can choose $m(x) = |x|^{-(p-2)}$. For $p > 2$ this is a superharmonic function and $\nabla \log m(x) = -x(p-2)/x'x$, so it yields the James–Stein estimator of Theorem 9.1.

The estimator $\delta^M(x) = x$ is a minimax rule, but it is not the only minimax rule. Any rule whose risk is bounded by p will also be minimax, so it is not too difficult to concoct shrinkage approaches that are minimax. This has been a topic of great interest, and the literature is vast. Our brief discussion is based on a review by Brandwein and Strawderman (1990), which is also a perfect entry point if you are interested in more.

A good starting point is the James–Stein estimator δ^{JS}, which is minimax, because its risk is bounded by p. Can we use the form of δ^{JS} to motivate a broader class of minimax estimators? This result is an example:

Theorem 9.3   *If h is a monotone increasing function such that $0 \leq h \leq 2(p-2)$, then the estimator*

$$\delta^M(x) = x\left[1 - \frac{h(x'x)}{x'x}\right]$$

is minimax.

Proof: In this proof we assume that $h(x'x)/x'x$ follows the conditions stated in Lemma 1. Then, by applying Lemma 1,

$$E_{x|\theta}\left[(x-\theta)'x\frac{h(x'x)}{x'x}\right] = (p-2)E_{x|\theta}\left[\frac{h(x'x)}{x'x}\right] + 2E_{x|\theta}\left[\frac{h'(x'x)}{x'x}\right]$$

$$\geq (p-2)E_{x|\theta}\left[\frac{h(x'x)}{x'x}\right]. \tag{9.19}$$

The above inequality follows from the fact that h is positive and increasing, which implies $h' \geq 0$. Moreover, by assumption,

$$0 \leq h(x'x) \leq 2(p-2),$$

which implies that for nonnull vectors x,

$$0 \leq \frac{h^2(x'x)}{x'x} \leq 2(p-2)\frac{h(x'x)}{x'x}.$$

Therefore,

$$E_{x|\theta}\left[\frac{h^2(x'x)}{x'x}\right] \leq 2(p-2)E_{x|\theta}\left[\frac{h(x'x)}{x'x}\right]. \tag{9.20}$$

By combining inequalities (9.19) and (9.20) above we conclude that the risk of this estimator is

$$R(\delta^M, \theta) = E_{x|\theta}\left[\left((x-\theta) - \frac{h(x'x)}{x'x}x'\right)'\left((x-\theta) - \frac{h(x'x)}{x'x}x'\right)\right]$$

$$= p + E_{x|\theta}\left[\frac{h^2(x'x)}{x'x}\right] - 2E_{x|\theta}\left[\frac{(x-\theta)'x}{x'x}h(x'x)\right]$$

$$\leq p + [2(p-2) - 2(p-2)]E_{x|\theta}\left[\frac{h(x'x)}{x'x}\right] = p.$$

In Section 7.7 we proved that x is minimax and it has constant risk p. Since the risk of δ^M is at most p, it is minimax.   □

The above lemma gives us a class of shrinkage estimators. The Bayes and generalized Bayes estimators may be found in this class. Consider a hierarchical model defined as follows: conditional on λ, $\boldsymbol{\theta}$ is distributed as $N(\mathbf{0}, (1 - \lambda)/\lambda \boldsymbol{I})$. Also, $\lambda \sim (1 - b)\lambda^{-b}$ for $0 \le b < 1$. The Bayes estimator is the posterior mean of $\boldsymbol{\theta}$ given by

$$\delta^*(x) = E[\boldsymbol{\theta}|x] = E[E[\boldsymbol{\theta}|x, \lambda]|x]$$

$$= E\left[\left(1 - \frac{1}{1 + [(1 - \lambda)/\lambda]}\right)x\Big|x\right]$$

$$= [1 - E[\lambda|x]]x. \tag{9.21}$$

The next theorem gives conditions under which the Bayes estimator given by (9.21) is minimax.

Theorem 9.4   *1. For $p \ge 5$, the proper Bayes estimator δ^* from (9.21) is minimax as long as $b \ge (6 - p)/2$.*

2. For $p \ge 3$, the generalized Bayes estimator δ^* from (9.21) is minimax if $(6 - p)/2 \le b < (p + 2)/2$.

Proof: One can show that the posterior mean of λ is

$$E[\lambda|x] = \frac{1}{x'x}\left[p + 2 - 2b - \frac{2\exp(-(x'x)/2)}{\int_0^1 \lambda^{(1/2)p-b}\exp(-(\lambda/2)x'x)d\lambda}\right] \tag{9.22}$$

$$= \frac{h(x'x)}{x'x}$$

where $h(x'x)$ is the term in square brackets on the right side of expression (9.22).

Observe that $h(x'x) \le p + 2 - 2b$. Moreover, it is monotone increasing because $\int_0^1 \lambda^{(1/2)p-b} \exp(-(\lambda/2)x'x)d\lambda$ is increasing. Application of Theorem 9.3 completes the proof.   □

We saw earlier that the Bayes rules, under very general conditions, can be written as $x + \nabla \log m(x)$, and that a sufficient condition for these rules to dominate δ is that m be superharmonic. It is therefore no surprise that we have the following theorem.

Theorem 9.5   *If $\pi(\boldsymbol{\theta})$ is superharmonic, then $x + \nabla \log m(x)$ is minimax.*

This theorem provides a broader intersection of the Bayes and minimax shrinkage approaches.

9.4   Shrinkage with different likelihood and losses

A Stein effect can also be observed when considering different sampling distributions as well as different loss functions. Brown (1975) discusses inadmissibility of the mean under whole families of loss functions.

On the likelihood side an important generalization is to the case of unknown variance. Stein (1981) discusses an extension of the results presented in this chapter using the unbiased estimation of risk technique. More broadly, *spherically symmetric* distributions are distributions with density $f((x - \theta)'(x - \theta))$ in $\Re^p$. When $p > 2$, $\delta(x) = x$ is inadmissible when estimating θ with any of these distributions. Formally we have:

Theorem 9.6   *Let $z = (x, y)$ in $\Re^p$, with distribution*

$$z \sim f((x - \theta)'(x - \theta) + y'y),$$

and $x \in \Re^q, y \in \Re^{p-q}$. The estimator

$$\delta_h(z) = (1 - h(x'x, y'y))x$$

dominates $\delta(x) = x$ under quadratic loss if there exist $\alpha, \beta > 0$ such that:

(i) $t^\alpha h(t, u)$ is a nondecreasing function of t for every u;

(ii) $u^{-\beta}h(t, u)$ is a nonincreasing function of u for every t; and

(iii) $0 \leq (t/u)h(t, u) \leq 2(q - 2)\alpha/(p - q - 2 + 4\beta)$.

For details on the proof see Robert (1994, pp. 67–68). Moreover, the Stein effect is robust in the class of spherically symmetric distributions with finite quadratic risk since the conditions on h do not depend on f.

In discrete data problems, there are, however, significant exceptions. For example, Alam (1979) and Brown (1981) show that the maximum likelihood estimator is admissible for estimating several binomial parameters under squared error loss. Also, the MLE is admissible for a vector of multinomial probabilities and a variety of other discrete problems.

9.5   Exercises

Problem 9.1   Consider a sample

$$x = (x_1, 0, 6, 5, 9, 0, 13, 0, 26, 1, 3, 0, 0, 4, 0, 34, 21, 14, 1, 9)'$$

where each $x_i \sim \text{Poi}(\lambda_i)$. Let δ^{EB} denote the 20-dimensional empirical Bayes decision rule for estimating the vector λ under squared error loss and let $\delta_1^{EB}(x)$ be the first coordinate, that is the estimate of λ_1.

Write a computer program to plot $\delta_1^{EB}(x)$ versus x_1 assuming the following prior distributions:

1. $\lambda_i \sim \text{Exp}(1)$, $i = 1, \ldots, 20$

2. $\lambda_i \sim \text{Exp}(\gamma)$, $i = 1, \ldots, 20$, $\gamma > 0$

3. $\lambda_i \sim \text{Exp}(\gamma)$, $i = 1, \ldots, 20$, γ either 1 or 10

4. $\lambda_i \sim (1 - \alpha)I_0 + \alpha\text{Exp}(\gamma)$, $\alpha \in (0, 1)$, $\gamma > 0$,

where $\text{Exp}(\gamma)$ is an exponential with mean $1/\gamma$.

Write a computer program to graph the risk functions corresponding to each of the choices above, as you vary λ_1 and fix the other coordinates $\lambda_i = x_i$, $i \geq 2$.

Problem 9.2   If x is a p-dimensional normal with mean θ and identity covariance matrix, and g is a continuous piecewise continuously differentiable function from $\Re^p$ to $\Re^p$ satisfying

$$\lim_{|x| \to 0} \frac{\partial g_i(x)}{\partial x_i} \exp^{(-0.25 x'x)} = 0,$$

then

$$E_{x|\theta}[(x + g(x) - \theta)'(x + g(x) - \theta)] = p + E_{x|\theta}[g(x)'g(x) + 2\nabla'g(x)],$$

where ∇ is the vector differential operator with coordinates

$$\nabla_i = \frac{\partial g_i}{\partial x_i}.$$

You can take this result for granted.

   Consider the estimator

$$\delta(x) = x + g(x)$$

where

$$g(x) = \begin{cases} -\dfrac{k-2}{x'_*x_*}x_i & \text{if} \quad |x_i| \leq z_{(k)} \\ -\dfrac{k-2}{x'_*x_*}z_{(k)}\text{sign}(x_i) & \text{if} \quad |x_i| > z_{(k)}, \end{cases}$$

where z is the vector of order statistics of x, and

$$x'_*x_* = \sum_i \min(x_i^2, z_{(k)}^2).$$

This estimator is discussed in Stein (1981).

Questions:

1. Why would anyone ever want to use this estimator?

2. Using the result above, show that the risk is

$$p - (k - 2)^2 E_{x|\theta} \left[\frac{1}{x'_* x_*} \right].$$

3. Find a vector θ for which this estimator has lower risk than the James–Stein estimator.

Problem 9.3   For $p = 10$, graph the risk of the two empirical Bayes estimators of Section 9.2.2 as a function of $\theta' \theta$.

Problem 9.4   Prove equation (9.4) using the assumptions of Theorem 9.1.

Problem 9.5   Construct an estimator with the following two properties:

1. δ is the limit of a sequence of Bayes rules as a hyperparamter gets closer to its extreme.

2. δ is not admissible.

You do not need to look very far.

10

Scoring rules

In Chapter 2 we studied coherence and explored relations between "acting rationally" and using probability to measure uncertainty about unknown events. De Finetti's "Dutch Book" theorem guaranteed that, if one wants to avoid a sure loss (that is, be coherent), then probability calculus ought to be used to represent uncertainty. The conclusion was that, regardless of one's beliefs, it is incoherent not to express such beliefs in the form of *some* probability distribution. In our discussion, the relationship between the agent's own knowledge and expertise, empirical evidence, and the probability distribution used to set fair betting odds was left unexplored. In this chapter we will focus on two related questions. The first is how to guarantee that probability assessors reveal their knowledge and expertise about unknowns in their announced probabilities. The second is how to evaluate, after events have occurred, whether their announced probabilities are "good."

To make our discussion concrete, we will focus on a simple situation, in which assessors have a clear interest in the quality of their announced probabilities. The prototypical example is forecasting. We will talk about weather forecasting, partly because it is intuitive, and partly because meteorologists were among the first to realize the importance of this problem. Weather forecasting is a good example to illustrate these concepts, but clearly not the only application of these ideas. In fact, any time you are trying to empirically compare different prediction algorithms or scientific theories you are in a situation similar to this.

In a landmark paper, Brier described the issue as follows:

> Verification of weather forecasts has been a controversial subject for more than a half century. There are a number of reasons why this problem has been so perplexing to meteorologists and others but one of the most important difficulties seems to be in reaching an agreement on

the specification of a scale of goodness of weather forecasts. Numerous systems have been proposed but one of the greatest arguments raised against forecast verification is that forecasts which may be the "best" according to the accepted system of arbitrary scores may not be the most useful forecasts. In attempting to resolve this difficulty the forecaster may often find himself in the position of choosing to ignore the verification system or to let it do the forecasting for him by "hedging" or "playing the system". This may lead the forecaster to forecast something other than what he thinks will occur, for it is often easier to analyze the effect of different possible forecasts on the verification score than it is to analyze the weather situation. It is generally agreed that this state of affairs is unsatisfactory, as one essential criterion for satisfactory verification is that the verification scheme should influence the forecaster in no undesirable way. (Brier 1950, p. 1)

We will study these questions formally by considering the measures used for the evaluation of forecasters. We will talk about scoring rules. In relation to our discussion this far, these can be thought of as loss functions for the decision problem of choosing a probability distribution. The process by which past data are used in reaching a prediction is not formally modeled—in contrast to previous chapters whose foci were decision functions. However, data will come into play in the evaluation of the forecasts.

Featured articles:

Brier, G. (1950). Verification of forecasts expressed in terms of probability, *Monthly Weather Review* **78**: 1–3.
Good, I. J. (1952). Rational decisions, *Journal of the Royal Statistical Society, Series B, Methodological* **14**: 107–114.

In the statistical literature, the earliest example we know of a scoring rule is Good (1952). Winkler (1969) discusses differences between using scoring rules for assessing probability forecasts and using them for elicitation. A general discussion of the role of scoring rules in elicitation can be found in Savage (1971) and Lindley (1982b). Additional references and discussion are in Bernardo and Smith (1994, Sec. 2.7).

10.1   Betting and forecasting

We begin with two simple illustrations in which announcing one's own personal probabilities will be the best thing to do for the forecaster.

The first example brings us back to the setting of Chapter 2. Suppose you are a bookmaker posting odds $q : (1 - q)$ on the occurrence of an event θ. Say your personal beliefs about θ are represented by π. Both π and q are points in [0, 1]. As the bookmaker, you get $-(1 - q)S$ if θ occurs, and qS if θ does not occur, where, as before, S is the stake and it can be positive or negative. Your expected gain, which in this case could also be positive or negative, is

$$-(1 - q)S\pi + qS(1 - \pi) = S(q - \pi).$$

Suppose you have to play by de Finetti's rules and post odds at which you are indifferent between taking bets with positive or negative stakes on θ. In order for you to be indifferent you need to set $q = \pi$. So in the betting game studied by de Finetti, a bookmaker interested in optimizing expected utility will choose betting odds that correspond to his or her personal probability. On the other hand, someone betting with you and holding beliefs π' about θ will have an expected gain of $S(\pi' - q)$ and will have an incentive to bet a positive stake on θ whenever $\pi' > q$.

The second example is a cartoon version of the problem of ranking forecasters. Commencement day is two days away and a departmental ceremony will be held outdoors. As one of the organizers you decide to seek the opinion of an expert about the possibility of rain. You can choose among two experts and you want to do so based on how well they predict whether it will rain tomorrow. Then you will decide on one of the experts and ignore the other, at least as far as the commencement day prediction goes. You are going to come up with a rule to rank them and you are going to tell them ahead of time which rule you will use. You are worried that if you do not specify the rule well enough, they may report to you something that is not their own best prediction, in an attempt to game the system—the same concern that was mentioned by Brier earlier.

So this is what you come up with. Let the event "rain tomorrow" be represented here by θ, and suppose that q is a forecaster's probability of rain for tomorrow. If θ occurs, you would like q to be close to 1; if θ does not occur you would like q to be close to 0. So you decide to assign to each forecaster the score $(q - \theta)^2$, and then choose the forecaster with the lowest score.

With this scoring rule, is it in a forecaster's best interest to announce his or her own personal beliefs about θ in playing this game? Let us call a forecaster's own probability of rain tomorrow π, and compute the expected score, which is

$$(q - 1)^2\pi + q^2(1 - \pi) = (q - \pi)^2 + \pi(1 - \pi).$$

To make this as small as possible a forecaster must choose $q = \pi$.

Incidentally, choosing one forecaster only and ignoring the other may not necessarily be the best approach from your point of view: there may be ways of combining the two to improve on both forecasts. A discussion of that would be too long a detour, but you can look at Genest and Zidek (1986) for a review.

10.2   Scoring rules

10.2.1   Definition

We are now ready to make these considerations more formal. Think of a situation where a forecaster needs to announce the probability of a certain event. To keep things from getting too technically complex we will assume that this event has a finite number J of possible mutually exclusive outcomes, represented by the vector $\boldsymbol{\theta} = (\theta_1, \ldots, \theta_J)$ of outcome indicators. When we say that θ_j occurred we mean that $\theta_j = 1$ and therefore $\theta_i = 0$ for $i \neq j$. The forecaster needs to announce probabilities for each of the J possibilities. The vector $\boldsymbol{q} = (q_1, \ldots, q_J)$ contains the announced probabilities for the corresponding indicator. Typically the action space for the choice of $\boldsymbol{q}$ will be the set $\mathcal{Q}$ of all probability distributions in the $J-1$ dimensional simplex. A scoring rule is a measure of the quality of the forecast $\boldsymbol{q}$ against the observed event outcome θ. Specifically:

Definition 10.1 (Scoring rule) *A scoring rule s for the probability distribution $\boldsymbol{q}$ is a function assigning a real number $s(\theta, \boldsymbol{q})$ to each combination $(\theta, \boldsymbol{q})$.*

We will often use the shorthand notation $s(\theta_j, \boldsymbol{q})$ to indicate $s(\theta, \boldsymbol{q})$ evaluated at $\theta_j = 1$ and $\theta_i = 0$ for $i \neq j$. This does not mean that the scoring rule only depends on one of the events in the partition. Note that after θ_j is observed, the score is computed based potentially on the entire vector of forecasts, not just the probability of θ_j. In the terminology of earlier lectures, $s(\theta_j, \boldsymbol{q})$ represents the loss when choosing $\boldsymbol{q}$ as the announced probability distribution, and θ_j turns out to be the true state of the world. We will say that a scoring rule is smooth if it is continuously differentiable in each q_j. This requires that small variations in the announced probabilities would produce only small variations in the score, which is often reasonable.

Because θ is unknown when the forecast is made, so is the score. We assume that the forecaster is a utility maximizer, and will choose $\boldsymbol{q}$ to minimize the expected score. This will be done based on his or her coherent personal probabilities, denoted by $\boldsymbol{\pi} = (\pi_1, \ldots, \pi_J)$. The expected loss from the point of view of the forecaster is

$$\mathcal{S}(\boldsymbol{q}) = \sum_{j=1}^{J} s(\theta_j, \boldsymbol{q}) \pi_j. \tag{10.1}$$

Note the different role played by the announced probabilities $\boldsymbol{q}$ that enter the score function, and the forecaster's beliefs that are weights in the computation of the expected score.

The fact that the forecaster seeks a Bayesian solution to this problem does not necessarily mean that the distribution $\boldsymbol{q}$ is based on Bayesian inference, only that the forecaster acts as a Bayesian in deciding which $\boldsymbol{q}$ will be best to report.

10.2.2   Proper scoring rules

In both of our examples in Section 10.1, the scoring rule was such that the decision maker's expected score was minimized by $\boldsymbol{\pi}$, the personal beliefs. Whenever this happens, no matter what the beliefs are, the scoring rule is called *proper*. More formally:

Definition 10.2 (Proper scoring rule)   *A scoring rule s is proper if, and only if, for each strictly positive π,*

$$\inf_{\mathbf{q} \in \mathcal{Q}} \mathcal{S}(\mathbf{q}) = \mathcal{S}(\pi). \tag{10.2}$$

If π is the only *solution that minimizes the expected score, then the scoring rule is called* strictly proper.

Winkler observes that this property is useful for both probability assessment and probability evaluation:

> In an ex ante sense, strictly proper scoring rules provide an incentive for careful and honest forecasting by the forecaster or forecast system. In an ex post sense, they reward accurate forecasts and penalize inferior forecasts. (Winkler 1969, p. 2)

Proper scoring rules can also be used as devices for eliciting one's subjective probability. Elicitation of expectations is studied by Savage (1971) who mathematically developed general forms for the scoring rules that are appropriate for eliciting one's expectation.

Other proper scoring rules are the logarithmic (Good 1952)

$$s(\theta_j, \boldsymbol{q}) = -\log q_j$$

and the spherical rule (Savage 1971)

$$s(\theta_j, \boldsymbol{q}) = -\frac{q_j}{\left(\sum_i q_i^2\right)^{1/2}}.$$

10.2.3   The quadratic scoring rules

Perhaps the best known proper scoring rule is the quadratic rule, already encountered in Section 10.1. The quadratic rule was introduced by Brier (1950) to provide a "verification score" for weather forecasts. A mathematical derivation of the quadratic scoring rule can be found in Savage (1971). A quadratic scoring rule is a function of the form

$$s(\theta_j, \boldsymbol{q}) = A \sum_{i=1}^{J} (q_i - 1_{\{i=j\}})^2 + B_j, \tag{10.3}$$

with $A > 0$. Here $1_{\{i=j\}}$ is 1 if $i = j$ and 0 otherwise.

Take a minute to convince yourself that the rule of the commencement party example of Section 10.1 is a special case of this, occurring when $J = \{1, 2\}$. Find the values of A and B_j implied by that example, and sort out why in the example there is only one term to s while here there are J.

Now we are going to show that the quadratic scoring rule is proper; that is, that $q = \pi$ is the Bayes decision. To do so, we have to minimize, with respect to q, the expected score

$$S(q) = \sum_{j=1}^{J} s(\theta_j, q)\pi_j = \sum_{j=1}^{J} \left\{ A \sum_{i=1}^{J} (q_i - 1_{\{i=j\}})^2 + B_j \right\} \pi_j \qquad (10.4)$$

subject to the constraint that the elements in q sum to one. The Lagrangian for this constrained minimization is

$$\Lambda(q) = S(q) + \lambda \left(\sum_{j=1}^{J} q_j - 1 \right).$$

Taking partial derivatives with respect to q_k we have

$$\frac{\partial \Lambda(q)}{\partial q_k} = \sum_{j=1}^{J} 2A(q_k - 1_{\{k=j\}})\pi_j + \lambda = 2A(q_k - \pi_k) + \lambda.$$

Setting $q_k = \pi_k$ and $\lambda = 0$ satisfies the first-order conditions and the constraint, since $\sum_j \pi_j = 1$. We should also verify that this gives a minimum. This is not difficult, but not insightful either, and we will not go over it here.

10.2.4   Scoring rules that are not proper

It is not difficult to construct scoring rules that look reasonable but are not proper. For example, go back to the commencement example, and say you state the score $|q - \theta|$. From the forecaster's point of view, the expected score is

$$(1 - q)\pi + q(1 - \pi) = \pi + (1 - 2\pi)q.$$

This is linear in q and is minimized by announcing $q = 1$ if $\pi > 0.5$ and $q = 0$ if $\pi < 0.5$. If π is exactly 0.5 then the expected score is flat and any q will do. You get very little information out of the forecaster with this scoring system. Problem 10.1 gives you a chance to work out the details in the more general cases.

Another temptation you should probably resist if you want to extract honest beliefs is that of including in the score the consequences of the forecaster's error on your own decision making. Say a company asks a geologist to forecast whether a large amount of oil is available (θ_1) or not (θ_2) in the region. Say q is the announced probability. After drilling, the geologist's forecast is evaluated by the company according to the following (modified quadratic) scoring rule:

$$s(\theta_1, q) = (q_1 - 1)^2$$
$$s(\theta_2, q) = 10(q_2 - 1)^2;$$

that is, the penalty for an error when the event is false is 10 times the penalty when the event is true. It may very well be that the losses to the company are different in the two cases, but that should affect the use the company makes of the probability obtained by the geologist, not the way the geologist's accuracy is rewarded. To see that note that the geologist will choose q that minimizes the expected score. Because $q_2 = 1 - q_1$

$$S(\boldsymbol{q}) = (q_1 - 1)^2 \pi + 10q_1^2(1 - \pi).$$

The minimum is attained when $q_1 = \pi/(10 - 9\pi)$. Clearly, this rule is not proper. The announced probability is always smaller than the geologist's own probability π when $\pi \in (0, 1)$.

One may observe at this point that as long as the announced q_1 is an invertible function of π the company is no worse off with the scoring rule above than it would be with a proper rule, as long as the forecasts are not taken at face value. This is somewhat general. Lindley (1982b) considers the case where the forecaster is scored based on the function $s(\theta_1, \boldsymbol{q})$ if θ_1 occurs, and $s(\theta_2, \boldsymbol{q})$ if θ_2 occurs. So the forecaster should minimize his or her expected score

$$S(\boldsymbol{q}) = \pi s(\theta_1, \boldsymbol{q}) + (1 - \pi)s(\theta_2, \boldsymbol{q}).$$

If the utility function has a first derivative, the forecaster should obtain $\boldsymbol{q}$ as a solution of the equation $dS/dq_1 = 0$ under the constraint $q_2 = 1 - q_1$. This implies $\pi s'(\theta_1, \boldsymbol{q}) + (1 - \pi)s'(\theta_2, \boldsymbol{q}) = 0$. Solving, we obtain

$$\pi(\boldsymbol{q}) = \frac{s'(\theta_2, \boldsymbol{q})}{s'(\theta_2, \boldsymbol{q}) - s'(\theta_1, \boldsymbol{q})} \tag{10.5}$$

which means that the probability π can be recovered via a transformation of the stated value $\boldsymbol{q}$. Can any value $\boldsymbol{q}$ lead to a probability? Lindley studied this problem and verified that if $\boldsymbol{q}$ solves the first-order equation above, then its transformation obeys the laws of probability. So while a proper scoring rule guarantees that the announced probabilities are the forecaster's own, other scoring rules may achieve the goal of extracting sufficient information to reconstruct the forecaster's probabilities.

10.3   Local scoring rules

A further restriction that one may impose on a scoring rule is the following. When event θ_i occurs, the forecaster is scored only on the basis of what was announced about θ_i, and not on the basis of what was announced for the events that did not occur. Such a scoring rule is called *local*. Formally we have the following.

Definition 10.3 (Local scoring rule)   *A scoring rule s is local if there exist functions $s_j(.), j \in J$, such that $s(\theta_j, \boldsymbol{q}) = s_j(q_j)$.*

For example, the logarithmic scoring rule $s(\theta_j, \boldsymbol{q}) = -\log q_j$ is local. The lower the probability assigned to the observed event, the higher the score. The rest of the announced probabilities are not taken into consideration. The quadratic scoring rule is not local. Say $J = 3, A = 1$, and $B_j = 0$. Suppose we observe $\theta_1 = 1$. The vectors $q = (0.5, 0.25, 0.25)$ and $q' = (0.5, 0.5, 0)$ both assigned a probability of 0.5 to the event θ_1, but q' gets a score of 0.5 while q gets a score of 0.375. This is because the quadratic rule penalizes a single error of size 0.5 more than it penalizes two errors of size 0.25. It can be debated whether this is appropriate or not. If you are tempted to replace the square in the quadratic rule with an absolute value, remember that it is not proper (see Problem 10.1). Another factor to consider in thinking about whether a local scoring rule is appropriate for a specific problem has to do with whether there is a natural ordering or a notion of closeness for the elements of the partition. A local rule gives no "partial credit" for near misses. If one is predicting the eye color of a newborn, who turns out to have green eyes, one may assess the prediction based only on the probability of green, and not worry about how the complement was distributed among, say, brown and blue. Things start getting trickier if we consider three events like "fair weather," "rain," and "snow" on a winter's day. Another example where a local rule may not be attractive is when the events in the partitions are the outcomes of a quantitative random variable, such as a count. Then the issue of near misses may be critical.

Naturally, having a local scoring rule gives us many fewer things to worry about and makes proving theorems easier. For example, the local property leads to a nice functional characterization of the scoring rule. Specifically, every smooth, proper, and local scoring rule can be written as $B_j + A \log q_j$ for $A < 0$. This means that if we are interested in choosing a scoring rule for assessing probability forecasters, and we find that smooth, proper, and local are reasonable requirements, we can restrict our choice to logarithmic scoring rules. This is usually considered a very strong point in favor of logarithmic scoring rules. We will come back to it later when discussing how to measure the information provided by an experiment.

Theorem 10.1 (Proper local scoring rules)   *If s is a smooth, proper, and local scoring rule for probability distributions q defined over Θ, then it must be of the form $s(\theta_j, q) = B_j + A \log q_j$, where $A < 0$ and the B_j are arbitrary constants.*

Proof: Assume that $\pi_j > 0$ for all j. Using the fact that s is local we can write

$$\inf_q \mathcal{S}(q) = \inf_q \sum_{j \in J} s(\theta_j, q) \pi_j = \inf_q \sum_{j \in J} s_j(q_j) \pi_j.$$

Because s is smooth, $\mathcal{S}$, which is a continuous function of s, will also be smooth. Therefore, in order for π to be a minimum, it has to satisfy the first-order conditions for the Lagrangian:

$$\Lambda(q) = \mathcal{S}(q) + \lambda \left(\sum_{j=1}^J q_j - 1 \right).$$

Differentiating,

$$\frac{\partial \Lambda(q)}{\partial q_j} = \pi_j s_j'(q_j) + \lambda \qquad j \in J. \tag{10.6}$$

Because s is proper, the minimum expected score is achieved when $q = \pi$. So, if π is a minimum, it must be that

$$s_j'(\pi_j) = -\frac{\lambda}{\pi_j} \qquad j \in J. \tag{10.7}$$

Integrating both sides of (10.7), we get that each s_j must be of the form $s_j(\pi_j) = -\lambda \log \pi_j + B_j$. In order to guarantee that the extremal found is a minimum we must have $\lambda > 0$. $\qquad\qquad\square$

The logarithmic scoring rule goes back to Good, who pointed out:

> A reasonable fee to pay to an expert who has estimated a probability as q is $A \log(2q)$ if the event occurs and $A \log(2 - 2q)$ if the event does not occur. If $q > 1/2$ the latter payment is really a fine. . . . This fee can easily be seen to have the desirable property that its expectation is minimized if $q = \pi$, the true probability, so that it is in the expert's own interest to give an objective estimate. (Good 1952, p. 112 with notational changes)

We changed Good's notation to match ours. It is interesting to see how Good specifies the constants B_j to set up a reward structure that could correspond to both a win and a loss, and what is his justification for this. It is also remarkable how Good uses the word "objective" for "true to one's own knowledge and beliefs." The reason for this is in what follows:

> It is also in his interest to collect as much evidence as possible. Note that no fee is paid if $q = 1/2$. The justification of this is that if a larger fee was paid the expert would have a positive expected gain by saying that $q = 1/2$ without looking at the evidence at all. (Good 1952, p. 112 with notational changes)

Imposing the condition that a scoring rule is local is not the only way to narrow the set of candidate proper scoring rules one may consider. Savage (1971) shows that the quadratic loss is the only proper scoring rule that satisfies the following conditions: first, the expected score S must be symmetric in π and q, so that reversing the role of beliefs and announced probabilities does not change the expected score. Second, S must depend on π and q only through their difference. The second condition highlights one of the rigidities of squared error loss: the pairs $(0.5, 0.6)$ and $(10^{-12}, 0.1 + 10^{-12})$ are considered equally, though in practice the two discrepancies may have very different implications.

We have not considered the effect of a scoring rule on how one collects and uses evidence, but it is an important topic, and we will return to this in Chapter 13, when exploring specific ways of measuring the information in a data set.

10.4   Calibration and refinement

10.4.1   The well-calibrated forecaster

In this section we discuss calibration and refinement, and their relation to scoring rules. Our discussion follows DeGroot and Fienberg (1982) and Seidenfeld (1985).

A set of probabilistic forecasts π is *well calibrated* if $\pi\%$ of all predictions reported at probability π are true. In other words, calibration looks at the agreement between relative frequency for the occurrence of an event (rain in the above context) and forecasts. Here we are using π for forecasts, and assume that $\pi = q$. To illustrate calibration, consider Table 10.1, adapted from Brier (1950). A forecaster is calibrated if the two columns in the table are close.

Requiring calibration in a forecaster is generally a reasonable idea, but calibration is not sufficient for the forecasts to be useful. DeGroot and Fienberg point out that:

> In practice, however, there are two reasons why a forecaster may not be well calibrated. First, his predictions can be observed for only a finite number of days. Second, and more importantly, there is no inherent reason why his predictions should bear any relation whatsoever to the actual occurrence of rain. (DeGroot and Fienberg 1983, p. 14)

If this comment sounds harsh, consider these two points: first, a calibration scoring rule based on calibration alone would, in a finite horizon, lead to forecasts whose only purpose is to game the system. Suppose a weather forecaster is scored at the end of the year based on how well calibrated he or she his. For example, we could take all days where the announced chance of rain was 10% and see whether or not the empirical frequency is close to 10%. Next we would do the same with 20% announced chance of rain and so on. The desire to be calibrated over a finite time period could induce the forecaster to make announcements that are radically at odds with his or her beliefs. Say that, towards the end of the year, the forecaster is finding that the empirical frequency for the 10% days is a bit too low—in the table it is 7%. If tomorrow promises to bring a flood of biblical proportions, it is to the forecaster's advantage to announce that the chance of rain is 10%. The forecaster's reputation may suffer, but the scoring rule will improve.

Table 10.1   Verification of a series of 85 forecasts expressed in terms of the probability of rain. Adapted from Brier (1950).

Forecast probability of rain	Observed proportion of rain cases
0.10	0.07
0.30	0.10
0.50	0.29
0.70	0.40
0.90	0.50

Second, even when faced with an infinite sequence of forecasts, a forecaster has no incentive to attempt at correlating the daily forecasts and the empirical observations. For example, a forecaster that invariably announces that the probability of rain is, say, 10%, because that is the likely overall average of the sequence, would likely be very well calibrated. These forecasts would not, however, be very useful.

So, suppose you have available two well-calibrated forecasters. Whose predictions would you choose? That is of course why we have scoring rules, but before we tie this discussion back to scoring rules, we describe a measure of forecasters' accuracy called *refinement*. Loosely speaking, refinement looks at the dispersion of the forecasts.

Consider a series of forecasts of the same type of event over time, as would be the case if we were to announce the probability of rain on the daily forecast. Consider a well-calibrated forecaster who announces discrete probabilities, so there is only a finite number of possibilities, collected in the set Π—for example, Π could be $\{0, 0.1, 0.2, \ldots, 1\}$. We can single out the days when the forecast is a particular π using the sequence of indicators $\epsilon_i^\pi = 1_{\{\pi_i = \pi\}}$. Consider a horizon consisting of the first n_k elements in the sequence. Let n_k^π denote the number of forecasts equal to π within the horizon. The proportion of days in the horizon with forecast π is

$$v_k(\pi) = n_k^\pi / n_k.$$

Let x_i be the indicator of rain in the ith day. Then

$$\bar{x}_k(\pi) = \sum_{i=1}^{k} \epsilon_i^\pi x_i / n_k^\pi$$

is the relative frequency of rain among those days in which the forecaster's prediction was π. The relative frequency of rainy days is

$$\mu = \sum_{\pi \in \Pi} \pi v_k(\pi).$$

We assume that $0 < \mu < 1$.

To simplify notation in what follows we drop the index k, but all functions are defined considering a finite horizon of k days. Before formally introducing refinement we need the following definition:

Definition 10.4 (Stochastic transformation)   *A stochastic transformation $h(\pi|p)$ is a function defined for all $\pi \in \Pi$ and $p \in \Pi$ such that*

$$h(\pi|p) \geq 0, \quad \forall \pi, p \in \Pi$$

$$\sum_{\pi \in \Pi} h(\pi|p) = 1, \quad \forall p \in \Pi.$$

Definition 10.5 (Refinement)   *Consider two well-calibrated forecasters A and B whose predictions are characterized by v_A and v_B, respectively. Forecaster A is at* least as refined as *[or, alternatively, forecaster A is* sufficient for*] forecaster B if there exists a stochastic transformation h such that*

$$\sum_{p \in \Pi} h(\pi|p)v_A(p) = v_B(\pi), \ \forall \ \pi \in \Pi$$

$$\sum_{p \in \Pi} h(\pi|p)pv_A(p) = \pi v_B(\pi), \ \forall \ \pi \in \Pi.$$

In other words, if forecaster A makes a prediction p, we can generate B's prediction frequencies by utilizing the conditional distribution $h(\pi|p)$.

Let us go back to the forecaster that always announces the overall average, that is

$$v(\pi) = \begin{cases} 1, & \text{if } \pi = \mu, \\ 0, & \text{if } \pi \neq \mu. \end{cases}$$

We need to assume that $\mu \in \Pi$—not so realistic if predictions are highly discrete, but convenient. This forecaster is the *least refined* forecaster. Alternatively, we say the forecaster exhibits *zero sharpness*. Any other forecaster is at least as refined as the least refined forecaster because we can set up the following stochastic transformation:

$$h(\mu|p) = 1, \text{ for } p \in \Pi$$

$$h(\pi|p) = 0, \text{ for } \pi \neq \mu$$

and apply Definition 10.5.

At another extreme of the spectrum of well-calibrated forecasters is the forecaster whose prediction each day is either 0 or 1:

$$v(\pi) = \begin{cases} \mu, & \text{if } \pi = 1 \\ (1 - \mu), & \text{if } \pi = 0 \\ 0, & \text{if } \pi \notin \{0, 1\}. \end{cases}$$

Because this forecaster is well calibrated, his or her predictions have to also be always correct. This is the *most refined forecaster* is said to have has *perfect sharpness*. To see this, consider any other forecaster B, and use Definition 10.5 with the stochastic transformation

$$h(\pi|1) = \frac{\pi}{\mu} v_B(\pi), \text{ for } \pi \in \Pi,$$

$$h(\pi|0) = \frac{1 - \pi}{1 - \mu} v_B(\pi), \text{ for } \pi \in \Pi.$$

DeGroot and Fienberg (1982, p. 302) provide a result that simplifies the construction of useful stochastic transformations.

Theorem 10.2 *Consider two well-calibrated forecasters A and B. Then A is at least as refined as B if and only if*

$$\sum_{i=0}^{j-1}(\pi_{(j)} - \pi_{(i)})[v_A(\pi_{(i)}) - v_B(\pi_{(i)})] \geq 0, \quad \forall j = 1, \ldots, J - 1,$$

where J is the number of elements in Π and $\pi_{(i)}$ is the ith smallest element of Π.

DeGroot and Fienberg (1982) present a more general result that compares forecasters who are not necessarily well calibrated using the concept of sufficiency as introduced by Blackwell (1951) and Blackwell (1953).

At this point we are finally ready to explore the relationship between proper scoring rules and the concepts of calibration and refinement. Let us go back to the quadratic scoring rule, or Brier score, which in the setting of this section is

$$BS = \sum_{\pi \in \Pi} v(\pi)\left[\bar{x}(\pi)(\pi - 1)^2 + (1 - \bar{x}(\pi))\pi^2\right].$$

This score can be rewritten as

$$BS = \sum_{\pi \in \Pi} v(\pi)\left[\pi - \bar{x}(\pi)\right]^2 + \sum_{\pi \in \Pi} v(\pi)\left[\bar{x}(\pi)(1 - \bar{x}(\pi))\right]. \tag{10.8}$$

The first summation on the right hand side of the above equation measures the distance between the forecasts and the relative frequency of rainy days; that is, it is a measure of the calibration of the forecaster. This component is zero if the forecast is well calibrated. The second summation term is a measure of the refinement of the forecaster and it shows that the score is improved with values of $\bar{x}(\pi)$ close to 0 or 1. It can be approximately interpreted as a weighted average of the bin-specific conditional variances, where bins are defined by the common forecast π. This decomposition illustrates how the squared error loss combines elements of calibration and forecasting, and is the counterpart of the bias/variance decomposition we have seen in equation (7.19) with regard to parameter estimation under squared error loss.

DeGroot and Fienberg (1983) generalize this beyond squared error loss: suppose that the forecaster's subjective prediction is π and that he or she is scored on the basis of a strictly proper scoring rule specified by functions s_1 and s_2, which are, respectively, decreasing and increasing functions of π. The forecaster's score is $s_1(\pi)$ if it rains and $s_2(\pi)$ otherwise. The next theorem tells us that strictly proper scoring rules can be partitioned into calibration and refinement terms.

Theorem 10.3 *Suppose that the forecaster's predictions are characterized by functions $\bar{x}(\pi)$ and $v(\pi)$ with a strictly proper scoring rule specified by functions $s_1(\pi)$*

*and $s_2(\pi)$. The forecaster's overall score is S which can be decomposed as $S = S_1 + S_2$
where*

$$S_1 = \sum_{\pi \in \Pi} v(\pi)\{\bar{x}(\pi)[s_1(\pi) - s_1(\bar{x}(\pi))] + [1 - \bar{x}(\pi)][s_2(\pi) - s_2(\bar{x}(\pi))]\}$$

$$S_2 = \sum_{\pi \in \Pi} v(\pi)\phi(\bar{x}(\pi))$$

with $\phi(p) = ps_1(p) + (1 - p)s_2(p), 0 \le p \le 1$, a strictly convex function.

Note that in the above decomposition, S_1 is a measure of the forecaster's calibration and achieves its minimum when $\bar{x}(\pi) = \pi$, $\forall \pi$; that is, when the forecaster is well calibrated. On the other hand, S_2 is a measure of the forecaster's refinement.

Although scoring rules can be used to compare probability assessments provided by competing forecasters, or models, in practice it is more common to separately assess calibration and refinement, or calibration and discrimination. Calibration is typically assessed by the bias term in BS, or the chi-square statistic built on the same two sets of frequencies. Discrimination is generally understood as the ability of a set of forecasts to separate rainy from dry days and is commonly quantified via the receiver operating characteristics (ROC) curve (Lusted 1971, McNeil *et al.* 1975). A ROC curve quantifies in general the ability of a quantitative or ordinal measurement and scores (possibly a probability) to classify an associate binary measurement.

For a biological example, consider a population of individuals, some of whom have a binary genetic marker $x = 1$ and some of whom do not. A forecaster, or prediction algorithm, gives you a measurement π that has cumulative distribution

$$F_x(\pi) = \sum_{\pi' \le \pi} v(\pi')$$

conditional on the true marker status x. In a specific binary decision problem we may be able to derive a cutoff point π_0 on the forecast and decide to "declare positive" all individuals for whom $\pi > \pi_0$. This declaration will lead to true positives but also some false positives. The fraction of true positives among the high-probability group, or sensitivity, is

$$\beta(\pi_0) = \frac{\sum_{\pi > \pi_0} v(\pi')\bar{x}(\pi)}{\sum_{\pi > \pi_0} v(\pi)}$$

while the fraction of true negatives among the low-probability group, or specificity, is

$$\alpha(\pi_0) = \frac{\sum_{\pi \le \pi_0} v(\pi)[1 - \bar{x}(\pi)]}{\sum_{\pi \le \pi_0} v(\pi)}.$$

Generally a higher cutoff will decrease the sensitivity and increase the specificity, similarly to what happens to the coordinates of the risk set as we vary the cutoff on the test statistic in Section 7.5.1.

The ROC curve is a graph of $\beta(\pi_0)$ versus $1 - \alpha(\pi_0)$ as we vary π_0. Formally the ROC curve is given by the equation

$$\beta = 1 - F_1[1 - F_0(1 - \alpha)].$$

The overall discrimination of a classifier is often summarized across all possible cutoffs by computing the area under the ROC curve, which turns out to be equal to the probability that a randomly selected person with the genetic marker has a forecast that is greater than a randomly selected person without the marker.

The literature on measuring the quality of probability assessments and the utility of predictive assays in medicine is extensive. Pepe (2003) provides a broad overview. Issues with comparing multiple predictions are explored in Pencina *et al.* (2008). An interesting example of evaluating probabilistic prediction in the context of specific medical decisions is given by Gail (2008).

10.4.2   Are Bayesians well calibrated?

A coherent probability forecaster expects to be calibrated in the long run. Consider a similar setting to the previous section and suppose in addition that, subsequent to each forecast π_i, the forecaster receives feedback in the form of the outcome x_i of the predicted event. Assume that π_{i+1} represents the posterior probability of rain given the prior history $x_1, \ldots, x_i$, that is

$$\pi_{i+1} = \pi(x_{i+1}|x_1, \ldots, x_i).$$

One complication now is that we can no longer assume that the forecasts live in a discrete set. One way to proceed is to select a subset, or subsequence, of days—for example, all days with forecast in a given interval—and compare forecasts π_i with the proportion of rainy day in the subset. To formalize, let ϵ_i be the indicator for whether day i is included in the set. Define

$$n_k = \sum_{i=1}^{k} \epsilon_i$$

to be the number of selected days within horizon k,

$$\bar{x}_k = n_k^{-1} \sum_{i=1}^{k} \epsilon_i x_i$$

to be the relative frequency of rainy days within the subset, and

$$\bar{\pi}_k = n_k^{-1} \sum_{i=1}^{k} \epsilon_i \pi_i$$

as the average forecast probability within the tested subsequence. In this setting we can define long-run calibration as follows:

Definition 10.6   *Given a subsequence such that $\epsilon_i = 1$ infinitely often, the forecaster is calibrated in the long run if*

$$\lim_{k \to \infty} (\bar{x}_k - \bar{\pi}_k) = 0$$

with probability 1.

The condition that $\epsilon_i = 1$ infinitely often is required so we never run out of events to compare and we can take the limit. To satisfy it we need to be careful in constructing subsets using a condition that will keep coming up.

This is a good point to remind ourselves that, from the point of view of the forecaster, the expected value of the forecaster's posterior is his or her prior. For example, at the beginning of the series

$$E_{x_1}[\pi_2 | x_1] = \sum_x \pi(x_2 | x_1 = x) m(x) = \pi(x_1) = \pi_1$$

and likewise at any stage in the process. This follows from the law of total probability, as long as m is the marginal distribution that is implied by the forecaster's likelihood and priors. So, as a corollary

$$E_{x_i}[\pi_{i+1} | x_i] - \pi_i = 0,$$

a property that makes the sequence of forecasts a so-called martingale. So if the data are generated from the very same stochastic model that is used in the updating step (that is, Bayes' rule) then the forecasts are expected to be stable. Of course this is a big if.

In our setting, a consequence of this is that in the long run, a coherent forecaster who updates probabilities according to a Bayes rule expects to be well calibrated almost surely in the long run. This can be formalized in various ways. Details of theorems and proofs are in Pratt (1962), Dawid (1982), and Seidenfeld (1985). The good side of this result is that any coherent forecaster is not only internally consistent, but also prepared to become consistent with evidence, at least with evidence of the type he or she expects will accumulate. For example, if the forecaster's probabilities are based on a parametric model whose parameters are unknown and are assigned a prior distribution, then, in the long run, if the data are indeed generated by the postulated model, the forecaster will learn model parameters no matter what the initial prior was, as long as a positive mass was assigned to the correct value. Also, any two coherent forecasters that agree on the model will eventually give very similar predictions.

Seidenfeld points out that this is a disappointment for those who may have hoped for calibration to provide an additional criterion for restricting the possible range of coherent probability specifications:

Subject to feedback, calibration in the long run is otiose. It gives no ground for validating one coherent opinion over another as each coherent forecaster is (almost) sure of his own long-run calibration. (Seidenfeld 1985, p. 274)

We are referring here to expected calibration by one's own model. In reality different forecasters will use different models and may never agree with each other or with evidence. Dawid is concerned by the challenge this poses to the foundations of Bayesian statistics:

Any application of the Theorem yields a statement of the form $\pi(A) = 1$, where A expresses some property of perfect calibration of the distribution π. In practice, however, it is rare for probability forecasts to be well calibrated (so far as can be judged from finite experience) and no realistic forecaster would believe too strongly in his own calibration performance. We have a paradox: an event can be distinguished . . . that is given subjective probability one, and yet is not regarded as "morally certain". How can the theory of coherence, which is founded on assumptions of rationality, allow for such irrational conclusions? (Dawid 1982, p. 607)

Whether this is indeed an irrational conclusion is a matter of debate, some of which can be enjoyed in the comments to Dawid's paper, as well as in his rejoinder.

Just as in our discussion of Savage's "small worlds," here we must keep in mind that the theory of rationality needs to be circumscribed to a small enough, reasonably realistic, microcosm, within which it is humanly possible to specify probabilities. For example, specifying a model for all possible future meteorological data in a broad sense, say including all new technological advances in measurement, climate change, and so forth, is beyond human possibility. So the "small world" may need to evolve with time, at the price of some reshaping of the probability model and a modicum of temporal incoherence. Morrie DeGroot used to say that he carried in his pocket an ϵ of probability for complete surprises. He was coherent up to that ϵ! His ϵ would come in handy in the event that a long-held model turned out not to be correct. Without it, or the willingness to understand coherence with some flexibility, we would be stuck for life with statistical models we choose here and now because they are useful, computable, or currently supported by a dominant scientific theory, but later become obsolete. How to learn statistical models from data, and how to try to do so rationally, is the matter of the next chapter.

10.5   Exercises

Problem 10.1   A forecaster must announce probabilities $q = (q_1, \ldots, q_J)$ for the events $\theta_1, \ldots, \theta_J$. These events form a partition: that is, one and only one of them will occur. The forecaster will be scored based on the scoring rule

$$s(\theta_j, \boldsymbol{q}) = \sum_{i=1}^{k} |q_i - 1_{i=j}|.$$

Here $1_{i=j}$ is 1 if $i = j$ and 0 otherwise. Let $\boldsymbol{\pi} = (\pi_1, \ldots, \pi_J)$ represent the forecaster's own probability for the events $\theta_1, \ldots, \theta_J$. Show that this scoring rule is not proper. That is, show that there exists a vector $\boldsymbol{q} \neq \boldsymbol{\pi}$ such that

$$\sum_{j=1}^{J} s(\theta_j, \boldsymbol{q})\pi_j > \sum_{j=1}^{J} s(\theta_j, \boldsymbol{\pi})\pi_j.$$

Because you are looking for a counterexample, it is okay to consider a simplified version of the problem, for example by picking a small J.

Problem 10.2   Show directly that the scoring rule $s(\theta_j, \boldsymbol{q}) = q_j$ is not proper. See Winkler (1969) for further discussion about the implications of this fact.

Problem 10.3   Suppose that θ_1 and θ_2 are indicators of disjoint events (that is, $\theta_1\theta_2 = 0$) and consider $\theta_3 = \theta_1 + \theta_2$ as the indicator of the union event. You have to announce probabilities q_{θ_1}, q_{θ_2}, and $q_{\theta_1+\theta_2}$ to these events, respectively. You will be scored according to the quadratic scoring rule: $s(\theta_1, \theta_2, \theta_3, q_{\theta_1}, q_{\theta_2}, q_{\theta_3}) = (q_{\theta_1} - \theta_1)^2 + (q_{\theta_2} - \theta_2)^2 + (q_{\theta_3} - \theta_3)^2$. Prove the additive law, that is $q_{\theta_3} = q_{\theta_1} + q_{\theta_2}$.
  Hint: Calculate the scores for the occurrence of $\theta_1(1 - \theta_2)$, $(1 - \theta_1)\theta_2$, and $(1 - \theta_1)(1 - \theta_2)$.

Problem 10.4   Show that the area under the ROC curve is equal to the probability that a randomly selected person with the genetic marker has an assay that is greater than a randomly selected person without the marker.

Problem 10.5   Prove Equation (10.8).

Problem 10.6   Consider a sequence of independent binary events with probability of success 0.4 Evaluate the two terms in equation (10.8) for the following four forecasters:

Charles: always says 0.4.
  Mary: randomly chooses between 0.3 and 0.5.
  Qing: says either 0.2 or 0.3; when he says 0.2 it never rains, when he says 0.3 it always rains.
  Ana: follows this table:

	Rain	No rain
$\pi = 0.3$	0.15	0.35
$\pi = 0.5$	0.25	0.25

Comment on the calibration and refinement of these forecasters.

11

Choosing models

In this chapter we discuss model choice. So far we postulated a fixed data-generating mechanism $f(x|\theta)$ without worrying about how f is chosen. From this perspective, we may think about model choice as the choice of which f to use. Depending on the application, f could be a simple parametric family, or a more elaborate model. George Box's famous comment that "all models are wrong but some models are useful" (Box 1979) highlights the importance of taking a pragmatic viewpoint in evaluating models, and to set criteria driven by the goals of the modeling. Decision theory would seem to be the perfect perspective to formalize Box's concise statement of principle.

A view we could take is that of Chapter 10. Forecasters are incarnations of predictive models that we can evaluate and compare based on utility functions such as scoring rules. If we do this, we neatly separate the information that was used by the forecasters to develop and tune the prediction models, from the information that we use to evaluate them. This separation is a luxury we do not always have. More often we would like to be able to entertain several approaches in parallel, and learn something about how well they do directly from the data that are used to develop them. Whether this is even possible is a matter of debate, and some hold, with good reasons, that model training and model assessment should be separate.

But let us say we give in to the temptation of training and evaluating models at the same time. An important question is whether we can talk about a model as a "state of the world" in the same way we did for parameters or future events. Box's comment that all models are wrong sounds like a negative answer. In the terminology of Savage, a model is perhaps like a "small world." Within the small world we apply a theory that explains how we should learn from data and make good decisions. But can the theory tell us whether the "small world" is right?

Both these considerations suggest that model choice may require a richer conceptual framework than that of statistical decision theory. However, we can still make a little bit of progress if we are willing to stipulate that a true model exists in our list of candidate models. This is not a real change of perspective conceptually: the "small world" is a little bit bigger, and the model is then simply another parameter, but the results are helpful in clarifying the underpinning of some popular model choice approaches.

In Section 11.1 we set up the general framework for decision problems in which the model is unknown, and look at the implications of model selection and prediction. Then, changing slightly the paradigm, we consider the situation in which only one model is being contemplated and the question arises as to whether or not the model may be adequate. We present a way in which decision theory can can be brought to bear for this problem in Section 11.2. We focus only on the Bayesian approach, though frequentist model selection approaches are also available.

We do not have a featured article for this chapter. Useful general readings are Box (1980), Bernardo and Smith (1994), and Clyde and George (2004).

11.1   The "true model" perspective

11.1.1   Model probabilities

Our discussion in most of this chapter is based on making our "small world" just a little bit bigger so that unknowns that are not normally considered part of it are now included. For example, we can imagine extending estimation of a population mean from a known to an unknown family of distributions, or extending a linear regression with two predictors to the bigger world in which any subset of five additional predictors could also be included in the model. We will assume we can make a list of possible models, and feel comfortable that one of the models is true—much as we could be confident that one real number or another is the true average height of a population. When we can do this, the model is part of the states of the world in the usual sense, and nothing differentiates it from what we normally call θ except habit. We can then apply all we know about decision theory, and handle the fact that the model is unknown in a goal-driven way. This approach takes very seriously the "some models are useful" part of Box's aphorism, but it ignores the "all models are wrong" part.

Formally, we can take our familiar $f(x|\theta)$ and consider it as a special case of a larger collection of possible data-generating mechanisms defined by $f(x|\theta_M, M)$, where M denotes the model. We have a list of models, called $\mathcal{M}$. To fix ideas, consider the original f to be a normal with mean θ and variance 1. If you do not trust the normal model but you are reasonably confident the distribution should be symmetric, and your primary concerns are occasional outliers, then you can consider the set of Student's t distributions with M degrees of freedom and median θ to be your new larger collection of models. If $\mathcal{M}$ is all positive integers, the normal case is approached at $M = \infty$. M is much like θ in that it is an unknown state of the world. The different denomination reflects the common usage of the normal as a fixed assumption (or model) in practical analyses.

In this example, the parameter θ can be interpreted as the population median across all models. However, the interpretation of θ, as well as its dimensionality, may change across models in more complex examples. This is why we need to define separate random variables θ_M for each M.

In general, we have a model index M, and as many parameter sets as there are models, potentially. These are all unknown. The axiomatic foundations tell us that, if the small world has not exploded on us yet, we should have a joint probability distribution on the whole set. If we have a finite list of models, so M is an integer between 1 and M_0, the prior is

$$\pi(\theta_1, \ldots, \theta_{M_0}, M). \tag{11.1}$$

This induces a joint distribution over the data, parameters, and models. Implicit in the specification of $f(x|\theta_M, M)$ is the idea that, conditional on M and θ_M, x is independent of all the $\theta_{M'}$ with $M' \neq M$. So the joint probability distribution on all unknowns can be written as

$$f(x|\theta_M, M)\pi(\theta_1, \ldots, \theta_{M_0}, M).$$

This can be a very complicated distribution to specify, though for special decision problems some simplifications take place. For example, if all models harbor a common parameter θ, and the loss function depends on the parameters only through θ, then we do not need to specify the horrendous-dimensional prior (11.1). We define $\theta_M = (\theta, \eta_M)$, where η_M are model-specific nuisance parameters. Then

$$\pi(\theta|x) = \frac{1}{m(x)} \sum_{M=1}^{M_0} \int_{H_1} \cdots \int_{H_{M_0}} f(x|\theta, \eta_M, M)\pi(\theta, \eta_1, \ldots, \eta_{M_0}, M)d\eta_1, \ldots, d\eta_{M_0}$$

$$= \frac{1}{m(x)} \sum_{M=1}^{M_0} \int_{H_M} f(x|\theta, \eta_M, M)\pi(\theta, \eta_M, M)d\eta_M, \tag{11.2}$$

where $\pi(\theta, \eta_M, M)$ is the model-specific prior defined by

$$\pi(\theta, \eta_M, M) = \int_{H_1} \cdots \int_{H_{M-1}} \int_{H_{M+1}} \cdots$$

$$\int_{H_{M_0}} \pi(\theta, \eta_1, \ldots, \eta_{M_0}, M)d\eta_1, \ldots, d\eta_{M-1}d\eta_{M+1} \ldots, d\eta_{M_0}. \tag{11.3}$$

To get equation (11.2) reorder each of the M_0 integrals so that the integral with respect to η_M is on the outside.

In this case, because of Theorem 7.1, we can operate directly from the distribution $\pi(\theta|x)$. Compared to working from prior (11.1), here we "only" need to specify model-specific priors $\pi(\theta, \eta_M, M)$.

Another important quantity is the marginal distribution of the data given the observed model. This will be the critical quantity to look at when the loss function depends on the model but not specifically on the model parameters. An integration similar to that leading to equation (11.2) will produce

$$f(x|M) = \int_{\Theta_M} f(x|\theta_M, M)\pi(\theta_M|M)d\theta_M \qquad (11.4)$$

which again depends only on the model-specific conditional prior $\pi(\theta_M|M)$. Then, conditional on the data x, the posterior model probabilities are given by

$$\pi(M|x) = \frac{f(x|M)\pi(M)}{m(x)}, \qquad (11.5)$$

where $\pi(M)$ is the prior model probability implied by (11.1). Moreover, the posterior predictive density for a new observation $\tilde{x}$ is given by

$$f(\tilde{x}|x) = \sum_{M=1}^{M_0} f(\tilde{x}|x, M)\pi(M|x); \qquad (11.6)$$

that is, a posterior weighted mixture of the conditional predictive distributions $f(\tilde{x}|x, M)$.

The approach outlined here is a very elegant way of handling uncertainty about which model to use: if the true model is unknown, use a set of models, and let the data weigh in about which models are more likely to be the true one. In the rest of this section we will look at the implications for model choice, prediction, and estimation. Two major difficulties with this approach are the specification of prior probabilities and the specification of a list of models large enough to contain the true model, but small enough that prior probabilities can be meaningful. There are also significant computational challenges. For more readings about the Bayesian perspective on model uncertainty see Madigan and Raftery (1994), Draper (1995), Clyde and George (2004), and references therein. Bernardo and Smith (1994) also discuss alternative perspectives where the assignment of probabilities to the model space is not logical, because it cannot be assumed that the true model is included in the list.

11.1.2   Model selection and Bayes factors

In model choice the goal is to choose a single "best" model according to some specified criterion. The simplest formulation is one in which we are interested in guessing the correct model, and consider all mistakes equally undesirable. Within the decision-theoretic framework the set of actions is $\mathcal{A} = \mathcal{M}$ and the loss function which is 0 for choosing the true model and 1 otherwise. To simplify our discussion we assume that $\mathcal{M}$ is a finite set, that is $\mathcal{M} = \{1, \ldots, M_0\}$. This loss function allows us to frame the discussion in a decision-theoretic way and get some clear-cut results, but it is not very much in the spirit of the Box aphorism: it really punts on the question of why the model is useful. In this setting the prior expected loss for action that declares M to be the true model is $1 - \pi(M)$ and the Bayes action a^* is to choose the model with highest prior probability. Similarly, after seeing observations x, the optimal decision is to choose the model with highest posterior probability (11.5). Many model selection procedures are motivated by the desire to approximate this property. However, often they are used in practice to select a model and then perform inference or prediction conditioning on the model. This practice is expedient and often necessary, but a more consistent decision-theoretic approach would be to specify the loss function directly in terms of the final use of the model. We will elaborate on this in the next section.

Given any two models M and M',

$$\frac{\pi(M|x)}{\pi(M'|x)} = \left[\frac{\pi(M)}{\pi(M')}\right]\left[\frac{f(x|M)}{f(x|M')}\right];$$

that is, the ratio between posterior probabilities for models M and M' is the product of the prior odds ratio and the Bayes factor. We discussed Bayes factors in the context of hypothesis testing in Chapter 7. As in hypothesis testing, the Bayes factor measures the relative support for M versus M' as provided by the data x. Because of its relation to posterior probabilities, choosing a model on the basis of the posterior probability is equivalent to choosing a model using Bayes factors.

In contrast to *model selection* where a true model is assumed and the utility function explicitly seeks to find it, in *model comparison* we are simply interested in quantifying the relative support that two models receive from the data. The literature on model comparison is extensive and Bayes factors play an important role. For an extensive discussion on Bayes factors in model comparison, see Kass and Raftery (1995). Alternatively, the Bayesian information criterion (BIC) (or Schwarz criterion) also provides a means for comparing models. The BIC is defined as

$$\text{BIC}_M = 2\log\sup_{\theta_M} f(x|\theta_M, M) - d\log n,$$

where d is the number of parameters in model M. A crude approximation to the Bayes factor for comparing models M and M' is

$$\exp\left[-\frac{1}{2}\left(\text{BIC}_M - \text{BIC}_{M'}\right)\right].$$

Spiegelhalter *et al.* (2002) propose an alternative criterion known as the deviance information criterion for model checking and comparison, which can be applied to complex settings such as generalized linear models and hierarchical models.

11.1.3 Model averaging for prediction and selection

Let us now bring in more explicitly the typical goals of a statistical analysis. We begin with prediction. Each model in our collection specifies a predictive density

$$f(\tilde{x}|x, \theta_M, M)$$

for a future observation $\tilde{x}$. If the loss function depends only on our actions and $\tilde{x}$, then model uncertainty is taken into account by model averaging. Similarly to the previous section, we first compute the distribution $f(\tilde{x}|x)$, integrating out both models and parameters, and then attack the decision problem by minimizing posterior expected loss. For squared error loss, point predictions can be expressed as

$$E[\tilde{x}|x] = \sum_{M=1}^{M_0} E[\tilde{x}|x, M]\pi(M|x), \qquad (11.7)$$

a weighted average of model-specific prediction, with weights given by posterior model probabilities. If, instead, we were to decide on a predictive distribution, using the negative log loss function of Section 10.3, the optimal predictive distribution would be $f(\tilde{x}|x)$, which can also be expressed as a weighted average of model-specific densities.

Consider now a slightly different case: the decision maker is uncertain about which model is true, but must make predictions based on a single model. This applies, for example, when the model predictions must be produced in a setting where the model averaging approach is not computationally feasible. Formally, the decision maker has to choose a model M in $\{1, \ldots, M_0\}$ and, subsequently, make a prediction for a future observation $\tilde{x}$ based on data x, and assuming model M is true. We formalize this decision problem and denote $\boldsymbol{a} = (a, b)$ as the action where we select model a and use it to make a prediction b. For simplicity, we consider squared error loss

$$L(\boldsymbol{a}, \tilde{x}) = (b - \tilde{x})^2.$$

This is a nested decision problem. The model affects the final loss through the constraints it imposes on the prediction b. There is some similarity between this and the multistage decision problems we will encounter in Part Three, though a key difference here is that there is no additional data acquisition between the choice of the model and the prediction.

For any model a, the optimal prediction rule is

$$\delta_a^*(x) = \arg\min_b \int_{\mathcal{X}} (b - \tilde{x})^2 f(\tilde{x}|x, a)\, d\tilde{x} = E[\tilde{x}|x, a].$$

Plugging this solution back into the posterior expected loss function, the optimal model a^* minimizes

$$\mathcal{L}(a, \delta_a^*) = \int_{\mathcal{X}} (\delta_a^*(x) - \tilde{x})^2 f(\tilde{x}|x)\, d\tilde{x}.$$

Working on the integral above we get

$$\int_{\mathcal{X}} (\delta_a^*(x) - \tilde{x})^2 f(\tilde{x}|x) \, d\tilde{x} = \int_{\mathcal{X}} (\delta_a^*(x) - \tilde{x})^2 \left[\sum_{M=1}^{M_0} \pi(M|x) f(\tilde{x}|x, M) \right] d\tilde{x}$$

$$= \sum_{M=1}^{M_0} \pi(M|x) \int_{\mathcal{X}} \left[(\delta_a^*(x) - \delta_M^*(x))^2 + (\delta_M^*(x) - \tilde{x})^2 \right] f(\tilde{x}|x, M) d\tilde{x}$$

$$= \sum_{M=1}^{M_0} \pi(M|x) \left[(\delta_a^*(x) - \delta_M^*(x))^2 + \mathrm{Var}[\tilde{x}|M, x] \right].$$

The above expression depends on model a only through the first element of the sum in square brackets. The optimal model minimizes the weighted difference between its own model-specific prediction and the predictions of the other possible models, with weights given by the posterior model probabilities.

We can also compare this to $\delta^*(x) = \sum_{M=1}^{M_0} \delta_M^*(x)\pi(M|x)$, the posterior averaged prediction. This is the global optimum when one is allowed to use predictions based on all models rather than being constrained to using a single one, and coincides with decision rule (11.7). Manipulating the first term of posterior expected loss above a little further we get

$$\sum_{M=1}^{M_0} \pi(M|x)(\delta_a^*(x) - \delta_M^*(x))^2 = (\delta_a^*(x) - \delta^*(x))^2 + \sum_{M=1}^{M_0} (\delta_M^*(x) - \delta^*(x))^2 \pi(M|x).$$

Thus, the best model gives a prediction rule δ_a^* closest to the posterior averaged prediction δ^*.

We now move to the case when the decision is about the entire prediction density. Let $\boldsymbol{a} = (a, b)$ denote the action where we choose model a and subsequently a density b as the predictive density for a future observation $\tilde{x}$. Let $L(\boldsymbol{a}, \tilde{x})$ denote the loss function. We assume that such a loss function is a proper scoring rule, so that the optimal choice for a predictive density is in fact the actual belief, that is $\delta_a^*(x) = f(\tilde{x}|x, a)$. Then the posterior expected loss of choosing model a and proceeding optimally is

$$\mathcal{L}(a, \delta_a^*) = \int_{\mathcal{X}} \left(\sum_{M=1}^{M_0} L(a, f(\cdot|x, a), \tilde{x}) f(\tilde{x}|x, M) \pi(M|x) \right) d\tilde{x}. \tag{11.8}$$

The optimal strategy is to choose the model that minimizes equation (11.8). San Martini and Spezzaferri (1984) consider the logarithmic loss function corresponding to the scoring rule of Section 10.3. In our context, this implies that model M is preferred to model M' iff

$$\int_{\mathcal{X}} \sum_{M''} \log \left(\frac{f(\tilde{x}|x, M)}{f(\tilde{x}|x, M')} \right) f(\tilde{x}|M'') \pi(M''|x) \, d\tilde{x} > 0.$$

San Martini and Spezzaferri (1984) further develop this choice criterion in the case of two nested linear regression models, and under additional assumptions regarding prior distributions observe that the criterion takes the form

$$\text{LR} - k(d_M - d_{M'}),\qquad(11.9)$$

where LR is the likelihood ratio statistic, d_M is the number of regression parameters in model M, and

$$k = \log\left(n\left[\frac{2e^{\text{LR}/n} - 1}{\text{LR}/n} - 1\right]^{\frac{2}{d_M - d_{M'}}}\right),$$

with n denoting the number of observations. Equation (11.9) resembles the comparison resulting from the AIC (Akaike 1973) for which $k = 2$ and the BIC (Schwartz 1978) where $k = \log(n)$. Poskitt (1987) provides another decision-theoretic development for a model selection criterion that resembles the BIC, assuming a continuous and bounded utility function.

11.2   Model elaborations

So far our approach to questioning our "small world" has been to make it bigger and deal with the new unknowns according to doctrine. In some settings, this approach also gives us guidance on how to make the world small again—by focusing back on a single model, perhaps different from the one we started out with. Here we briefly consider a different perspective, which is based on looking at small perturbations (or "elaborations") of the small world that are designed to explore whether bigger worlds are likely to change our behavior substantially, without actually having to build a complete probabilistic representation for those.

In more statistical language, we are interested in model criticism—we plan by default to consider a specific model and we wish to get a sense for whether this choice may be inadequate. In the vast majority of applications this task is addressed by a combination of significance testing, typically for goodness of fit of the model, and exploratory data analysis, for example examination of residuals. For a great discussion and an entry point to the extensive literature see Box (1980). In this paper, Box describes scientific learning as "an *iterative process* consisting of Criticism and Estimation," and holds that "sampling theory is needed for exploration and ultimate criticism of entertained models in the light of data, while Bayes' theory is needed for estimation." An interesting related discussion is in Gelman *et al.* (1996).

While we do find much wisdom in Box's comment, in this section, we look at how traditional Bayesian decision theory can be harnessed to criticize a model, and also we revisit traditional criticism metrics from a decision perspective. In our discussion, model M will be the model initially proposed and the question is whether or not the decision maker should embark on more complex modeling before carrying out the decision analysis. A simple approach, reflecting some statistical practice, is to set up, for probing purposes, a second model M' and compare it to M. Bernardo and

Smith (1994) propose to choose an appropriate loss function for model evaluation, and look at the change in posterior expected loss as an indication of the worthiness of M' compared to M. A related approach is described in Carota *et al.* (1996) who use model elaboration to estimate the change in utility resulting from a larger model in a neighborhood of M.

If the current model holds, the joint density of the observed data x and unobserved parameters θ is

$$f(x, \theta | M) = f(x | \theta, M)\pi(\theta | M).$$

To evaluate model M we embed it in a class of models $\mathcal{M}$, called *model elaboration* (see Box and Tiao 1973, Smith 1983, West 1992). For example, suppose that in the current model data are exponential. To elaborate on this model we may consider $\mathcal{M}$ to be the family of Weibull distributions. The parameter ϕ will index models in $\mathcal{M}$ so that

$$f(x, \theta, \phi | \mathcal{M}) = f(y | \theta, \phi, \mathcal{M})\pi(\theta | \phi, \mathcal{M})\pi(\phi | \mathcal{M}).$$

The idea of an elaboration is that the original model M is still a member of $\mathcal{M}$ for some specific value ϕ_M of the elaboration parameter. In the model criticism situation, the prior distribution $\pi(\phi | \mathcal{M})$ is concentrated around ϕ_M, to reflect the initial assumption that M is the default model. One way to think of this prior is that it provides a formalization of DeGroot's pocket ϵ from Section 10.4.2.

To illustrate, let $\bar{y} | \theta, M \sim N(\theta, \sigma^2/n)$, and $\theta | M \sim N(\mu_0, \tau_0^2)$ where (σ^2, τ_0^2) are both known. A useful elaboration $\mathcal{M}$ is defined by

$$\bar{y} | \theta, \phi, \mathcal{M} \sim N(\theta, \sigma^2/n\phi) \tag{11.10}$$

$$\theta | \phi, \mathcal{M} \sim N(\mu_0, \tau_0^2/\phi)$$

$$\phi | \mathcal{M} \sim Gamma(\nu/2, 1/2\tau^2).$$

The elaboration parameter ϕ corresponds to a variance inflation factor, and $\phi_M = 1$. West (1985) and Efron (1986) show how this generalizes to the case where M is an exponential family.

For another illustration, a general way to connect two nonnested models M and M' defined by densities $f(x | M)$ and $g(x | M')$ is the elaboration

$$x | \phi, \mathcal{H} \sim c(\phi) f(x | M)^{\phi} g(x | M')^{1-\phi}$$

where $c(\phi)$ is an appropriate normalization function. See Cox (1962) for further discussion.

Given any elaboration, model criticism can be carried out by comparing the original and elaborated posterior expected losses, by comparing the posteriors $\pi(\theta | x, M)$ to $\pi(\theta | x, \phi, \mathcal{M})$, or, lastly, by comparing $\pi(\phi | x, \mathcal{M})$ to $\pi(\phi | \mathcal{M})$. Carota *et al.* (1996) consider the latter for defining a criticism measure, and define a loss function for capturing the distance between the two distributions. Even though this loss is not the actual terminal loss of the problem we started out with, if the data change the marginal posterior of ϕ by a large amount, it is likely that model M will perform poorly for the purpose of the original loss as well. To simplify notation, we drop $\mathcal{M}$ in the above distributions.

Motivated by the logarithmic loss function of Section 10.3, we define the diagnostic measure as

$$\Delta = E_{\phi|x}\left(\log\left(\frac{\pi(\phi|x)}{\pi(\phi)}\right)\right). \tag{11.11}$$

Δ is the Kullback–Leibler divergence between the prior and posterior distributions of ϕ. In a decision problem where the goal is to choose a probability distribution on ϕ, and the utility function is logarithmic, it measures the change in expected utility attributable to observing the data. We will return to this point in Chapter 13 in our discussion of the Lindley information.

Low values of Δ indicate agreement between prior and posterior distributions of the model elaboration parameter ϕ validating the current model M. Interpreting high values of Δ is trickier, though, and may require the investigation of $\pi(\phi|x)$. If the value of Δ is large and $\pi(\phi|x)$ is peaked around ϕ_0, the value for which the elaborated model is equal to M, then model M is adequate. Otherwise, it indicates that model M is inappropriate.

Direct evaluation of equation (11.11) is often difficult. As an alternative, the function Δ can be computed either by an approximation of the prior and the posterior, leading to analytical expressions for Δ, or by using a Monte Carlo approach (Müller and Parmigiani 1996). Another possibility is to consider a linearized diagnostic measure Δ_L which approximates Δ when the prior on ϕ is peaked around ϕ_M. To derive the linearized diagnostic Δ_L, observe that

$$\Delta = E_{\phi|x}\left(\log\left(\frac{f(x|\phi)}{m(x)}\right)\right).$$

Now, expanding $\log f(x|\phi)$ about ϕ_M we have

$$\log f(x|\phi) = \log f(x|\phi_M) + (\phi - \phi_M)\left.\frac{\partial}{\partial\phi}\log f(x|\phi)\right|_{\phi=\phi_M} + R(\phi)$$

for some remainder function $R(.)$. The linearized version Δ_L is defined as

$$\Delta_L = \log\frac{f(x|\phi_M)}{m(x)} + E_{\phi|x}(\phi - \phi_M)\left.\frac{\partial}{\partial\phi}\log f(x|\phi)\right|_{\phi=\phi_M}. \tag{11.12}$$

Δ_L combined three elements all of which are relevant model criticism statistics in their own right: the Savage density ratio defined by

$$\frac{f(x|\phi_M)}{m(x)};$$

the posterior expected value of $\phi - \phi_M$; and the marginal score function $(\partial/\partial\phi)\log f(x|\phi)$. The Savage density ratio is equivalent, under certain conditions, to the Bayes factor for the null hypothesis that $\phi = \phi_M$ against the family of alternatives defined by the elaboration. However, in the diagnostic context, the Bayes factor is a sufficient summary of the data only when the loss function is assigning the same penalty to all incorrect models or when the elaboration is binary. The diagnostic approach differs from a model choice analysis based on a Bayes factor as both Δ and Δ_L incorporate a penalty for the severity of the departure in the utility function.

11.3 Exercises

Problem 11.1 In the context of Section 11.2 suppose that M is such that $\bar{x}|M \sim N(\theta_0, 1/n)$ where θ_0 is a known mean. Consider the elaborated model $\mathcal{M}$ given by $\bar{x}|\phi, \mathcal{M} \sim N(\theta_0 + \phi, 1/n)$, and $\phi|\mathcal{M} \sim N(0, \tau^2)$. Show that the diagnostic Δ is

$$\Delta = \frac{n^2\tau^2}{2(n\tau^2 + 1)^2} (\bar{y} - \theta_0)^2 - \frac{n\tau^2}{2(n\tau^2 + 1)} + \frac{1}{2} \log(n\tau^2 + 1)$$

and the linearized diagnostic Δ_L is

$$\Delta_L = \frac{n^2\tau^2}{2(n\tau^2 + 1)} (\bar{y} - \theta_0)^2 + \frac{1}{2} \log(n\tau^2 + 1).$$

Comment on the strength and limitations of these as metrics for model criticism.

Part Three
Optimal Design

12

Dynamic programming

In this chapter we begin our discussion of *multistage decision problems*, where multiple decisions have to be made over time and with varying degrees of information. The salient aspect of multistage decision making is that, like in a chess game, decisions made now affect the worthiness of options available in the future and sometimes also the information available when making future decisions. In statistical practice, multistage problems can be used to provide a decision-theoretic foundation to the design of experiments, in which early decisions are concerned with which data to collect, and later decisions with how to use the information obtained. Chapters 13, 14, and 15 consider multistage statistical decisions in some detail. In this chapter we will be concerned with the general principles underlying multistage decisions for expected utility maximizers.

Finite multistage problems can be represented by *decision trees*. A decision tree is a graphical representation that allows us to visualize a large and complex decision problem by breaking it into smaller and simpler decision problems. In this chapter, we illustrate the use of decision trees in the travel insurance example of Section 7.3 and then present a general solution approach to two-stage (Section 12.3.1) and multistage (Section 12.3.2) decision trees. Examples are provided in Sections 12.4.1 through 12.5.2. While conceptually general and powerful, the techniques we will present are not of easy implementation: we will discuss some of the computational issues in Section 12.7.

The main principle used in solving multistage decision trees is called *backwards induction* and it emerged in the 1950s primarily through the work of Bellman (1957) on *dynamic programming*.

Featured book (chapter 3):

Bellman, R. E. (1957). *Dynamic programming*, Princeton University Press.

Our discussion is based on Raiffa and Schlaifer (1961). Additional useful references are DeGroot (1970), Lindley (1985), French (1988), Bather (2000), and Bernardo and Smith (2000).

12.1   History

Industrial process control has been one of the initial motivating applications for dynamic programming algorithms. Nemhauser illustrates the idea as follows:

> For example, consider a chemical process consisting of a heater, reactor and distillation tower connected in series. It is desired to determine the optimal temperature in the heater, the optimal reaction rate, and the optimal number of trays in the distillation tower. All of these decisions are interdependent. However, whatever temperature and reactor rate are chosen, the number of trays must be optimal with respect to the output from the reactor. Using this principle, we may say that the optimal number of trays is determined as a *function* of the reactor output. Since we do not know the optimal temperature or reaction rate yet, *the optimal number of trays and return from the tower must be found for all feasible reactor outputs.*
>
> Continuing sequentially, we may say that, whatever temperature is chosen, the reactor rate and number of trays must be optimal with respect to the heater output. To choose the best reaction rate as a function of the heater output, we must account for the dependence of the distillation tower of the reactor output. But we already know the optimal return from the tower as a function of the reactor output. Hence, the optimal reaction rate can be determined as a function of the reactor input, by optimizing the reactor together with the optimal return from the tower as a function of the reactor output.
>
> In making decisions sequentially as a function of the preceding decisions, the first step is to determine the number of trays as a function of the reactor output. Then, the optimal reaction rate is established as a function of the input to the reactor. Finally, the optimal temperature is determined as a function of the input to the heater. Finding a *decision function*, we can optimize the chemical process one stage at a time. (Nemhauser 1966, pp. 6–7)

The technique just described to solve multidimensional decision problems is called *dynamic programming*, and is based on the principle of *backwards induction*. Backwards induction has its roots in the work of Arrow *et al.* (1949) on optimal stopping problems and that of Wald (1950) on sequential decision theory. In optimal stopping problems one has the option to collect data sequentially. At each decision node two options are available: either continue sampling, or stop sampling and take a terminal action. We will discuss this class of problems in some detail in Chapter 15. Richard Bellman's research on a large class of sequential problems in the early 1950s led to the first book on the subject (Bellman 1957) (see also Bellman and Dreyfus 1962). Bellman also coined the term dynamic programming:

> The problems we treat are programming problems, to use a terminology now popular. The adjective "dynamic", however, indicates that we are interested in processes in which time plays a significant role, and in which the order of operations may be crucial. (Bellman 1957, p. xi)

A far more colorful description of the political motivation behind this choice is reported in Bellman's autobiography (Bellman 1984), as well as in Dreyfus (2002).

The expression backward induction comes from the fact that the sequence of decisions is solved by reversing their order in time, as illustrated with the chemical example by Nemhauser (1966). Lindley explains the inductive technique as follows:

> For the expected utility required is that of going on and then doing the best possible from then onwards. Consequently in order to find the best decision *now* ... it is necessary to know the best decision *in the future*. In other words the natural time order of working from the present to the future is not of any use because the present optimum involves the future optimum. The only method is to work backwards in time: from the optimum future behaviour to deduce the optimum present behaviour, and so on back into the past. ... The *whole* of the future must be considered in deciding whether to go on. (Lindley 1961, pp. 42–43)

The dynamic programming method allows us to conceptualize and solve problems that would be far less tractable if each possible decision function, which depends on data and decisions that accumulate sequentially, had to be considered explicitly:

> In the conventional formulation, we consider the entire multi-stage decision process as essentially one stage, at the expense of vastly increasing the dimension of the problem. Thus, if we have an N-stage process where M decisions are to be made at each stage, the classical approach envisages an MN-dimensional single-stage process.... [I]n place of determining the optimal sequence of decisions from some *fixed* state of the system, we wish to determine the optimal decision to be made at any state of the system. ... The mathematical advantage of this formulation lies first of all in the fact that it reduces the dimension of the process to its proper level, namely the dimension of the decision which confronts one at any particular stage. This makes the problem analytically more tractable and computationally vastly simpler. Secondly, ... it furnishes us with a type of approximation which has a unique mathematical property, that of monotonicity of convergence, and is well suited to applications, namely, "approximation in policy space". (Bellman 1957, p. xi)

Roughly speaking, the technique allows one to transform a multistage decision problem into a series of one-stage decision problems and thus make decisions one at a time. This relies on the *principle of optimality* stated by Bellman:

> An optimal policy has the property that whatever the initial state and initial decision are, the remaining decisions must constitute an optimal policy with regard to the state resulting from the first decision. (Bellman 1957, p. 82)

This ultimately allows for computational advantages as explained by Nemhauser:

> We may say that a problem with N decision variables can be transformed into N subproblems, each containing only one decision variable. As a rule of thumb, the computations increase exponentially with the number of variables, but only linearly with the number of subproblems. Thus there can be great computational savings. Of this savings makes the difference between an insolvable problem and one requiring only a small amount of computer time. (Nemhauser 1966, p. 6)

Dynamic programming algorithms have found applications in engineering, economics, medicine, and most recently computational biology, where they are used to optimally align similar biological sequences (Ewens and Grant 2001).

12.2   The travel insurance example revisited

We begin our discussion of dynamic programming techniques by revisiting the travel insurance example of Section 7.3. In that section we considered how to optimally use the information about the medical test. We are now going to consider the sequential decision problem in which, in the first stage, you have to decide whether or not to take the test, and in the second stage you have to decide whether or not to buy insurance. This is an example of a *two-stage sequential decision problem*. Two-stage or, more generally, multistage decision problems can be represented by decision trees by introducing additional decision nodes and chance nodes. Figure 12.1 shows the decision tree for this sequential problem.

> in describing the tree from left to right the natural order of the events in time has been followed, so that at any point of the tree the past lies to our left and can be studied by pursuing the branches down to the trunk, and the future lies to our right along the branches springing from that point and leading to the tips of the tree. (Lindley 1985, p. 141)

In this example we use the loss function in the original formulation given in Table 7.6. However, as previously discussed, the Bayesian solution would be the same should we use regret losses instead, and this still applies in the multistage case.

We are going to work within the expected utility principle, and also adhere to the before/after axiom across all stages of decisions. So we condition on information as

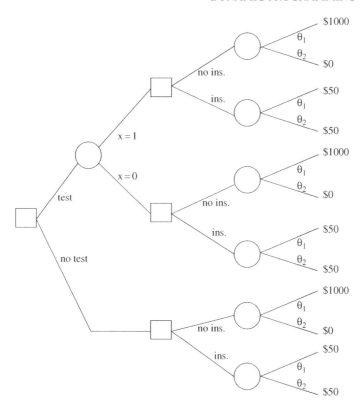

Figure 12.1   Decision tree for the two-stage travel insurance example.

it accrues, and we revise our probability distribution on the states of the world using the Bayes rule at each stage. In this example, to calculate the Bayes strategy we need to obtain the posterior probabilities of becoming ill given each of the possible values of x, that is

$$\pi(\theta|x) = \frac{f(x|\theta)\pi(\theta)}{m(x)}$$

for $\theta = \theta_1, \theta_2$ and $x = 0, 1$. Here

$$m(x) = \sum_{k=1}^{2} f(x|\theta_k)\pi(\theta_k)$$

is the marginal distribution of x. We get

$$m(x = 1) = 0.250$$
$$\pi(\theta_1|x = 0) = 0.004$$
$$\pi(\theta_1|x = 1) = 0.108.$$

Hence, we can calculate the posterior expected losses for rules δ_1 and δ_2 given $x = 0$ as

$$\delta_1: \quad \text{Expected loss} = 1000 \times 0.004 + 0 \times 0.996 = 4$$

$$\delta_2: \quad \text{Expected loss} = 50 \times 0.004 + 50 \times 0.996 = 50$$

and given $x = 1$ as

$$\delta_1: \quad \text{Expected loss} = 50 \times 0.108 + 50 \times 0.892 = 50$$

$$\delta_2: \quad \text{Expected loss} = 1000 \times 0.108 + 0 \times 0.892 = 108.$$

The expected losses for rules δ_0 and δ_3 remain as they were, because those rules do not depend on the data. Thus, if the test is performed and it turns out positive, then the optimal decision is to buy the insurance and the expected loss of this action is 50. If the test is negative, then the optimal decision is not to buy the insurance with expected loss 4. Incidentally, we knew this from Section 7.3, because we had calculated the Bayes rule by minimizing the Bayes risk. Here we verified that we get the same result by minimizing the posterior expected loss for each point in the sample space.

Now we can turn to the problem of whether or not to get tested. When no test is performed, the optimal solution is not to buy the insurance, and that has an expected loss of 30. When a test is performed, we calculate the expected losses associated with the decision in the first stage from this perspective: we can evaluate what happens if the test is positive and we proceed according to the optimal strategy thereafter. Similarly, we can evaluate what happens if the test is negative and we proceed according to the optimal strategy thereafter. So what we expect to happen is a weighted average of the two optimal expected losses, each conditional on one of the possible outcomes of x. The weights are the probabilities of the outcomes of x at the present time. Accordingly, the expected loss if the test is chosen is

$$50 \times m(x = 1) + 4 \times m(x = 0) = 50 \times 0.25 + 4 \times 0.75 = 15.5.$$

Comparing this to 30, the value we get if we do not test, we conclude that the optimal decision is to test. This is reassuring: the test is free (so far) and the information provided by the test would contribute to our decision, so it is logical that the optimum at the first stage is to acquire the information. Overall, the optimal sequential strategy is as follows. You should take the test. If the test is positive, then buy the medical insurance. If the test is negative, then you should not buy the medical insurance. The complete analysis of this decision problem is summarized in Figure 12.2.

Lindley comments on the rationale behind solving a decision tree:

> Like a real tree, a decision tree contains parts that act together, or cohere. Our method solves the easier problems that occur at the different parts of the tree and then uses the rules of probability to make them cohere . . . it is coherence that is the principal novelty and the major tool in what

we do. . . . The solution of the common subproblem, plus coherence fit-
ting together of the parts, produces the solution to the complex problem.
The subproblems are like bricks, the concrete provides the coherence.
(Lindley 1985, pp. 139,146)

In our calculations so far we ignored the costs of testing. How would the solution
change if the cost of testing was, say, $10? All terminal branches that stem out from
the action "test" in Figure 12.2 would have the added cost of $10. This would imply
that the expected loss for the action "test" would be 25.5 while the expected loss for
the action "no test" would still be 30.0. Thus, the optimal decision would still be to
take the test. The difference between the expected loss of 30 if no test information
is available and that of 15.5 if the test is available is the largest price you should

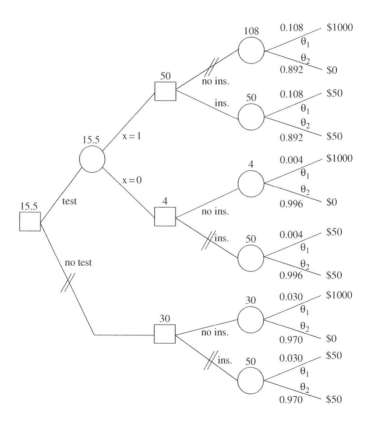

Figure 12.2   Solved decision tree for the two-stage travel insurance example.
Losses are given at the end of the branches. Above each chance node is the expected
loss, while above each decision node is the maximum expected utility. Alongside
the branches stemming from each chance node are the probabilities of the states of
nature. Actions that are not optimal are crossed out by a double line.

be willing to pay for the test. In the context of this specific decision this difference measures the worthiness of the information provided by the test—a theme that will return in Chapter 13.

Here we worked out the optimal sequential strategy in stages, by solving the insurance problem first, and then nesting the solution into the diagnostic test problem. Alternatively, we could have laid out all the possible sequential strategies. In this case there are six:

Stage 1	Stage 2
Do not test	Do not buy the insurance
Do not test	Buy the insurance
Test	Do not buy the insurance
Test	Buy the insurance if $x = 1$. Otherwise, do not
Test	Buy the insurance if $x = 0$. Otherwise, do not
Test	Buy the insurance

We could alternatively figure out the expected loss of each of these six options, and pick the one with the lowest value. It turns out that the answer would be the same. This is an instance of the equivalence between the *normal* form of the analysis—which lists all possible decision rules as we just did—and the *extensive* form of analysis, represented by the tree. We elaborate on this equivalence in the next section.

12.3   Dynamic programming

12.3.1   Two-stage finite decision problems

A general two-stage finite decision problem can be represented by the decision tree of Figure 12.3. As in the travel insurance example, the tree represents the sequential problem with decision nodes shown in a chronological order.

To formalize the solution we first introduce some additional notation. Actions will carry a superscript in brackets indicating the stage at which they are available. So $a_1^{(s)}, \ldots, a_{I_s}^{(s)}$ are the actions available at stage s. In Figure 12.3, s is either 1 or 2. For each action at stage 1 we have a set of possible observations that will potentially guide the decision at stage 2. For action $a_i^{(1)}$ these are indicated by $x_{i1}, \ldots, x_{iJ}$. For each action at stage 2 we have the same set of possible states of the world $\theta_1, \ldots, \theta_K$, and for each combination of actions and states of the world, as usual, an outcome z. If actions $a_{i_1}^{(1)}$ and $a_{i_2}^{(2)}$ are chosen and θ_k is the true state of the world, the outcome is $z_{i_1 i_2 k}$. A warning about notation: for the remainder of the book, we are going to abandon the loss notation in favor of the utility notation used in the discussion of foundation. Finally, in the above, we assume an equal number of possible states of the world and outcomes to simplify notation. A more general formulation is given in Section 12.3.2.

In our travel insurance example we used *backwards induction* in that we worked from the outermost branches of the decision tree (the right side of Figure 12.3) back to the root node on the left. We proceeded from the terminal branches to the root by

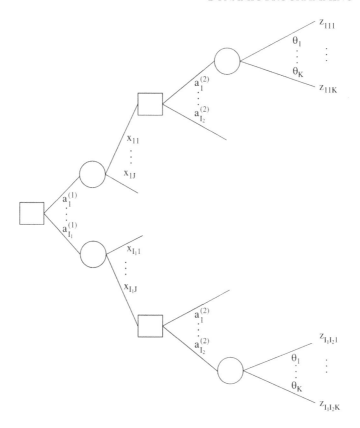

Figure 12.3   A general two-stage decision tree with finite states of nature and data.

alternating the calculation of expected utilities at the random nodes and maximization of expected utilities at the decision nodes. This can be formalized for a general two-stage decision tree as follows.

At the second stage, given that we chose action $a_{i_1}^{(1)}$ in the first stage, and that outcome x_{i_1j} was observed, we choose a terminal action $a^{*(2)}$ to achieve

$$\max_{1 \le i_2 \le I_2} \sum_{k=1}^{K} \pi(\theta_k|x_{i_1j})\, u\,(z_{i_1 i_2 k}). \tag{12.1}$$

At stage 2 there only remains uncertainty about the state of nature θ. As usual, this is addressed by taking the expected value of the utilities with respect to the posterior distribution. We then choose the action that maximizes the expected utility. Expression (12.1) depends on i_1 and j, so this maximization defines a function $\delta^{*(2)}(a_{i_1}^{(1)}, x_{i_1j})$ which provides us with a recipe for how to optimally proceed in every possible scenario. An important difference with Chapter 7 is the dependence on the first-stage decision, typically necessary because different actions may be available for $\delta^{(2)}$ to choose from, depending on what was chosen earlier.

Equipped with $\delta^{*(2)}$, we then step back to stage 1 and choose an action $a^{*(1)}$ to achieve

$$\max_{1 \leq i_1 \leq I_1} \left\{ \sum_{j=1}^{J} \left[\max_{1 \leq i_2 \leq I_2} \sum_{k=1}^{K} \pi(\theta_k | x_{i_1 j}) u(z_{i_1 i_2 k}) \right] m(x_{i_1 j}) \right\}. \tag{12.2}$$

The inner maximum is expression (12.1) and we have it from the previous step. The outer summation in (12.2) computes the expected utility associated with choosing the i_1th action at stage 1, and then proceeding optimally in stage 2. When the stage 1 decision is made, the maximum expected utilities ensuing from the available actions are uncertain because the outcome of x is not known. We address this by taking the expected value of the maximum utilities calculated in (12.1) with respect to the marginal distribution of x, for each action $a_{i_1}^{(1)}$. The result is the outer summation in (12.2). The optimal solution at the first stage is then the action that maximizes the outer summation. At the end of the whole process, an optimal sequential decision rule is available in the form of a pair $(a^{*(1)}, \delta^{*(2)})$.

To map this procedure back to our discussion of optimal decision functions in Chapter 7, imagine we are minimizing the negative of the function u in expression (12.2). The innermost maximization is the familiar minimization of posterior expected loss. The optimal $a^{*(2)}$ depends on x. In the terminology of Chapter 7 this optimization will define a formal Bayes rule with respect to the loss function given by the negative of the utility. The summation with respect to j computes an average of the posterior expected losses with respect to the marginal distribution of the observations, and so it effectively is a Bayes risk r, so long as the loss is bounded. In the two-stage setting we get a Bayes risk for every stage 1 action $a_{i_1}^{(1)}$, and then choose the action with the lowest Bayes risk. We now have a plan for what to do at stage 1, and for what to do at stage 2 in response to any of the potential experimental outcomes that result from the stage 1 decision.

In this multistage decision, the likelihood principle operates at stage 2 but not at stage 1. At stage 2 the data are known and the alternative results that never occurred have become irrelevant. At stage 1 the data are unknown, and the distribution of possible experimental results is essential for making the experimental design decision.

Another important connection between this discussion and that of Chapter 7 concerns the equivalence of normal and extensive forms of analysis. Using the Bayes rule we can rewrite the outer summation in expression (12.2) as

$$\sum_{j=1}^{J} \left[\max_{1 \leq i_2 \leq I_2} \sum_{k=1}^{K} \pi(\theta_k | x_{i_j}) u(z_{i_1 i_2 k}) \right] m(x_{i_1 j}) = \sum_{j=1}^{J} \left[\max_{1 \leq i_2 \leq I_2} \sum_{k=1}^{K} \pi(\theta_k) f(x_{i_1 j} | \theta_k) u(z_{i_1 i_2 k}) \right].$$

The function δ^* that is defined by maximizing the inner sum pointwise for each i and j will also maximize the average expected utility, so we can rewrite the right hand side as

$$\max_{\delta} \left[\sum_{j=1}^{J} \sum_{k=1}^{K} \pi(\theta_k) f(x_{i_1 j} | \theta_k) u(z_{i_1 i_2 k}) \right].$$

Here δ is not explicitly represented in the expression in square brackets. Each choice of δ specifies how subscript i_2 is determined as a function of i_1 and j. Reversing the order of summations we get

$$\max_{\delta} \left\{ \sum_{k=1}^{K} \pi(\theta_k) \left[\sum_{j=1}^{J} f(x_{i_1 j} | \theta_k) \, u \, (z_{i_1 i_2 k}) \right] \right\}.$$

Lastly, inserting this into expression (12.2) we obtain an equivalent representation as

$$\max_{1 \le i_1 \le I_1} \max_{\delta} \left\{ \sum_{k=1}^{K} \pi(\theta_k) \left[\sum_{j=1}^{J} f(x_{i_1 j} | \theta_k) \, u \, (z_{i_1 i_2 k}) \right] \right\}. \qquad (12.3)$$

This equation gives the solution in normal form: rather than alternating expectations and maximizations, the two stages are maximized jointly with respect to pairs $(a^{(1)}, \delta^{(2)})$. The inner summation gives the expected utility of a decision rule, given the state of nature, over repeated experiments. The outer summation gives the expected utility of a decision rule averaging over the states of nature. A similar equivalence was brought up earlier on in Chapter 7 when we discussed the relationship between the posterior expected loss (in the extensive form) and the Bayes risk (in the normal form of analysis).

12.3.2   More than two stages

We are now going to extend these concepts to more general multistage problems, and consider the decision tree of Figure 12.4. As before, stages are chronological from left to right, information may become available between stages, and each decision is dependent on decisions made in earlier stages. We will assume that the decision tree is bounded, in the sense that there is a finite number of both stages and decisions at each stage.

First we need to set the notation. As before, S is the number of decision stages. Now $a_0^{(s)}, \ldots, a_{I_s}^{(s)}$ are the decisions available at the sth stage, with $s = 1, \ldots, S$. At each stage, the action $a_0^{(s)}$ is different from the rest in that, if $a_0^{(s)}$ is taken, no further stages take place, and the decision problem terminates. Formally $a_0^{(s)}$ maps states to outcomes in the standard way. For each stopping action we have a set of relevant states of the world that constitute the domain of the action. At stage s, the possible states are $\theta_{01}^{(s)}, \ldots, \theta_{0K_s}^{(s)}$. For each continuation action $a_i^{(s)}, i > 0, 1 \le s < S$, we observe a random variable $x_i^{(s)}$ with possible values $x_{i1}^{(s)}, \ldots, x_{iJ_s}^{(s)}$. If stage s is reached, the decision may depend on all the decisions that were made, and all the information that was accrued in preceding stages. We concisely refer to this information as the history, and use the notation $\mathcal{H}_{s-1}$ where

$$\mathcal{H}_{s-1} = \{ a_{i_1}^{(1)}, \ldots, a_{i_{s-1}}^{(s-1)}, x_{i_1 j_1}^{(1)}, \ldots, x_{i_{s-1} j_{s-1}}^{(s-1)} \}, \qquad s = 2, \ldots, S.$$

For completeness of notation, the (empty) history prior to stage 1 is denoted by $\mathcal{H}_0$. Finally, at the last stage S, the set of states of the world constituting the domain of action $a_{i_S}^{(S)}$ is $\theta_{i_S 1}^{(S)}, \ldots, \theta_{i_S K_S}^{(S)}$.

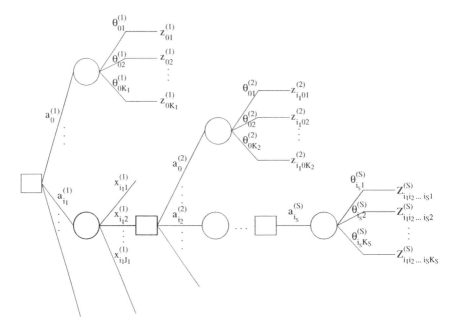

Figure 12.4   Decision tree for a generic multistage decision problem.

It is straightforward to extend this setting to the case in which J_s also depends on i without any conceptual modification. We avoid it here for notational simplicity.

Dynamic programming proceeds as follows. We start by solving the decision problem from the last stage, stage S, by maximizing expected utility for every possible history. Then, conditional on the optimal choice made at stage S, we solve the problem at stage $S - 1$, by maximizing again the expected maximum utility. This procedure is repeated until we solve the decision problem at the first stage. Algorithmically, we can describe this recursive procedure as follows:

I. At stage S:

    (i) For every possible history $\mathcal{H}_{S-1}$, compute the expected utility of the actions $a_i^{(S)}$, $i = 0, \ldots, I_S$, available at stage S, using

$$\mathcal{U}^S(a_i^{(S)}) = \sum_{k=1}^{K_S} u_{ik}^{(S)}(\mathcal{H}_{S-1}) \pi(\theta_{ik}^{(S)} | \mathcal{H}_{S-1}) \qquad (12.4)$$

    where $u_{ik}^{(S)}(\mathcal{H}_{S-1}) = u(z_{i_1 \ldots i_{S-1} ik}^{(S)})$.

    (ii) Obtain the optimal action, that is

$$a^{*(S)}(\mathcal{H}_{S-1}) = \arg \max_i \mathcal{U}^S(a_i^{(S)}). \qquad (12.5)$$

This is a function of $\mathcal{H}_{s-1}$ because both the posterior distribution of θ and the utility of the outcomes depend on the past history of decisions and observations.

II. For stages $S - 1$ through 1, repeat:

(i) For every possible history $\mathcal{H}_{s-1}$, compute the expected utility of actions $a_i^{(s)}$, $i = 0, \dots, I_s$, available at stage s, using

$$\mathcal{U}^s(a_i^{(s)}) = \sum_{j=1}^{J_s} u_{ij}^{(s)}(\mathcal{H}_{s-1})\, m\,(x_{ij}^{(s)}|\mathcal{H}_{s-1}) \qquad i > 0 \qquad (12.6)$$

$$\mathcal{U}^s(a_0^{(s)}) = \sum_{k=1}^{K_s} u_{0k}^{(s)}(\mathcal{H}_{s-1})\pi\,(\theta_{0k}^{(s)}|\mathcal{H}_{s-1})$$

where now $u_{ij}^{(s)}$ are the expected utilities associated with the optimal continuation from stage $s+1$ on, given that that $a_i^{(s)}$ is chosen and $x_{ij}^{(s)}$ occurs. If we indicate by $\{\mathcal{H}_{s-1}, a_i^{(s)}, x_{ij}^{(s)}\}$ the resulting history, then we have

$$u_{ij}^{(s)} = \mathcal{U}^{s+1}\big(a^{*(s+1)}(\{\mathcal{H}_{s-1}, a_i^{(s)}, x_{ij}^{(s)}\})\big) \qquad i > 0. \qquad (12.7)$$

A special case is the utility of the decision to stop, which is

$$u_{0k}^{(s)} = u(z_{i_1 \dots i_{s-1} 0k}^{(s)}).$$

(ii) Obtain the optimal action, that is

$$a^{*(s)}(\mathcal{H}_{s-1}) = \arg\max_i\ \mathcal{U}^s(a_i^{(s)}). \qquad (12.8)$$

(iii) Move to stage $s - 1$, or stop if $s = 1$.

This algorithm identifies a sequence of functions $a^{*(s)}(\mathcal{H}_{s-1})$, $s = 1, \dots, S$, that defines the optimal sequential solution.

In a large class of problems we only need to allow z to depend on the unknown θ and the terminal action. This still covers problems in which the first $S - 1$ stages are information gathering and free, while the last stage is a statistical decision involving θ. See also Problem 12.4. We now illustrate this general algorithm in two artificial examples, and then move to applications.

12.4   Trading off immediate gains and information

12.4.1   The secretary problem

For an application of the dynamic programming technique we will consider a version of a famous problem usually referred to in the literature under a variety of politically incorrect denominations such as the "secretary problem" or the "beauty-contest problem." Lindley (1961), DeGroot (1970), and Bernardo and Smith (1994) all discuss this example, and there is a large literature including versions that can get quite complicated. The reason for all this interest is that this example, while relatively simple, captures one of the most fascinating sides of sequential analysis: that is, the ability of optimal solutions to negotiate trade-offs between immediate gains and information-gathering activities that may lead to greater future gains.

We decided to make up yet another story for this example. After successful interviews in S different companies you are the top candidate in each one of them. One by one, and sequentially, you are going to receive an offer from each company. Once you receive an offer, you may accept it or you may decline it and wait for the next offer. You cannot consider more than one offer at the time, and if you decide to decline an offer, you cannot go back and accept it later. You have no information about whether the offers that you have not seen are better or worse than the ones you have seen: the offers come in random order. The information you do have about each offer is its relative rank among the offers you previously received. This is reasonable if the companies are more or less equally desirable to you, but the conditions of the offers vary. If you refuse all previous offers, and you reach the Sth stage, you take the last offer.

When should you accept an offer? How early should you do it and how high should the rank be? If you accept too soon, then it is possible that you will miss the opportunity to take better future offers. On the other hand, if you wait too long, you may decline offers that turn out to be better than the one you ultimately accept. Waiting gives you more information, but fewer opportunities to use it.

Figure 12.5 has a decision tree representation of two consecutive stages of the problem. At stage s, you examined s offers. You can choose decision $a_1^{(s)}$, to wait for

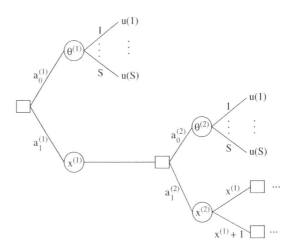

Figure 12.5 Decision tree representing two stages of the sequential problem of Section 12.4.1. At any stage s, if action $a_0^{(s)}$ is chosen, the unknown rank $\theta^{(s)}$ takes a value between 1 and S. If action $a_1^{(s)}$ is chosen, a relative rank between 1 and s is observed.

the next offer, or $a_0^{(s)}$, to accept the current offer and stop the process. If you stop at s, your utility will depend on the unknown rank $\theta^{(s)}$ of offer s: $\theta^{(s)}$ is a number between 1 and S. At s you make a decision based on the observed rank $x^{(s)}$ of offer s among the s already received. This is a number between 1 and s. In summary, $\theta^{(s)}$ and $x^{(s)}$ refer to the same offer. The first is the unknown rank if the offer is accepted. The second is the known rank being used to make the decision. If you continue until stage S, the true relative rank of each offer will become known also.

The last assumption we need is a utility function. We take $u(\theta)$ be the utility of selecting the job that has rank θ among all S options, and assume $u(1) \geq u(2) \geq \cdots \geq u(S)$. This utility will be the same irrespective of the stage at which the decision is reached.

We are now able to evaluate the expected utility of both accepting the offer (stopping) and declining it (continuing) at stage s. If we are at stage s we must have rejected all previous offers. Also the relative ranks of the older offers are now irrelevant. Therefore the only part of the history $\mathcal{H}_s$ that matters is that the sth offer has rank $x^{(s)}$.

Stopping. Consider the probability $\pi(\theta^{(s)}|x^{(s)})$ that the sth offer, with observed rank $x^{(s)}$, has, in fact, rank $\theta^{(s)}$ among all S offers. As we do not have any a priori information about the offers that have not yet been made, we can evaluate $\pi(\theta^{(s)}|x^{(s)})$ as the probability that, in a random sample of s companies taken from a population with S companies, $x^{(s)} - 1$ are from the highest-ranking $\theta^{(s)} - 1$ companies, one has rank $\theta^{(s)}$, and the remaining $s - x^{(s)}$ are from the lowest-ranking $S - \theta^{(s)}$ companies. Thus, we have

$$\pi(\theta^{(s)}|x^{(s)}) = \binom{\theta^{(s)} - 1}{x^{(s)} - 1}\binom{S - \theta^{(s)}}{s - x^{(s)}} \bigg/ \binom{S}{s}, \quad \text{for } x^{(s)} \leq \theta^{(s)} \leq S - s + x^{(s)}. \quad (12.9)$$

Let $\mathcal{U}(a_0^{(s)})$ denote the expected utility of decision $a_0^{(s)}$; that is, of accepting the sth offer. Equation (12.9) implies that

$$\mathcal{U}(a_0^{(s)}) = \sum_{\theta = x^{(s)}}^{S - s + x^{(s)}} u(\theta) \binom{\theta - 1}{x^{(s)} - 1}\binom{S - \theta}{s - x^{(s)}} \bigg/ \binom{S}{s} \qquad s = 1, \ldots, S. \quad (12.10)$$

Waiting. On the other hand, we need to consider the expected utility of taking decision $a_1^{(s)}$ and continuing optimally after s. Say again the sth offer has rank $x^{(s)}$, and define $\mathcal{U}^*(a_1^{(s)}) = b(s, x^{(s)})$. If you decide to wait for the next offer, the probability that the next offer will have an observed rank $x^{(s+1)}$, given that the sth offer has rank $x^{(s)}$, is $1/(s + 1)$, as the offers arrive at random and all the $s + 1$ values of $x^{(s+1)}$ are equally probable. Thus, the expected utility of waiting for the next offer and continuing optimally is

$$\mathcal{U}^*(a_1^{(s)}) = b(s, x^{(s)}) = \frac{1}{s + 1} \sum_{x=1}^{s+1} b(s + 1, x). \quad (12.11)$$

Also, at stage S we must stop, so the following relation must be satisfied:

$$b(S, x^{(S)}) = \mathcal{U}^*(a_1^{(S)}) = \mathcal{U}(a_0^{(S)}) \qquad x^{(S)} = 1, \ldots, S. \tag{12.12}$$

Because we know the right hand side from equation (12.10), we can use equation (12.11) to recursively step back and determine the expected utilities of continuing. The optimal solution at stage s is, depending on $x^{(s)}$, to wait for the next offer if $\mathcal{U}^*(a_1^{(s)}) > \mathcal{U}(a_0^{(s)})$ or accept the current offer if $\mathcal{U}^*(a_1^{(s)}) = \mathcal{U}(a_0^{(s)})$.

A simple utility specification that allows for a more explicit solution is one where your only goal is to get the best offer, with all other ranks being equally disliked. Formally, $u(1) = 1$ and $u(\theta) = 0$ for $\theta > 1$. It then follows from equation (12.10) that, for any s,

$$\mathcal{U}(a_0^{(s)}) = \begin{cases} s/S & \text{if } x = 1 \\ 0 & \text{if } x > 1. \end{cases} \tag{12.13}$$

This implies that $\mathcal{U}^*(a_1^{(s)}) > \mathcal{U}(a_0^{(s)})$ whenever $x^{(s)} > 1$. Therefore, you should wait for the next offer if the rank of the sth offer is not the best so far. So far this is a bit obvious: if your current offer is not the best so far, it cannot be the best overall, and you might as well continue. But what should you do if the current offer has observed rank 1?

The largest expected utility achievable at stage s is the largest of the expected utilities of the two possible decisions. Writing the expected utility of continuing using equation (12.11), we have

$$\mathcal{U}^*(a_1^{(s)}) = \max\left(\mathcal{U}(a_0^{(s)}), \frac{1}{s+1} \sum_{x=1}^{s+1} b(s+1, x)\right). \tag{12.14}$$

Let $v(s) = (1/(s+1)) \sum_{x=1}^{s+1} b(s+1, x)$. One can show (see Problem 12.5) that

$$v(s) = \frac{1}{s+1}\mathcal{U}^*(a_1^{(s+1)}) + \frac{s}{s+1}v(s+1). \tag{12.15}$$

At the last stage, from equations (12.12) and (12.14) we obtain that $\mathcal{U}^*(a_1^{(S)}) = 1$ and $v(S) = 0$. By backward induction on s it can be verified that (see Problem 12.5)

$$v(s) = \frac{s}{S}\left(\frac{1}{S-1} + \frac{1}{S-2} + \cdots + \frac{1}{s}\right). \tag{12.16}$$

Therefore, from equations (12.14) and (12.16) at stage s, with $x^{(s)} = 1$

$$\mathcal{U}^*(a_1^{(s)}) = \max\left(\frac{s}{S}, \frac{s}{S}\left(\frac{1}{S-1} + \cdots + \frac{1}{s}\right)\right) \tag{12.17}$$

Let s^* be the smallest positive integer so that $(1/(S-1) + \cdots + 1/s) \leq 1$. The optimal procedure is to wait until s^* offers are made. If the sth offer is the best so far far, accept it. Otherwise, wait until you reach the best offer so far and accept it. If you reach stage S you have to accept the offer irrespective of the rank. If S is large, $s^* \approx S/e$. So you should first just wait and observe approximately $1/e \approx 36\%$ of the offers. At that point the information-gathering stage ends and you accept the first offer that ranks above all previous offers.

Lindley (1961) provides a solution to this problem when the utility is linear in the rank, that is when $u(\theta) = S - \theta$.

12.4.2   The prophet inequality

This is an example discussed in Bather (2000). It has points in common with the one in the previous section. What it adds is a simple example of a so-called "prophet inequality"—a bound on the expected utility that can be expected if information that accrues sequentially was instead revealed in its entirety at the beginning.

You have the choice of one out of a set of S options. The values of the options are represented by nonnegative and independent random quantities $\theta^{(1)}, \ldots, \theta^{(S)}$ with known probability distributions. Say you decided to invest a fixed amount of money for a year, and you are trying to choose among investment options with random returns. Let μ_s denote the expected return of option s, that is $\mu_s = E[\theta^{(s)}]$, $s = 1, \ldots, S$. Assume that these means are finite, and assume that returns are in the utility scale.

Let us first consider two extreme situations. If you are asked to choose a single option now, your expected utility is maximized by choosing the option with largest mean, and that maximum is $g_S = \max(\mu_1, \ldots, \mu_S)$. At the other extreme, if you are allowed to communicate with a prophet that knows all the true returns, then your expected utility is $h_S = E[\max(\theta^{(1)}, \ldots, \theta^{(S)})]$. Now, consider the sequential case in which you can ask the prophet about one option at the time. Suppose that the values of $\theta^{(1)}, \ldots, \theta^{(S)}$ are revealed sequentially in this order. At each stage, you are only allowed to ask about the next option if you reject the ones examined so far. Cruel perhaps, but it makes for another interesting trade-off between learning and immediate gains.

At stage s, decisions can be $a_1^{(s)}$, wait for the next offer, or $a_0^{(s)}$, accept the offer and stop bothering the prophet. If you stop, your utility will be the value $\theta^{(s)}$ of the option you decided to accept. On the other hand, the expected utility of continuing $\mathcal{U}^{(s)}(a_1^{(s)})$. At the last stage, when there is only one option to be revealed, the expectation of the best utility that can be reached is $\mathcal{U}^{*(S)}(a_0^{(S)}) = \mu_S$, because $a_0^{(S)}$ is the only decision available. So when folding back to stage $S - 1$, you either accept option $\theta^{(S-1)}$ or wait to see $\theta^{(S)}$. That is, the maximum expected utility is $\mathcal{U}^{*(S-1)} = E[\max(\theta^{(S-1)}, \mu_S)] = E[\max(\theta^{(S-1)}, \mathcal{U}^{*(S)})]$. With backward induction, we find that the expected maximum utility at stage s is

$$\mathcal{U}^{*(s)} = E[\max(\theta^{(s)}, \mathcal{U}^{*(s+1)})], \quad \text{for } s = 1, \ldots, S - 1. \tag{12.18}$$

To simplify notation, let w_s denote the expected maximum utility when there are s options to consider, that is $w_s = \mathcal{U}^{*(S-s+1)}$, $s = 1, \ldots, S$. What is the advantage of obtaining full information of the options over the sequential procedure? We will prove the prophet's inequality

$$h_S \leq 2w_S. \tag{12.19}$$

This inequality is explained by Bather as follows:

> [I]t shows that a sequential decision maker can always expect to gain at least half as much as a prophet who is able to foresee the future, and make a choice with full information on all the values available. (Bather 2000, p. 111)

Proof: To prove the inequality, first observe that $w_1 = \mu_S$. Then, since $w_2 = E[\max(\theta^{(S-1)}, w_1)]$, it follows that $w_2 \geq E[\theta^{(S-1)}] = \mu_{S-1}$ and $w_2 \geq w_1$. Recursively, we can prove that the sequence w_s is increasing, that is

$$w_1 \leq w_2 \leq \cdots \leq w_S. \tag{12.20}$$

Define $y_1 = \theta^{(1)}$ and $y_r = \max(\theta^{(r)}, w_{r-1})$ for $r = 2, \ldots, S$. Moreover, define $w_0 \equiv 0$ and $z_r = \max(\theta^{(r)} - w_{r-1}, 0)$. This implies that $y_r = w_{r-1} + z_r$. From (12.20), $w_{r-1} \leq w_{S-1}$ for $r = 1, \ldots, S$. Thus,

$$y_r \leq w_{S-1} + z_r, \text{ for } r = 1, \ldots, S. \tag{12.21}$$

For any r, it follows from our definitions that $\theta^{(r)} \leq y_r$. Thus,

$$\max(\theta^{(1)}, \ldots, \theta^{(S)}) \leq \max(y_1, \ldots, y_S)$$
$$\text{using equation (12.21),}$$
$$\leq w_{S-1} + \max(z_1, \ldots, z_S). \tag{12.22}$$

By definition, $z_r \geq 0$. Thus, $\max(z_1, \ldots, z_S) \leq z_1 + \cdots + z_S$. By combining the latter inequality with that in (12.22), and taking expectations, we obtain

$$E[\max(\theta^{(1)}, \ldots, \theta^{(S)})] \leq w_{S-1} + E[z_1 + \cdots + z_S]$$
$$\Longleftrightarrow h_S \leq w_{S-1} + \sum_{r=1}^{S} E[z_r] = w_{S-1} + \sum_{r=1}^{S} E[y_r - w_{r-1}]$$
$$\Longleftrightarrow h_S \leq w_{S-1} + \sum_{r=1}^{S} (w_r - w_{r-1})$$
$$\Longleftrightarrow h_S \leq w_{S-1} + w_S \leq 2w_S, \tag{12.23}$$

which completes the proof.   □

12.5   Sequential clinical trials

Our next set of illustrations, while still highly stylized, bring us far closer to real applications, particularly in the area of clinical trials. Clinical trials are scientific experiments on real patients, and involve the conflicting goals of learning about treatments, so that a broad range of patients may benefit from medical advances, and ensuring a good outcome for the patients enrolled in the trial. For a more general discussion on Bayesian sequential analysis of clinical trials see, for example, Berry (1987) and Berger and Berry (1988). While we will not examine this controversy in detail here, in the sequential setting the Bayesian and frequentist approaches can differ in important ways. Lewis and Berry (1994) discuss the value of sequential decision-theoretic approaches over conventional approaches. More specifically, they compare performances of Bayesian decision-theoretic designs to classical group-sequential designs of Pocock (1977) and O'Brien and Fleming (1979). This comparison is within the hypothesis testing framework in which the parameters associated with the utility function are chosen to yield classical type I and type II error rates comparable to those derived under the Pocock and O'Brien and Fleming designs. While the power functions of the Bayesian sequential decision-theoretic designs are similar to those of the classical designs, they usually have smaller-mean sample sizes.

12.5.1   Two-armed bandit problems

A decision maker can take a fixed number n of observations sequentially. At each stage, he or she can choose to observe either a random variable x with density $f_x(.|\theta_0)$ or a random variable y with $f_y(.|\theta_1)$. The decision problem is to find a sequential procedure that maximizes the expected value of the sum of the observations. This is an example of a *two-armed bandit problem* discussed in DeGroot (1970) and Berry and Fristedt (1985). For a clinical trial connection, imagine the two arms being two treatments for a particular illness, and the random variables being measures of well-being of the patients treated. The goal is to maximize the patient's well-being, but some early experimentation is necessary on both treatments to establish how to best do so.

We can formalize the solution to this problem using dynamic programming. At stages $s = 0, \ldots, n$, the decision is either $a_0^{(s)}$, observe x, or $a_1^{(s)}$, observe y. Let π denote the joint prior distribution for $\theta = (\theta_0, \theta_1)$ and consider the maximum expected sum of the n observations given by

$$V^{(n)}(\pi) = \max\left(\mathcal{U}(a_0^{(n)}), \mathcal{U}(a_1^{(n)})\right) \qquad (12.24)$$

where $\mathcal{U}(a_i^{(n)})$, $i = 0, 1$, is calculated under distribution π.

Suppose the first observation is taken from x. The joint posterior distribution of (θ_0, θ_1) is π_x and the expected sum of the remaining $(n - 1)$ observations is given by $V^{(n-1)}(\pi_x)$. Then, the expected sum of all n observations is $\mathcal{U}(a_0^{(n)}) = E_x[x + V^{(n-1)}(\pi_x)]$. Similarly, if the first observation comes from y, the expected sum of all n observations is $\mathcal{U}(a_1^{(n)}) = E_y[y + V^{(n-1)}(\pi_y)]$. Thus, the optimal procedure $a^{*(n)}$ has expected utility

$$V^{(n)}(\pi) = \max \left\{ E_x \left[x + V^{(n-1)}(\pi_x) \right], E_y \left[y + V^{(n-1)}(\pi_y) \right] \right\}. \tag{12.25}$$

With the initial condition that $V^{(0)}(\pi) \equiv 0$, one can solve $V^{(s)}(\pi), s = 1, \ldots, n$, and find the sequential optimal procedure by induction.

12.5.2   Adaptive designs for binary outcomes

To continue in a similar vein, consider a clinical trial that compares two treatments, say A and B. Patients arrive sequentially, and each can only receive one of the two therapies. Therapy is chosen on the basis of the information available up until that point on the treatments' efficacy. Let n denote the total number of patients in the trial. At the end of the trial, the best-looking therapy is assigned to $(N - n)$ additional patients. The total number N of patients involved is called the patient horizon. A natural question in this setting is how to optimally allocate patients in the trial to maximize the number of patients who respond positively to the treatment over the patient horizon N. This problem is solved with dynamic programming.

Let us say that response to treatment is a binary event, and call a positive response simply "response." Let θ_A and θ_B denote the population proportion of responses under treatments A and B, respectively. For a prior distribution, assume that θ_A and θ_B are independent and with a uniform distribution on $(0, 1)$. Let n_A denote the number of patients assigned to treatment A and r_A denote the number of responses among those n_A patients. Similarly, define n_B and r_B for treatment B.

To illustrate the required calculations, assume $n = 4$ and $N = 100$. When using dynamic programming in the last stage, we need to consider all possibilities for which $n_A + n_B = n = 4$. Consider for example $n_A = 3$ and $n_B = 1$. Under this configuration, r_A takes values in $0, \ldots, 3$ while r_B takes values 0 or 1. Consider $n_A = 3, r_A = 2, n_B = 1, r_B = 0$. The posterior distribution of θ_A is $Beta(3, 2)$ and the predictive probability of response is $3/5$, while θ_B is $Beta(1, 2)$ with predictive probability $1/3$. Because $3/5 > 1/3$ the remaining $N - n = 100 - 4 = 96$ patients are assigned to A with an expected number of responses given by $96 \times 3/5 = 57.6$. Similar calculations can be carried out for all other combinations of treatment allocations and experimental outcomes.

Next, take one step back and consider the cases in which $n_A + n_B = 3$. Take $n_A = 2, n_B = 1$ for an example. Now, r_A takes values $0, 1$, or 2 while r_B takes values 0 or 1. Consider the case $n_A = 2, r_A = 1, n_B = 1, r_B = 0$. The current posterior distributions are $Beta(2, 2)$ for θ_A and $Beta(1, 2)$ for θ_B. If we are allowed one additional observation in the trial before we get to $n = 4$, to calculate the expected number of future responses we need to consider two possibilities: assigning the last patient to treatment A or to treatment B.

Choosing A moves the process to the case $n_A = 3, n_B = 1$. If the patient responds, $r_A = 2, r_B = 0$ with predictive probability $2/4 = 1/2$. Otherwise, with probability $1/2$, the process moves to the case $r_A = 1, r_B = 0$. The maximal number of responses is 57.6 (as calculated above) with probability $1/2$ and 38.4 (this calculation is omitted and left as an exercise) with probability $1/2$. Thus, the expected number of future responses with A is $(1+57.6) \times 1/2 + 38.4 \times 1/2 = 48.5$. Similarly, when treatment B is chosen for the next patient $n_A = 2, n_B = 2$. If the patient responds to treatment B, which happens with probability $1/3$, the process moves to $r_A = 1, r_B = 1$; otherwise it moves to $r_A = 1, r_B = 0$. One can show that the expected number of successes under treatment B is 48.3. Because $48.5 > 48.3$, the optimal decision is to assign the next patient to treatment A.

After calculating the maximal expected number of responses for all cells under which $n_A + n_B = 3$, we turn to the case with $n_A + n_B = 2$ and so on until $n_A + n_B = 0$. Figure 12.6 shows the optimal decisions when $n = 4$ and $N = 100$ for each combination of n_A, r_A, n_B, r_B. Each separated block of cells corresponds to a pair (n_A, n_B). Within each block, each individual cell is a combination of $r_A = 0, \ldots, n_A$ and $r_B = 0, \ldots, n_B$. In the figure, empty square boxes denote cases for which A is the optimal decision while full square boxes denote cases for which B is optimal. An asterisk represents cases where both treatments are optimal. For instance, the left bottom corner of the figure has $n_A = r_A = 0, n_B = r_B = 0$, for which both treatments are optimal. Then, in the top left block of cells when $n_A = r_A = 0$ and $n_B = 4$, the optimal treatment is A for $r_B = 0, 1$. Otherwise, B is optimal. Figure 12.7 shows the optimal design under two different prior choices on θ_A. These priors were chosen to have mean 0.5, but different variances.

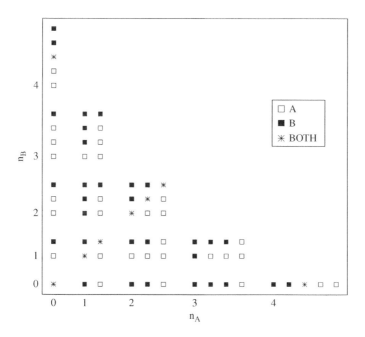

Figure 12.6   Optimal decisions when $n = 4$ and $N = 100$ given available data as given by $n_A, r_A, n_B,$ and r_B.

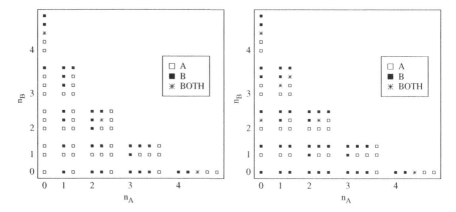

Figure 12.7   Sensitivity of the optimal decisions under two different prior choices for θ_A when $n = 4$ and $N = 100$. Designs shown in the left panel assume that $\theta_A \sim Beta(0.5, 0.5)$. In the right panel we assume $\theta_A \sim Beta(2, 2)$.

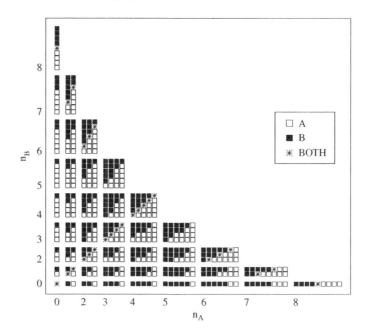

Figure 12.8   Optimal decisions when $n = 8$ and $N = 100$ given available data as given by n_A, r_A, n_B, and r_B.

In Figure 12.8 we also show the optimal design when $n = 8$ and $N = 100$ assuming independent uniform priors for θ_A and θ_B, while in Figure 12.9 we give an example of the sensitivity of the design to prior choices on θ_A. Figures in this section are similar to those presented in Berry and Stangl (1996, chapter 1).

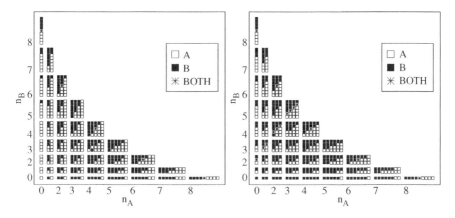

Figure 12.9   Sensitivity of the optimal decisions under two different prior choices for θ_A when $n = 8, N = 100$. Designs shown in the left panel assume that $\theta_A \sim Beta(0.5, 0.5)$. In the right panel we assume $\theta_A \sim Beta(2, 2)$.

In the example we presented, the allocation of patients to treatments is adaptive, in that it depends on the results of previous patients assigned to the same treatments. This is an example of an *adaptive design*. Berry and Eick (1995) discuss the role of adaptive designs in clinical trials and compare some adaptive strategies to balanced randomization which has a traditional and dominant role in randomized clinical trials. They conclude that the Bayesian decision-theoretic adaptive design described in the above example (assuming that θ_A and θ_B are independent and with uniform distributions) is better for any choice of patient horizon N. However, when the patient horizon is very large, this strategy does not perform much better than balanced randomization. In fact, balanced randomization is a good solution when the aim of the design is maximizing effective treatment in the whole population. However, if the condition being treated is rare then learning with the aim of extrapolating to a future population is much less important, because the future population may not be large. In these cases, using an adaptive procedure is more critical.

12.6   Variable selection in multiple regression

We now move to an application to multiple regression, drawn from Lindley (1968a). Let y denote a response variable, while $\boldsymbol{x} = (x_1, \ldots, x_p)'$ is vector of explanatory variables. The relationship between y and $\boldsymbol{x}$ is governed by

$$E[y|\boldsymbol{\theta}, \boldsymbol{x}] = \boldsymbol{x}'\boldsymbol{\theta}$$
$$\text{Var}[y|\boldsymbol{\theta}, \boldsymbol{x}] = \sigma^2;$$

that is, we have a linear multiple regression model with constant variance. To simplify our discussion we assume that σ^2 is known. We make two additional assumptions. First, we assume that θ and x are independent, that is

$$f(x, \theta) = f(x)\pi(\theta).$$

This means that learning the value of the explanatory variables alone does not carry any information about the regression coefficients. Second, we assume that y has known density $f(y|x, \theta)$, so that we do not learn anything new about the likelihood function from new observations of either x or y.

The decision maker has to predict a future value of the dependent variable y. To help with the prediction he or she can observe some subset (or all) of the p explanatory variables. So, the questions of interest are: (i) which explanatory variables should the decision maker choose; and (ii) having observed the explanatory variables of his or her choice, what is the prediction for y?

Let I denote the subset of integers in the set $\{1, \ldots, p\}$ and J denote the complementary subset. Let x^I denote a vector with components $x_i, i \in I$. Our decision space consists of elements $a = (I, g(.))$ where g is a function from $\Re^s$ to $\Re$; that is, it consists of a subset of explanatory variables and a prediction function. Let us assume that our loss function is

$$u(a(\theta)) = -(y - g(x^I))^2 - c_I;$$

that is, we have a quadratic utility for the prediction $g(x^I)$ for y with a cost c_I for observing explanatory variables with indexes in I. This allows different variables to have different costs.

Figure 12.10 shows the decision tree for this problem. At the last stage, we consider the expected utility, for fixed x^I, averaging over y, that is we compute

$$E_{y|x^I}[-(y - g(x^I))^2] - c_I. \tag{12.26}$$

Next, we select the prediction $g(x^I)$ that maximizes equation (12.26). Under the quadratic utility function, the optimal prediction is given by

$$g(x^I) = E[y|x^I]. \tag{12.27}$$

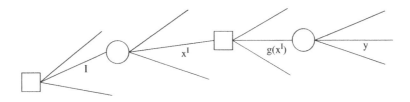

Figure 12.10   Decision tree for selecting variables with the goal of prediction. Figure adapted from Lindley (1968a).

The unknowns in this prediction problem include both the parameters and the unobserved variables. Thus

$$E[y|x^J] = \int_{\mathcal{X}} \int_{\Theta} E[y|x^J, x^J, \theta]\pi(\theta)f(x^J|x^J)d\theta dx^J$$

$$= \int_{\mathcal{X}} \int_{\Theta} x'\theta\pi(\theta)f(x^J|x^J)d\theta dx^J$$

$$= E[x|x^J]'E[\theta];$$

that is, to obtain the optimal prediction we first estimate the unobserved explanatory variables x^J and combine those with the observed x^J and the estimated regression parameters.

Folding back the decision tree, by averaging over x^J for fixed I, we obtain

$$E[(y - E[x|x^J]'E[\theta])^2] = \sigma^2 + E[x'\theta - E[x|x^J]'E[\theta]]^2$$

$$= \sigma^2 + tr(\text{Var}[\theta]\text{Var}[x]) + E[x]'\text{Var}[\theta]E[x]$$

$$+ E[\theta]'\text{Var}[x^J]E[\theta]$$

where $\text{Var}[\theta]$ and $\text{Var}[x]$ are the covariance matrices for θ and x, respectively, and $\text{Var}[x^J] = E[(x - E[x|x^J])(x - E[x|x^J])')]$, that is the covariance matrix of the whole predictor vector, once the subset x^J is fixed.

Thus, the expected utility for fixed I is

$$-(\sigma^2 + tr(\text{Var}[\theta]\text{Var}[x]) + E[x]'\text{Var}[\theta]E[x] + E[\theta]'\text{Var}[x^J]E[\theta] + c_I). \quad (12.28)$$

The optimal solution at the first stage of the decision tree is obtained by maximizing equation (12.28) over I. Because only the last two elements of equation (12.28) depend on I, this corresponds to choosing I to reach

$$\min_I \{E[\theta]'\text{Var}[x^J]E[\theta] + c_I\}.$$

The solution will depend both on how well the included and excluded x predict y, and on how well the included x^J predict the excluded x^J.

As a special case, suppose that the costs of each observation $x_i, i \in I$, are additive so that $c_I = \sum_{i \in I} c_i$. Moreover, assume that the explanatory variables are independent. This implies

$$\min_I \{E[\theta]'\text{Var}[x^J]E[\theta] + c_I\} = \min_I \left\{ \sum_{j \in J}(E[\theta_j])^2\text{Var}[x_j] + \sum_{i \in I} c_i \right\}.$$

It follows from the above equation that x_i should be observed if and only if

$$(E[\theta_i])^2\text{Var}[x_i] > c_i;$$

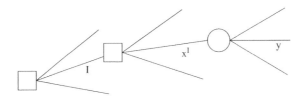

Figure 12.11   Decision tree for selecting variables with the goal of controlling the response. Figure adapted from Lindley (1968a).

that is, if either its variation is large or the squared expected value of the regression parameter is large compared to the cost of observation.

Let us now change the decision problem and assume that we can set the explanatory variables as we want, and the goal is to bring the value of the response variable towards a preassigned value y_0. The control of the response variable depends on the selection of explanatory variables and on the choice of a setting x_0^I. Our decision space consists of elements $a = (I, x^I)$; that is, it consists of selecting a subset of explanatory variables and the values assigned to these variables. Let us assume that our utility function is

$$u(a(\boldsymbol{\theta})) = -(y - y_0)^2 - c(\boldsymbol{x}^I).$$

This is no longer a two-stage sequential decision problem, because no new data are accrued between the choice of the subset and the choice of the setting. As seen in Figure 12.11 this is a nested decision problem, and it can be solved by solving for the setting given I, and plugging the solution back into the expected utility to solve for I. In a sense, this is a special case of dynamic programming where we skip one of the expectations. Using additional assumptions on the distributions of x and on the form of the cost function, in a special case Lindley (1968a) shows that if $y_0 = 0$ and the cost function does not depend on $\boldsymbol{x}^I$, then an explanatory variable x_i is chosen for controlling if and only if

$$E[\theta_i^2]\mathrm{Var}[x_i] > c_i.$$

This result parallels that seen earlier. However, a point of contrast is that the decision depends on the variance, rather than the mean, of the regression parameter, as a result of the differences in goals for the decision making. For a detailed development see Lindley (1968a).

12.7   Computing

Implementation of the fully sequential decision-theoretic approach is challenging in applications. In addition to the general difficulties that apply with the elicitation of priors and utilities in decision problems, the implementation of dynamic programming is limited by its computational complexity that grows exponentially with the maximum number of stages S. In this section we review a couple of examples that give a sense of the challenges and possible approaches.

Suppose one wants to design a clinical trial to estimate the effect θ of an experimental treatment over a placebo. Prior information about this parameter is summarized by $\pi(\theta)$ and the decisions are made sequentially. There is a maximum of S times, called monitoring times, when we can examine the data accrued that far, and take action. Possible actions are a_1 (stop the trial and conclude that the treatment is better), a_2 (stop the trial and conclude that the placebo is preferred), or a_3 (continue the trial). At the final monitoring time S, the decision is only between actions a_1 and a_2 with utility functions

$$u^S(a_1(\theta)) = -k_1^{(S)}(\theta - b_1)_+$$
$$u^S(a_2(\theta)) = -k_2^{(S)}(b_2 - \theta)_+$$

with $y_+ = \max(0, y)$. This utility builds in an indifference zone (b_2, b_1) within which the effect of the treatment is considered similar to that of placebo and, thus, either is acceptable. Low θ are good (say, they imply a low risk) so the experimental treatment is preferred when $\theta < b_2$, while placebo is preferred when $\theta > b_1$. Also, suppose a constant cost C for any additional observation.

As S increases, backwards induction becomes increasingly computationally difficult. Considering the two-sided sequential decision problem described above, Carlin *et al.* (1998) propose an approach to reduce the computational complexity associated with dynamic programming. Their *forward sampling algorithm* can be used to find optimal stopping boundaries within a class of decision problems characterized by decision rules of the form

$$E[\theta|x^{(1)}, \ldots, x^{(s)}] \le \gamma_{s,L} \quad \text{choose } a_1$$
$$E[\theta|x^{(1)}, \ldots, x^{(s)}] > \gamma_{s,U} \quad \text{choose } a_2$$
$$\gamma_{s,L} < E[\theta|x^{(1)}, \ldots, x^{(s)}] \le \gamma_{s,U} \quad \text{choose } a_3$$

for $s = 1, \ldots, S$ and with $\gamma_{S,L} = \gamma_{S,U} = \gamma_S$.

To illustrate how their method works, suppose that $S = 2$ and that a cost C is associated with carrying out each stage of the trial. To determine a sequential decision rule of the type considered here, we need to specify a total of $(2S - 1)$ decision boundaries. Let

$$\gamma = (\gamma_{S-2,L}, \gamma_{S-2,U}, \gamma_{S-1,L}, \gamma_{S-1,U}, \gamma_S)'$$

denote the vector of these boundaries. To find an optimal γ we first generate a Monte Carlo sample of size M from $f(\theta, x^{(S-1)}, x^{(S)}) = \pi(\theta)f(x^{(S-1)}|\theta)f(x^{(S)}|\theta)$. Let

$(\theta_m, x_m^{(S-1)}, x_m^{(S)})$ denote a generic element in the sample. For a fixed γ, the utilities achieved with this rule are calculated using the following algorithm:

$$
\begin{aligned}
&\text{If} && E[\theta] \le \gamma_{S-2,L} && u_m = -k_1^{(S-2)}(\theta_m - b_1) \\
&\text{else if} && E[\theta] > \gamma_{S-2,U} && u_m = -k_2^{(S-2)}(b_2 - \theta_m) \\
&\text{else if} && E[\theta|x_m^{(S-1)}] \le \gamma_{S-1,L} && u_m = -k_1^{(S-1)}(\theta_m - b_1) - C \\
&\text{else if} && E[\theta|x_m^{(S-1)}] > \gamma_{S-1,U} && u_m = -k_2^{(S-1)}(b_2 - \theta_m) - C \\
&\text{else if} && E[\theta|x_m^{(S-1)}, x_m^{(S)}] \le \gamma_S && u_m = -k_1^{(S)}(\theta_m - b_1) - 2C \\
&\text{else} && && u_m = -k_2^{(S)}(b_2 - \theta_m) - 2C.
\end{aligned}
$$

A Monte Carlo estimate of the posterior expected utility incurred with γ is given by

$$
\bar{u} = \frac{1}{M} \sum_{m=1}^{M} u_m.
$$

This algorithm provides a way of evaluating the expected utility of any strategy, and can be embedded into a $(2S - 1)$-dimensional optimization to numerically search for the optimal decision boundary γ^*. With this procedure Carlin et al. (1998) replace the initial decision problem of deciding among actions a_1, a_2, a_3 by a problem in which one needs to find optimal thresholds that define the stopping times in the sequential procedure. The strategy depends on the history of actions and observations only through the current posterior mean. The main advantage of the forward sampling algorithm is that it grows only linearly in S, while the backward induction algorithm would grow exponentially with S. A disadvantage is that it requires a potentially difficult maximization over a continuous space of dimension $(2S - 1)$, while dynamic programming is built upon simple one-dimensional discrete maximizations over the set of actions.

Kadane and Vlachos (2002) consider a *hybrid algorithm* which combines dynamic programming with forward sampling. Their hybrid algorithm works backwards for S_0 stages and provides values of the expected utility for the optimal continuation. Then, the algorithm is completed by forward sampling from the start of the trial up to the stage S_0 when backward induction starts. While reducing the optimization problem with forward sampling, the hybrid algorithm allows for a larger number of stages in sequential problems previously intractable with dynamic programming. The trade-off between accuracy in the search for optimal strategies and computing time is controlled by the maximum number of backward induction steps S_0.

Müller et al. (2007a) propose a combination of forward simulation, to approximate the integral expressions, and a reduction of the allowable action space and sample space, to avoid problems related to an increasing number of possible trajectories in the backward induction. They also assume that at each stage, the choice of an action depends on the data portion of the history set $\mathcal{H}_{s-1}$ only through a function T_s of $\mathcal{H}_{s-1}$. This function could potentially be any summary statistic, such as the posterior mean as in Carlin, Kadane & Gelfand (1998). At each stage s, both T_{s-1} and $a^{(s-1)}$

are discretized over a finite grid. Evaluation of the expected utility uses forward sampling. M samples $(\theta_m, \boldsymbol{x}_m)$ are generated from the distribution $f(x^{(1)}, \ldots, x^{(S-1)} | \theta) \pi(\theta)$. This is done once for all, and stretching out the data generation to its maximum number of stages, irrespective of whether it is optimal to stop earlier or not. The value of the summary statistic $T_{s,m}$ is then recorded at stage 1 through $S - 1$ for every generated data sequence. The coarsening of data and decisions to a grid simplifies the computation of the expected utilities, which are now simple sums over the set of indexes m that lead to trajectories that belong to a given cell in the history. As the grid size is kept constant, the method avoids the increasing number of pathways with the traditional backward induction. This *constrained backward induction* is an approximate method and its successful implementation depends on the choice of the summary statistic T_s, the number of points on the grid for T_s, and the number M of simulations. An application of this procedure to a sequential dose-finding trial is described in Berry *et al.* (2001).

Brockwell and Kadane (2003) also apply discretization of well-chosen statistics to simplify the history $\mathcal{H}_s$. Their examples focus on statistics of dimension up to three and a maximum of 50 stages. They note, however, that sequential problems involving statistics with higher dimension may be still intractable with the gridding method. More broadly, computational sequential analysis is still an open area, where novel approaches and progress could greatly contribute to a more widespread application of the beautiful underlying ideas.

12.8   Exercises

Problem 12.1   (French 1988) Four well-shuffled cards, the kings of clubs, diamonds, hearts, and spades, are laid face down on the table. A man is offered the following choice. Either: a randomly selected card will be turned over. If it is red, he will win £100; if it is black, he will lose £100. Or: a randomly selected card will be turned over. He may then choose to pay £35 and call the bet off or he may continue. If he continues, one of the remaining three cards will be randomly selected and turned over. If it is red, he will win £100; if it is black, he will lose £100. Draw a decision tree for the problem. If the decision maker is risk neutral, what should he do? Suppose instead that his utility function for sums of money in this range is $u(x) = \log(1 + x/200)$. What should he do in this case?

Problem 12.2   (French 1988) A builder is offered two plots of land for house building at £20 000 each. If the land does not suffer from subsidence, then he would expect to make £10 000 net profit on each plot, when a house is built on it. However, if there is subsidence, the land is only worth £2000 so he would make an £18 000 loss. He believes that the chance that both plots will suffer from subsidence is 0.2, the chance that one will suffer from subsidence is 0.3, and the chance that neither will is 0.5. He has to decide whether to buy the two plots or not. Alternatively, he may buy one, test it for subsidence, and then decide whether to buy the other plot. Assuming that the test is a perfect predictor of subsidence and that it costs £200 for the test, what should he do? Assume that his preferences are determined purely by money and that he is risk neutral.

Problem 12.3   (French 1988) You have decided to buy a car, and have eliminated possibilities until you are left with a straight choice between a brand new car costing £4000 and a second-hand car costing £2750. You must have a car regularly for work. So, if the car that you buy proves unreliable, you will have to hire a car while it is being repaired. The guarantee on the new car covers both repair costs and hire charges for a period of two years. There is no guarantee on the second-hand car. You have considered the second-hand-car market and noticed that cars tend to be either very good buys or very bad buys, few cars falling between the two extremes. With this in mind you consider only two possibilities.

   (i) The second-hand car is a very good buy and will involve you in £750 expenditure on car repairs and car hire over the next two years.

  (ii) The second-hand car is a very bad buy and will cost you £1750 over the next two years.

You also believe that the probability of its being a good buy is only 0.25. However, you may ask the AA for advice at negligible financial cost, but risking a probability of 0.3 that the second-hand car will be sold while they are arranging a road test. The road test will give a satisfactory or unsatisfactory report with the following probabilities:

Probability of AA report being:	Conditional on the car being:	
	a very bad buy	a very good buy
Satisfactory	0.1	0.9
Unsatisfactory	0.9	0.1

If the second-hand car is sold before you buy it, you must buy the new car. Alternatively you may ask your own garage to examine the second-hand car. They can do so immediately, thus not risking the loss of the option to buy, and will also do so at negligible cost, but you do have to trust them not to take a "back-hander" from the second-hand-car salesman. As a result, you evaluate their reliability as:

Probability of garage report being:	Conditional on the car being:	
	a very bad buy	a very good buy
Satisfactory	0.5	1
Unsatisfactory	0.5	0

You can arrange at most one of the tests. What should you do, assuming that you are risk neutral for monetary gains and losses and that no other attribute is of importance to you?

Problem 12.4   Write the algorithm of Section 12.3.2 for the case in which z only depends on the unknown θ and the terminal action, and in which observation of each of the random variables in stages $1, \ldots, S - 1$ has cost C.

Problem 12.5   Referring to Section 12.4.1:

1. Prove equation (12.15).

2. Use backwards induction to prove equation (12.16).

Problem 12.6   Consider the example discussed in Section 12.4.2. Suppose that $\theta^{(1)}, \ldots, \theta^{(S)}$ are independent and identically distributed with a uniform distribution on $[0, 1]$. Prove that

$$h_S = \frac{S}{S+1} \quad \text{and} \quad w_S = \frac{1}{2}(1 + w_{S-1}^2) \geq \frac{S+1}{S+3}.$$

13

Changes in utility as information

In previous chapters we discussed situations in which a decision maker, before making a decision, has the opportunity to observe data x. We now turn to the questions of whether this observation should be made and how worthwhile it is likely to be. Observing x could give information about the state of nature and, in this way, lead to a better decision; that is, a decision with higher expected utility. In this chapter we develop this idea more formally and present a general approach for assessing the value of information. Specifically, the value of information is quantified as the expected change in utility from observing x, compared to the "status quo" of not observing any additional data. This approach permits us to measure the information provided by an experiment on a metric that is tied to the decision problem at hand. Our discussion will follow Raiffa and Schlaifer (1961) and DeGroot (1984).

In many areas of science, data are collected to accumulate knowledge that will eventually contribute to many decisions. In that context the connection outlined above between information and a specific decision is not always useful. Motivated by this, we will also explore an idea of Lindley (1956) for measuring the information in a data set, which tries to capture, in a decision-theoretic way, "generic learning" rather than specific usefulness in a given problem.

Featured articles:

Lindley, D. V. (1956). On a measure of the information provided by an experiment, *Annals of Mathematical Statistics* **27**: 986–1005.

DeGroot, M. H. (1984). Changes in utility as information, *Theory and Decision* **17**: 287–303.

Useful references are Raiffa and Schlaifer (1961) and Raiffa (1970).

Decision Theory: Principles and Approaches   G. Parmigiani, L. Y. T. Inoue
© 2009 John Wiley & Sons, Ltd

13.1   Measuring the value of information

13.1.1   The value function

The first step in quantifying the change in utility resulting from a change in knowledge is to describe the value of knowledge in absolute terms. This can be done in the context of a specific decision problem, for a given utility specification, as follows. As in previous chapters, let a^* denote the Bayes action, that is a^* maximizes the expected utility

$$\mathcal{U}_\pi(a) = \int_\Theta u(a(\theta))\pi(\theta)d\theta. \tag{13.1}$$

As usual, a is an action, u is the utility function, θ is a parameter with possible values in a parameter space Θ, and π is the decision maker's prior distribution. Expectation (13.1) is taken with respect to π. We need to keep track of this fact in our discussion, and we do so using the subscript π on $\mathcal{U}$. The amount of utility we expect to achieve if we decide to make an immediate choice without experimentation, assuming we choose the best action under prior π, is

$$V(\pi) = \sup_a \mathcal{U}_\pi(a). \tag{13.2}$$

This represents, in absolute terms, the "value" to the decision maker of solving the problem as well as possible, given the initial state of knowledge. We illustrate $V(\pi)$ using three stylized statistical decision problems:

1. Consider an estimation problem where a represents a point estimate of θ and the utility is

   $$u(a(\theta)) = -(\theta - a)^2,$$

   that is the negative of the familiar squared error loss. The optimal decision is $a^* = E[\theta]$ as we discussed in Chapter 7. Computing the expectation in (13.2) gives

   $$V(\pi) = -\int_\Theta (\theta - a^*)^2 \pi(\theta)d\theta = -\int_\Theta (\theta - E[\theta])^2 \pi(\theta)d\theta = -\text{Var}[\theta].$$

   The value of solving the problem as well as possible given the knowledge represented by π is the negative of the variance of θ.

2. Next, consider the discrete parameter space $\Theta = \{1, 2, \ldots\}$ and imagine that the estimation problem is such that we gain something only if our estimate is exactly right. The corresponding utility is

   $$u(a(\theta)) = I_{\{a=\theta\}},$$

   where, again, a represents a point estimate of θ. The expected utility is maximized by $a^* = \text{mode}(\theta) \equiv \theta^0$ (use your favorite tie-breaking rule if there is more than one mode) and

$$V(\pi) = \pi(\theta^0).$$

The value V is now the largest mass of the prior distribution.

3. Last, consider forecasting θ when forecasts are expressed as a probability distribution, as is done, for example, for weather reports. Now a is the whole probability distribution on θ. This is the problem we discussed in detail in Chapter 10 when we talked about scoring rules. Take the utility to be

$$u(a(\theta)) = \log(a(\theta)).$$

This utility is a proper local scoring rule, which implies that the optimal decision is $a^*(\theta) = \pi(\theta)$. Then,

$$V(\pi) = \int_\Theta \log(\pi(\theta))\pi(\theta)d\theta.$$

The negative of this quantity is also known as the entropy (Cover and Thomas 1991) of the probability distribution $\pi(\theta)$. A high value of V means low entropy, which corresponds to low variability of θ.

In all these examples the value V of the decision problem is directly related to a measure of the strength of the decision maker's knowledge about θ, as reflected by the prior π. The specific aspects of the prior that determine V depend on the characteristics of the problem. All three of the above problems are statistical decisions. In Section 13.2.2 we discuss in detail an example using a two-stage decision tree similar to those of Chapter 12, in which the same principles are applied to a practical decision problem.

The next question for us to consider is how V changes as the knowledge embodied in π changes. To think of this concretely, consider example 2 above, and suppose priors π_1 and π_2 are as follows:

θ	π_1	π_2
-1	0.5	0.1
0	0.4	0.4
1	0.1	0.5

For a decision maker with prior π_1, the optimal choice is -1 and $V(\pi_1) = 0.5$. For one with prior π_2, the optimal choice is 1 and $V(\pi_1) = 0.5$. For a third with prior $\alpha\pi_1 + (1-\alpha)\pi_2$ with, say, $\alpha = 0.5$, the optimal choice is 0, with $V(\alpha\pi_1 + (1-\alpha)\pi_2) = 0.4$, a smaller value. The third decision maker is less certain about θ than any of the previous two. In fact, the prior $\alpha\pi_1 + (1 - \alpha)\pi_2$ is hedging bets between π_1 and π_2 by taking a weighted average. It therefore embodies more variability than either one of the other two. This is why the third decision maker expects to gain less than the others from having to make an immediate decision, and, as we will see later, may be more inclined to experiment.

You can check that no matter how you choose α in $(0, 1)$, you cannot get V to be above 0.5, so the third decision maker expects to gain less no matter what α is. This is an instance of a very general inequality which we consider next.

Theorem 13.1   *The function $V(\pi)$ is convex; that is, for any two distributions π_1 and π_2 on θ and $0 < \alpha < 1$*

$$V(\alpha\pi_1 + (1 - \alpha)\pi_2) \leq \alpha V(\pi_1) + (1 - \alpha)V(\pi_2). \tag{13.3}$$

Proof: The main tool here is a well-known inequality from calculus, which says that $\sup\{f_1(x) + f_2(x)\} \leq \sup f_1(x) + \sup f_2(x)$. Applying this to the left hand side of (13.3) we get

$$V(\alpha\pi_1 + (1 - \alpha)\pi_2)$$

$$= \sup_{a \in \mathcal{A}} \int_{\Theta} u(a(\theta))\big[\alpha\pi_1(\theta) + (1 - \alpha)\pi_2(\theta)\big]d\theta$$

$$= \sup_{a \in \mathcal{A}} \left[\alpha \int_{\Theta} u(a(\theta))\pi_1(\theta)d\theta + (1 - \alpha) \int_{\Theta} u(a(\theta))\pi_2(\theta)d\theta\right]$$

$$\leq \alpha \left[\sup_{a \in \mathcal{A}} \int_{\Theta} u(a(\theta))\pi_1(\theta)d\theta\right] + (1 - \alpha) \left[\sup_{a \in \mathcal{A}} \int_{\Theta} u(a(\theta))\pi_2(\theta)d\theta\right]$$

$$= \alpha V(\pi_1) + (1 - \alpha)V(\pi_2).$$

$\square$

We will use this theorem later when measuring the expected change in V that results from observing additional data.

13.1.2   Information from a perfect experiment

So far we quantified the value of solving the decision problem at hand based on current knowledge, as captured by π. At the opposite extreme we can consider the value of solving the decision problem after having observed an ideal experiment that reveals the value of θ exactly. We denote experiments in general by $\mathcal{E}$, and this ideal experiment by $\mathcal{E}^\infty$. We define a_θ as an action that would maximize the decision's maker utility if θ was known. Formally, for a given θ,

$$a_\theta = \arg\sup_{a \in \mathcal{A}} u(a(\theta)). \tag{13.4}$$

Because θ is unknown, so is a_θ. For any given θ, the difference

$$\mathcal{V}_\theta(\mathcal{E}^\infty) = u(a_\theta(\theta)) - u(a^*(\theta)) \tag{13.5}$$

measures the gap between the best that can be achieved under the current state of knowledge and the best that can be achieved if θ is known exactly. This is called *conditional value of perfect information*, since it depends on the specific θ that is chosen. What would the value be to the decision maker of knowing θ exactly? Because different θ will change the utility by different amounts, it makes sense to answer this question by computing the expectation of (13.5) with respect to the prior π, that is

$$\mathcal{V}(\mathcal{E}^\infty) = E_\theta[\mathcal{V}_\theta(\mathcal{E}^\infty)], \tag{13.6}$$

which is called *expected value of perfect information*. Using equations (13.2), (13.4), and (13.5) we can, equivalently, rewrite

$$\mathcal{V}(\mathcal{E}^\infty) = E_\theta\left[\sup_a u(a(\theta))\right] - V(\pi). \tag{13.7}$$

The first term on the right hand side of equation (13.7) is the prior expectation of the utility of the optimal action given perfect information on θ. Section 13.2.2 works through an example in detail.

13.1.3   Information from a statistical experiment

From a statistical viewpoint, interesting questions arise when one can perform an experiment $\mathcal{E}$ which may be short of ideal, but still potentially useful. Say $\mathcal{E}$ consists of observing a random variable x, with possible values in the sample space $\mathcal{X}$, and whose probability density function (or probability mass) is $f(x|\theta)$. The tree in Figure 13.1 represents two decisions: whether to experiment and what action to take. We can solve this two-stage decision problem using dynamic programming, as we described in Chapter 12. The solution will tell us what action to take, whether we should perform the experiment before reaching that decision, and, if so, how the results should affect the subsequent choice of an action.

Suppose the experiment has cost c. The largest value of c such that the experiment should be performed can be thought of as the value of the experiment for this decision problem. From this viewpoint, an experiment has value only if it may affect subsequent decisions. How large this value is depends on the utilities assigned to the final outcome of the decision process, on the strength of the dependence between x and θ, and on the probability of observing the various outcomes of x. We now make this more formal.

The "No Experiment" branch of the tree was discussed in Section 13.1.1. In summary, if we do not experiment, we choose an action a that maximizes the expected utility (13.1). The amount of utility we expect to gain is given by (13.2). When we have the option to perform an experiment, two questions can be considered:

1. After observing x, how much did we learn about θ?

2. How much do we expect to learn from x prior to observing it?

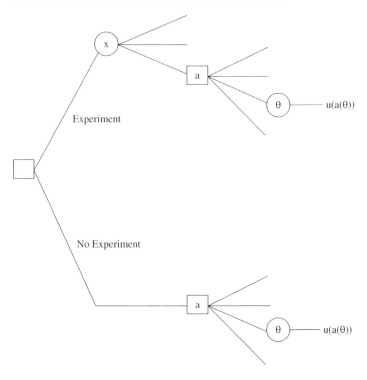

Figure 13.1   Decision tree for a generic statistical decision problem with discrete x. The decision node on the left represents the choice of whether or not to experiment, while the ones on the right represent the terminal choice of an action.

Question 2 is the relevant one for solving the decision tree. Question 1 is interesting retrospectively, as a quantification of observed information: answering question 1 would provide a measure of the *observed information* about θ provided by observing x. To simplify the notation, let π_x denote the posterior probability density function (or posterior probability mass in the discrete case) of θ, that is $\pi_x = \pi(\theta|x)$. A possible answer to question 1 is to consider the observed change in expected utility, that is

$$V(\pi_x) - V(\pi). \tag{13.8}$$

However, it is possible that the posterior distribution of θ will leave the decision maker with more uncertainty about θ or, more generally, with a less favorable scenario. Thus, with definition (13.8), the observed information can be both positive and negative. Also, it can be negative in situations where observing x was of great practical value, because it revealed that we knew far less about θ than we thought we did.

Alternatively, DeGroot (1984) proposed to define the observed information as the expected difference, calculated with respect to the posterior distribution of θ, between the expected utility of the Bayes decision and the expected utility of the decision a^* that would be chosen if observation x had not been available. Formally

$$\mathcal{V}_x(\mathcal{E}) = V(\pi_x) - \mathcal{U}_{\pi_x}(a^*), \tag{13.9}$$

where a^* represents a Bayes decision with respect to the prior distribution π. The observed information is always nonnegative, as $V(\pi_x) = \max_a \mathcal{U}_{\pi_x}(a)$. Expression (13.9) is also known as the conditional value of sample information (Raiffa and Schlaifer 1961) or conditional value of imperfect information (Clemen and Reilly 2001).

Let us explore further the definition of observed information in two of the stylized cases we considered earlier.

1. We have seen that in an estimation problem with utility function $u(a(\theta)) = -(\theta - a)^2$, we have $V(\pi) = -\text{Var}(\theta)$. The observed information (using equation (13.9)) is

$$\begin{aligned}
\mathcal{V}_x(\mathcal{E}) &= V(\pi_x) - \mathcal{U}_{\pi_x}(a^*) \\
&= -\text{Var}[\theta|x] - \mathcal{U}_{\pi_x}\big(E[\theta]\big) \\
&= -\text{Var}[\theta|x] - \left(-\int_\Theta (\theta - E[\theta])^2 \pi_x(\theta)d\theta\right) \\
&= -\text{Var}[\theta|x] + \big(\text{Var}[\theta|x] + (E[\theta|x] - E[\theta])^2\big) \\
&= \big(E[\theta|x] - E[\theta]\big)^2,
\end{aligned}$$

that is the square of the change in mean from the prior to the posterior.

2. Take now $u(a(\theta)) = I_{\{a=\theta\}}$, and let θ^0 and θ_x^0 be the modes for π and π_x, respectively. As we saw, $V(\pi) = \pi(\theta^0)$. Then

$$\mathcal{V}_x(\mathcal{E}) = V(\pi_x) - \mathcal{U}_{\pi_x}(a^*) = \pi_x(\theta_x^0) - \pi_x(\theta^0),$$

the difference in the posterior probabilities of the posterior and prior modes. This case is illustrated in Figure 13.2.

We can now move to our second question: how much do we expect to learn from x prior to observing it? The approach we will follow is to compute the expectation of the observed information, with respect to the marginal distribution of x. Formally

$$\mathcal{V}(\mathcal{E}) = E_x[\mathcal{V}_x(\mathcal{E})] = E_x[V(\pi_x) - \mathcal{U}_{\pi_x}(a^*)]. \tag{13.10}$$

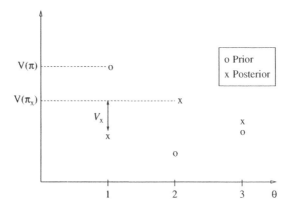

Figure 13.2   Observed information for utility $u(a(\theta)) = I_{\{a=\theta\}}$, when $\Theta = \{1,2,3\}$. Observing x shifts the probability mass and also increases the dispersion, decreasing the point mass at the mode. In this case the value V of the decision problem is higher before the experiment, when available evidence indicated $\theta = 1$ as a very likely outcome, than it is afterwards. However, the observed information is positive because the experiment suggested that the decision $a = \theta = 1$, which we would have chosen in its absence, is not likely to be a good one.

An important simplification of this expression can be achieved by expanding the expectation in the second term of the right hand side:

$$
\begin{aligned}
E_x[\mathcal{U}_{\pi_x}(a^*)] &= \int_{\mathcal{X}} \mathcal{U}_{\pi_x}(a^*)m(x)dx \\
&= \int_{\mathcal{X}} \int_{\Theta} u(a^*(\theta))\pi(\theta|x)m(x)d\theta dx \\
&= \int_{\mathcal{X}} \int_{\Theta} u(a^*(\theta))\frac{f(x|\theta)\pi(\theta)}{m(x)}m(x)d\theta dx \\
&= \int_{\Theta} u(a^*(\theta))\pi(\theta) \int_{\mathcal{X}} f(x|\theta)dxd\theta \\
&= \int_{\Theta} u(a^*(\theta))\pi(\theta)d\theta \\
&= \mathcal{U}_{\pi}(a^*) \\
&= V(\pi), \qquad\qquad\qquad\qquad\qquad (13.11)
\end{aligned}
$$

where $m(x)$ is the marginal distribution of x. Replacing the above into the definition of $\mathcal{V}(\mathcal{E})$, we get

$$
\mathcal{V}(\mathcal{E}) = E_x[V(\pi_x)] - V(\pi). \qquad\qquad\qquad (13.12)
$$

This expression is also known as the expected value of sample information (Raiffa and Schlaifer 1961). Note that this is the same as the expectation of (13.8).

Now, because $E_x[\pi_x] = \pi$, we can write

$$\mathcal{V}(\mathcal{E}) = E_x[V(\pi_x)] - V(E_x[\pi_x]). \tag{13.13}$$

From the convexity of V (Theorem 13.1) and Jensen's inequality, we can then derive the following:

Theorem 13.2 $\mathcal{V}(\mathcal{E}) \geq 0$.

This inequality means that no matter what decision problem one faces, expected utility cannot be decreased by taking into account free information. This very general connection between rationality and knowledge was first brought out by Raiffa and Schlaifer (1961), and Good (1967), whose short note is very enjoyable to read. Good also points out that in the discrete case considered in his note, the inequality is strict unless there is a dominating action: that is, an action that would be chosen no matter how much experimentation is done. Later, analysis of Ramsey's personal notes revealed that he had already noted Theorem 13.2 in the 1920s (Ramsey 1991).

We now turn to the situation where the decision maker can observe two random variables, x_1 and x_2, potentially in sequence. We define $\mathcal{E}_1$ and $\mathcal{E}_2$ as the experiments corresponding to observation of x_1 and x_2 respectively, and define $\mathcal{E}_{12}$ as the experiment of observing both. Let $\pi_{x_1 x_2}$ be the posterior after both random variables are observed. Let a_0^* be an optimal action when the prior is π and a_1^* be an optimal action when the prior is π_{x_1}. The observed information if both random variables are observed at the same time is, similarly to expression (13.9),

$$\mathcal{V}_{x_1 x_2}(\mathcal{E}_{12}) = V(\pi_{x_1 x_2}) - \mathcal{U}_{\pi_{x_1 x_2}}(a_0^*). \tag{13.14}$$

If instead we first observe x_1 and revise our prior to π_{x_1}, the additional observed information from observing that x_2 can be defined is

$$\mathcal{V}_{x_1 x_2}(\mathcal{E}_2|\mathcal{E}_1) = V(\pi_{x_1 x_2}) - \mathcal{U}_{\pi_{x_1 x_2}}(a_1^*). \tag{13.15}$$

In both cases the final results depend on both x_1 and x_2, but in the second case, the starting point includes knowledge of x_1.

Taking the expectation of (13.14) we get that the expected information from $\mathcal{E}_{12}$ is

$$\mathcal{V}(\mathcal{E}_{12}) = E_{x_1 x_2}[V(\pi_{x_1 x_2})] - V(\pi). \tag{13.16}$$

Next, taking the expectation of (13.15) we get that the expected conditional information is

$$\begin{aligned}
\mathcal{V}(\mathcal{E}_2|\mathcal{E}_1) &= E_{x_1 x_2}\left[V(\pi_{x_1 x_2}) - \mathcal{U}_{\pi_{x_1 x_2}}(a_1^*)\right] \\
&= E_{x_1 x_2}\left[V(\pi_{x_1 x_2})\right] - E_{x_1}\left[V(\pi_{x_1})\right].
\end{aligned} \tag{13.17}$$

This measures the expected value of observing x_2 conditional on having already observed x_1. The following theorem gives an additive decomposition of the expected information of the two experiments.

Theorem 13.3

$$\mathcal{V}(\mathcal{E}_{12}) = \mathcal{V}(\mathcal{E}_1) + \mathcal{V}(\mathcal{E}_2|\mathcal{E}_1). \tag{13.18}$$

Proof:

$$
\begin{aligned}
\mathcal{V}(\mathcal{E}_{12}) &= E_{x_1 x_2}[V(\pi_{x_1 x_2})] - V(\pi) \\
&= E_{x_1 x_2}[V(\pi_{x_1 x_2})] - E_{x_1}[V(\pi_{x_1})] + E_{x_1}[V(\pi_{x_1})] - V(\pi) \\
&= \mathcal{V}(\mathcal{E}_2|\mathcal{E}_1) + \mathcal{V}(\mathcal{E}_1).
\end{aligned}
$$

$\square$

It is important that additivity is in $\mathcal{V}(\mathcal{E}_2|\mathcal{E}_1)$ and $\mathcal{V}(\mathcal{E}_1)$, rather than $\mathcal{V}(\mathcal{E}_2)$ and $\mathcal{V}(\mathcal{E}_1)$. This reflects the fact that we accumulate knowledge incrementally and that the value of new information depends on what we already know about an unknown. Also, in general, $\mathcal{V}(\mathcal{E}_2|\mathcal{E}_1)$ and $\mathcal{V}(\mathcal{E}_1)$ will not be the same even if x_1 and x_2 are exchangeable.

13.1.4   The distribution of information

The observed information $\mathcal{V}_x$ depends on the observed sample and it is unknown prior to experimentation. So far we focused on its expectation, but it can be useful to study the distribution of $\mathcal{V}_x$, both for model validation and design of experiments.

We illustrate this using an example based on normal data. Suppose that a random quantity x is drawn from a $N(\theta, \sigma^2)$ distribution, and that the prior distribution for θ is $N(\mu_0, \tau_0^2)$. The posterior distribution of θ given x is a $N(\mu_x, \tau_x^2)$ distribution where, as given in Appendix A.4,

$$\mu_x = \frac{\sigma^2}{\sigma^2 + \tau_0^2} \mu_0 + \frac{\tau_0^2}{\sigma^2 + \tau_0^2} x \quad \text{and} \quad \tau_x^2 = \frac{\sigma^2 \tau_0^2}{\sigma^2 + \tau_0^2}$$

and the marginal distribution of x is $N(\mu_0, \sigma^2 + \tau_0^2)$. Earlier we saw that with the utility function $u(a(\theta)) = -(\theta - a)^2$, we have

$$\mathcal{V}_x(\mathcal{E}) = (E[\theta|x] - E[\theta])^2.$$

Thus, in this example, $\mathcal{V}_x(\mathcal{E}) = (\mu_x - \mu_0)^2$. The difference in means is

$$\mu_x - \mu_0 = \frac{\sigma^2}{\sigma^2 + \tau_0^2} \mu_0 + \frac{\tau_0^2}{\sigma^2 + \tau_0^2} x - \mu_0 = \frac{\tau_0^2}{\sigma^2 + \tau_0^2}(x - \mu_0).$$

Since $(x - \mu_0) \sim N\left(0, \sigma^2 + \tau_0^2\right)$, we have

$$\mu_x - \mu_0 \sim N\left(0, \frac{\tau_0^4}{\sigma^2 + \tau_0^2}\right).$$

Therefore,

$$\frac{\sigma^2 + \tau_0^2}{\tau_0^4}(\mu_x - \mu_0)^2 \sim \chi_1^2.$$

Knowing the distribution of information gives us a more detailed view of what we can anticipate to learn from an observation. Features of this distribution could be used to discriminate among experiments that have the same expected utility. Also, after we make the observation, comparing the observed information to the distribution expected a priori can provide an informal overall diagnostic of the joint specification of utility and probability model. Extreme outliers would lead to questioning whether the model we specified was appropriate for the experiment.

13.2   Examples

13.2.1   Tasting grapes

This is a simple example taken almost verbatim from Savage (1954). While obviously a bit contrived, it is useful to make things concrete, and give you numbers to play around with. A decision maker is considering whether to buy some grapes and, if so, in what quantity. To his or her taste, grapes can be classified as of poor, fair, or excellent quality. The unknown θ represents the quality of the grapes and can take values $1, 2$, or 3, indicating increasing quality. The decision maker's personal probabilities for the quality of the grapes are stated in Table 13.1.

The decision maker can buy $0, 1, 2$, or 3 pounds of grapes. This defines the basic or terminal acts in this example. The utilities of each act according to the quality of the grapes are stated in Table 13.2. Buying 1 pound of grapes maximizes the expected utility $\mathcal{U}_\pi(a)$. Thus, $a^* = 1$ is the Bayes action with value $V(\pi) = 1$.

Table 13.1   Prior probabilities for the quality of the grapes.

Quality	1	2	3
Prior probability	1/4	1/2	1/4

Table 13.2   Utilities associated with each action and quality of the grapes.

a		θ		$\mathcal{U}_\pi(a)$
	1	2	3	
0	0	0	0	0
1	-1	1	3	1
2	-3	0	5	1/2
3	-6	-2	6	-1

Table 13.3   Joint probabilities (multiplied by 128) of quality θ and outcome x.

x	θ		
	1	2	3
1	15	5	1
2	10	15	2
3	4	24	4
4	2	15	10
5	1	5	15

Table 13.4   Expected utility of action a (in pounds of grapes) given outcome x. For each x, the highest expected utility is in italics.

a	x				
	1	2	3	4	5
0	*0/21*	*0/27*	0/32	0/27	0/21
1	−7/21	*11/27*	*32/32*	43/27	49/21
2	−40/21	−20/27	8/32	*44/27*	72/21
3	−94/21	−78/27	−48/32	18/27	*74/21*

Suppose the decision maker has the option of tasting some grapes. How much should he or she pay for making this observation? Suppose that there are five possible outcomes of observation x, with low values of x suggesting low quality. Table 13.3 shows the joint distribution of x and θ.

Using Tables 13.2 and 13.3 it can be shown that the conditional expectation of the utility of each action given each possible outcome is as given in Table 13.4. The highest value of the expected utility $V(\pi_x)$, for each x, is shown in italics. Averaging with respect to the marginal distribtion of x from Table 13.3, $E_x[V(\pi_x)] = 161/128 \approx 1.26$. The decision maker would pay, if necessary, up to $\mathcal{V}(\mathcal{E}) = E_x[V(\pi_x)] - V(\pi) \approx 1.26 - 1.00 = 0.26$ utilities for tasting the grapes before buying.

13.2.2   Medical testing

The next example is a classic two-stage decision tree. We will use it to illustrate in a simple setting the value of information analysis, and to show how the value of information varies as a function of prior knowledge about θ. Return to the travel insurance example of Section 7.3. Remember you are about to take a trip overseas. You are not sure about the status of your vaccination against a certain mild disease that is common in the country you plan to visit, and need to decide whether to buy health insurance for the trip. We will assume that you will be exposed to the disease,

Table 13.5   Hypothetical monetary consequences of buying or not buying an insurance plan for the trip to an exotic country.

Actions	Events	
	θ_1: ill	θ_2: not ill
Insurance	−50	−50
No insurance	−1000	0

but you are uncertain about whether your present immunization will work. Based on aggregate data on tourists like yourself, the chance of developing the disease during the trip is about 3% overall. Treatment and hospital abroad would normally cost you, say, 1000 dollars. There is also a definite loss in quality of life in going all the way to a foreign country and being grounded at a local hospital instead of making the most out of your experience, but we are going to ignore this aspect here. On the other hand, if you buy a travel insurance plan, which you can do for 50 dollars, all your expenses will be covered. This is a classical gamble versus sure outcome situation.

Table 13.5 presents the outcomes for this decision problem. We are going to analyze this decision problem assuming both a risk-neutral and a risk-averse utility function for money. In the risk-neutral case we can simply look at the monetary outcome, so when we talk about "utility" we refer to the risk-averse case. We will start by working out the risk-neutral case first and come back to the risk-averse case later.

If you are risk neutral, based on the overall rate of disease of 3% in tourists, you should not buy the insurance plan because this action has an expected loss of 30 dollars, which is better than a loss of 50 dollars for the insurance plan. Actually, you would still choose not to buy the insurance for any value of $\pi(\theta_1)$ less than 0.05, as the observed expected utility (or expected return) is

$$V(\pi) = \begin{cases} -1000\,\pi(\theta_1) & \text{if } \pi(\theta_1) \leq 0.05 \\ -50 & \text{if } \pi(\theta_1) \geq 0.05. \end{cases} \tag{13.19}$$

How much money could you save if you knew exactly what will happen? That amount is the value of perfect information. The optimal decision under perfect information is to buy insurance if you know you will become ill and not to buy it if you know you will stay healthy. In the notation of Section 13.1.2

$$a_\theta(\theta) = \begin{cases} \text{insurance} & \text{if } \theta = \theta_1 \\ \text{no insurance} & \text{if } \theta = \theta_2. \end{cases}$$

The returns that go with the two cases are

$$u(a_\theta(\theta)) = \begin{cases} -50 & \text{if } \theta = \theta_1 \\ 0 & \text{if } \theta = \theta_2 \end{cases}$$

so that the expected return under perfect information is $-50\,\pi(\theta_1)$. In our example, $-50 \times 0.03 = -1.5$. The difference between this amount and the expected return under current information is the expected value of perfect information. For a general prior, this works out to be

$$\mathcal{V}(\mathcal{E}^\infty) = -50\,\pi(\theta_1) - V(\pi) = \begin{cases} 950\,\pi(\theta_1) & \text{if } \pi(\theta_1) < 0.05 \\ 50\,(1 - \pi(\theta_1)) & \text{if } \pi(\theta_1) \geq 0.05. \end{cases} \quad (13.20)$$

An alternative way of thinking about the value of perfect information as a function of the initial prior knowledge on θ is presented in Figure 13.3. The expected utility of not buying the insurance plan is $-1000\,\pi(\theta_1)$, while the expected utility of buying insurance is -50. The expected value of perfect information for a given $\pi(\theta_1)$ is the difference between the expected return with perfect information and the maximum between the expected returns with and without insurance. The expected value of perfect information increases for $\pi(\theta_1) < 0.05$ and it decreases for $\pi(\theta_1) \geq 0.05$.

When $\pi(\theta_1) = 0.03$, the expected value of perfect information is $950 \times 0.03 = 28.5$. In other words, you would be willing to spend up to 28.5 dollars for an infallible medical test that can predict exactly whether you are immune or not. In practice, it

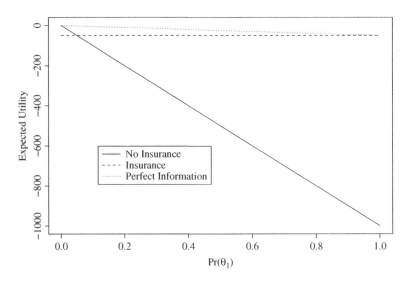

Figure 13.3   The expected utility of no insurance (solid line) and insurance (dashed line) as a function of the prior $\pi(\theta_1)$. The dotted line, a weighted average of the two, is the expected utility of perfect information, that is $E_\theta[u(a_\theta(\theta))]$. The gap between the dotted line and the maximum of the lower two lines is the expected value of perfect information. As a function of the prior, this is smallest at the extremes, where we think we know a lot, and greatest at the point of indifference between the two actions, where information is most valued.

is rare to have access to perfect information, but this calculation gives you an upper bound for the value of the information given by a reliable although not infallible, test. We turn to this case now.

Suppose you have the option of undergoing a medical test that can inform you about whether your immunization is likely to work. After the test, your individual chances of illness will be different from the overall 3%. This test costs c dollars, so from what we know about the value of perfect information, c has to be less than or equal to 28.5 for you to consider this possibility. The result x of the test will be either 1 (for high risk or bad immunization) or 0 (for low risk or good immunization). From past experience, the test is correct in 90% of the subjects with the bad immunization and 77% of subjects with good immunization. In medical terms, these figures represent the sensitivity and specificity of the test (Lusted 1968). Formally they translate into $f(x = 1|\theta_1) = 0.9$ and $f(x = 1|\theta_2) = 0.23$. Figure 13.4 shows the solution of this decision problem using a decision tree, assuming that $c = 10$. The optimal strategy

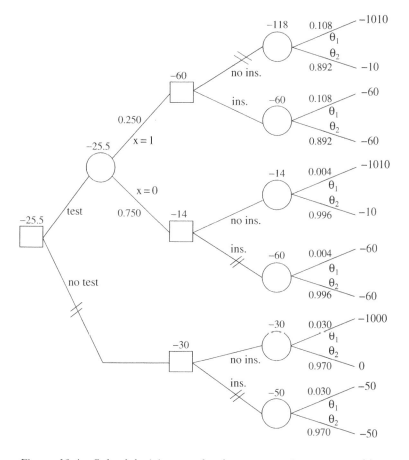

Figure 13.4   Solved decision tree for the two-stage insurance problem.

is to buy the test. If $x = 1$, then the best strategy is buying the insurance. Otherwise, it is best not to buy it. If no test was made, then not buying the insurance would be best, as observed earlier.

The test result x can alter your course of action and therefore it provides valuable information. The expected value of the information in x is defined by equation (13.12). Not considering cost for the moment, if $x = 1$ the optimal decision is to buy the insurance and that has conditional expected utility of -50; if $x = 0$ the optimal decision is not to buy insurance, which has conditional expected utility of -4. So in our example the first term in expression (13.12) is

$$E_x[V(\pi_x)] = -50 \times m(X = 1) - 4 \times m(X = 0)$$
$$= -50 \times 0.25 - 4 \times 0.75$$
$$= -15.5,$$

and thus the expected value of the information provided by the test is $V(\mathcal{E}) = -15.5 - (-30) = 14.5$. This difference exceeds the cost, which is 10 dollar, which is why according to the tree it is optimal to buy the test.

We now explore, as we had done in Figure 13.3, how the value of information changes as a function of the prior information on θ. We begin by evaluating $V(\pi_x)$ for each of the two possible experimental outcomes. From (13.19)

$$V(\pi_{x=1}) = \begin{cases} -1000\,\pi(\theta_1|x = 1) & \text{if } \pi(\theta_1|x = 1) \le 0.05 \\ -50 & \text{if } \pi(\theta_1|x = 1) \ge 0.05 \end{cases}$$

and

$$V(\pi_{x=0}) = \begin{cases} -1000\,\pi(\theta_1|x = 0) & \text{if } \pi(\theta_1|x = 0) \le 0.05 \\ -50 & \text{if } \pi(\theta_1|x = 0) \ge 0.05. \end{cases}$$

Replacing Bayes' formula for $\pi(\theta_1|x)$ and solving the inequalities, we can rewrite these as

$$V(\pi_{x=1}) = \begin{cases} -1000\,\pi(\theta_1|x = 1) & \text{if } \pi(\theta_1) \le 0.013 \\ -50 & \text{if } \pi(\theta_1) \ge 0.013 \end{cases}$$

and

$$V(\pi_{x=0}) = \begin{cases} -1000\,\pi(\theta_1|x = 0) & \text{if } \pi(\theta_1) \le 0.288 \\ -50 & \text{if } \pi(\theta_1) \ge 0.288. \end{cases}$$

Averaging these two functions with respect to the marginal distribution of x we get

$$E_x[V(\pi_x)] = \begin{cases} -1000\,\pi(\theta_1) & \text{if } \pi(\theta_1) \le 0.013 \\ -11.5 - 133.5\,\pi(\theta_1) & \text{if } 0.013 \le \pi(\theta_1) \le 0.288 \\ -50 & \text{if } \pi(\theta_1) \ge 0.288. \end{cases} \qquad (13.21)$$

The intermediate range in this expression is the set of priors such that the optimal solution can be altered by the results of the test. By contrast, values outside this interval are too far from the indifference point $\pi(\theta_1) = 0.05$ for the test to make a difference: the two results of the test lead to posterior probabilities of illness that are both on the same side of 0.05, and the best decision is the same.

Subtracting (13.19) from (13.21) gives the expected value of information

$$
\mathcal{V}(\mathcal{E}) = E_x[V(\pi_x)] - V(\pi) = \begin{cases} 0 & \text{if } \pi(\theta_1) \leq 0.013 \\ -11.5 + 866.5\,\pi(\theta_1) & \text{if } 0.013 \leq \pi(\theta_1) \leq 0.05 \\ 38.5 - 133.5\,\pi(\theta_1) & \text{if } 0.05 \leq \pi(\theta_1) \leq 0.288 \\ 0 & \text{if } \pi(\theta_1) \geq 0.288. \end{cases}
$$

$$(13.22)$$

Figure 13.5 graphs $\mathcal{V}(\mathcal{E})$ and $\mathcal{V}(\mathcal{E}^{\infty})$ versus the prior $\pi(\theta_1)$. At the extremes, evidence from the test is not sufficient to change a strong initial opinion about becoming or not becoming ill, and the test is not expected to contribute valuable information. As was the case with perfect information, the value of the test is largest at the point of indifference $\pi(\theta_1) = 0.05$.

As a final exploration of this example, let us imagine that you are slightly averse to risk, as opposed to risk neutral. Specifically, your utility function for a change in wealth of z dollars is of the form

$$
u(z) = a - b\,e^{-\lambda z}, \qquad b, \lambda > 0. \tag{13.23}
$$

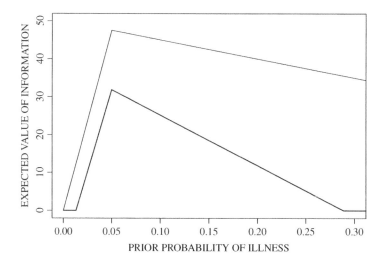

Figure 13.5   Expected value of information versus prior probability of illness. The thin line is the expected value of perfect information, and corresponds to the vertical gap between the dotted line and the maximum of the lower two lines in Figure 13.3. The thicker line is the expected value of the information provided by the medical test.

In the terminology of Chapter 4 this is a constantly risk-averse utility. To elicit your utility you consider a lottery where you win $1010 with probability 0.5, and nothing otherwise, and believe that you are indifferent between $450 and the lottery ticket. Then you set $u(0) = 0, u(1010) = 1$, and $u(450) = 0.5$. Solving for the parameters of your utility function gives $a = b = 2.813\,721\,8$ and $\lambda = 0.000\,434\,779$. Using equation (13.23) we can extrapolate the utilities to the negative monetary outcomes in the decision tree of Figure 13.4. Those are

Outcomes in dollars	-10	-50	-60	-1000	-1010
Utility	-0.012	-0.062	-0.074	-1.532	-1.551

Solving the tree again, the best decision in the absence of testing continues to be do not buy the insurance plan. In the original case, not buying the insurance plan was the best decision as long as $\pi(\theta_1)$ was less than 0.05. Now such a cutoff has diminished to about 0.04 (give or take rounding error) as a result of your aversion to the risk of spending a large sum for the hospital bill. Finally, the overall optimal strategy is still to buy the test and buy insurance only if the test is positive.

The story about the trip abroad is fictional, but it captures the essential features of medical diagnosis and evaluation of the utility of a medical test. Real applications are common in the medical decision-making literature (Lusted 1968, Barry *et al.* 1986, Sox 1996, Parmigiani 2002). The approach we just described for quantification of the value of information in medical diagnosis relies on knowing probabilities such as $\pi(\theta)$ and $f(x|\theta)$ with certainty. In reality these will be estimates and are also uncertain. While for a utility-maximizing decision maker it is appropriate to average out this uncertainty, there are broader uses of a value of information analysis, such as supporting decision making by a range of patients and physicians who are not prepared to solve a decision tree every time they order a test, or to support regulatory decisions. With this in mind, it is interesting to represent uncertainty about model inputs and explore the implication of such uncertainty on the conclusions. One approach to doing this is probabilistic sensitivity analysis (Doubilet *et al.* 1985, Critchfield and Willard 1986), which consists of drawing a random set of inputs from a distribution that reflects the uncertainty about them (for example, is consistent with confidence intervals reported in the studies that were used to determine those quantities), and evaluating the value of information for each.

Figure 13.6 shows the results of probabilistic sensitivity analysis applied to a decision analysis of the value of genetic testing for a cancer-causing gene called BRCA1 in guiding preventative surgery decisions. Details of the underlying modeling assumptions are discussed by Parmigiani *et al.* (1998). Figure 13.6 reminds us of Figure 13.5 with additional noise. There is little variability in the range of priors for which the test has value, while there is more variability in the value of information to the left of the mode. As was the case in our simpler example above, the portions of the curve on either side of the mode are affected by different input values.

In Figure 13.6 we show variability in the value of information as inputs vary. In general, uncertainty about inputs translates into uncertainty about optimal decisions

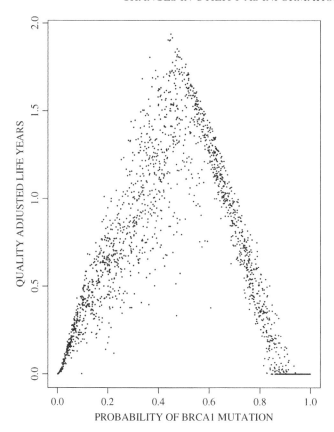

Figure 13.6   Results of a probabilistic sensitivity analysis of the additional expected quality adjusted life resulting from genetic testing for the cancer-causing gene BRCA1. Figure from Parmigiani et al. (1998).

at all stages of a multistage problem. Because of this, in a probabilistic sensitivity analysis of the value of diagnostic information it can be interesting to separately quantify and convey the uncertainty deriving directly from randomness in the input, and the uncertainty deriving from randomness in the optimal future decisions. In clinical settings, this distinction can become critical, because it has implications for the appropriateness of a treatment. A detailed discussion of this issue is in Parmigiani (2004).

13.2.3   Hypothesis testing

We now turn to a more statistical application of value of information analysis, and work out in detail an example of observed information in hypothesis testing. Assume that $x_1, \ldots, x_9$ form a random sample from a $N(\theta, 1)$ distribution. Let the prior for θ be $N(0, 4)$ and the goal be that of testing the null hypothesis $H_0 : \theta < 0$ versus the

Table 13.6   Utility function for the hypotheses testing example.

Actions	States of nature	
	H_0	H_1
$a_0 = $ accept H_0	0	-1
$a_1 = $ accept H_1	-3	0

alternative hypothesis $H_1 : \theta \geq 0$. Set $\pi(H_0) = \pi(H_1) = 1/2$. Let a_i denote the action of accepting hypothesis H_i, $i = 0, 1$, and assume that the utility function is that shown in Table 13.6.

The expected utilities for each decision are:

- for decision a_0:

$$\mathcal{U}_\pi(a_0) = u(a_0(H_0))\pi(H_0) + u(a_0(H_1))\pi(H_1)$$
$$= 0\,\pi(H_0) - 1\,\pi(H_1) = -1/2,$$

- for decision a_1:

$$\mathcal{U}_\pi(a_1) = u(a_1(H_0))\pi(H_0) + u(a_1(H_1))\pi(H_1)$$
$$= -3\,\pi(H_0) + 0\,\pi(H_1) = -3/2,$$

with $V(\pi) = \sup_{a \in \mathcal{A}} \mathcal{U}_\pi(a) = -1/2$. Based on the initial prior, the best decision is action a_0. The two hypotheses are a priori equally likely, but the consequences of a wrong decision are worse for a_1 than they are for a_0.

Using standard conjugate prior algebra, we can derive the posterior distribution of θ, which is

$$\pi(\theta|x^n) = N\left(\frac{36}{37}\bar{x}, \frac{4}{37}\right)$$

which in turn can be used to compute the posterior probabilities of H_0 and H_1 as

$$\pi(H_0|x^n) = \pi(\theta < 0|x^n) = \Phi\left(\frac{-18\bar{x}}{\sqrt{37}}\right)$$
$$\pi(H_1|x^n) = \pi(\theta \geq 0|x^n) = 1 - \Phi\left(\frac{-18\bar{x}}{\sqrt{37}}\right)$$

where $\Phi(x)$ is the cumulative distribution function of the standard normal distribution. After observing $x^n = (x_1, \ldots, x_9)$, the expected utilities of the two alternative decisions are:

- for decision a_0:

$$\mathcal{U}_{\pi_x}(a_0) = u(a_0(H_0))\pi_x(H_0) + u(a_0(H_1))\pi_x(H_1)$$
$$= 0\,\pi_x(H_0) - 1\,\pi_x(H_1)$$
$$= \Phi\left(\frac{-18\bar{x}}{\sqrt{37}}\right) - 1,$$

- for decision a_1:

$$\mathcal{U}_{\pi_x}(a_1) = u(a_1(H_0))\pi_x(H_0) + u(a_1(H_1))\pi_x(H_1)$$
$$= -3\,\pi_x(H_0) + 0\,\pi_x(H_1)$$
$$= -3\,\Phi\left(\frac{-18\bar{x}}{\sqrt{37}}\right)$$

and a_0 is optimal if $\mathcal{U}_{\pi_x}(a_0) > \mathcal{U}_{\pi_x}(a_1)$, that is if

$$\Phi\left(\frac{-18\bar{x}}{\sqrt{37}}\right) - 1 > -3\,\Phi\left(\frac{-18\bar{x}}{\sqrt{37}}\right)$$

or, equivalently, if $\bar{x} < 0.2279$. Therefore,

$$V(\pi_x) = \max\{\mathcal{U}_{\pi_x}(a_0), \mathcal{U}_{\pi_x}(a_1)\}$$
$$= \begin{cases} \mathcal{U}_{\pi_x}(a_0) = \Phi\left(-18\bar{x}/\sqrt{37}\right) - 1 & \text{if } \bar{x} < 0.2279 \\ \mathcal{U}_{\pi_x}(a_1) = -3\,\Phi\left(-18\bar{x}/\sqrt{37}\right) & \text{if } \bar{x} \geq 0.2279. \end{cases}$$

Also,

$$\mathcal{U}_{\pi_x}(a^*) = \mathcal{U}_{\pi_x}(a_0) = \pi_x(H_0)u(a_0(H_0)) + \pi_x(H_1)u(a_0(H_1))$$
$$= 0\,\pi_x(H_0) - 1\,\pi_x(H_1) = -\pi_x(H_1)$$
$$= -1 + \Phi\left(\frac{-18\bar{x}}{\sqrt{37}}\right).$$

Therefore, the observed information is

$$V_x(\mathcal{E}) = V(\pi_x) - \mathcal{U}_{\pi_x}(a_0) = \begin{cases} 0 & \text{if } \bar{x} < 0.2279 \\ 1 - 4\,\Phi(-18\bar{x}/\sqrt{37}) & \text{if } \bar{x} \geq 0.2279. \end{cases}$$

In words, small values of $\bar{x}$ result in the same decision that would have been made a priori, and does not contribute to decision making. Values larger than 0.2279 lead to a reversal of the decision and are informative. For very large values the difference in posterior expected utility of the two decisions approaches one. The resulting observed information is shown in Figure 13.7.

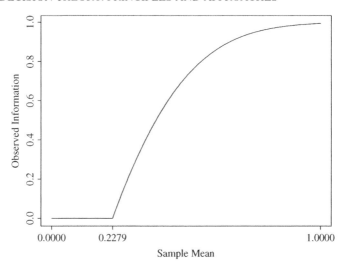

Figure 13.7   Observed information for the hypothesis testing example, as a function of the sample mean.

Finally, we can now evaluate the expected value of information. Using the fact that $\bar{x} \sim N(0, 37/9)$, we have

$$E_x[\mathcal{V}_x(\mathcal{E})] = \int_{0.2279}^{\infty} \left[1 - 4 \, \Phi \left(-\frac{18\bar{x}}{\sqrt{37}} \right) \right] m(\bar{x}) d\bar{x}$$

$$= Pr(\bar{x} > 0.2279) - 4 \int_{0.2279}^{\infty} \Phi \left(-\frac{18\bar{x}}{\sqrt{37}} \right) m(\bar{x}) d\bar{x}$$

$$= 0.416.$$

In expectation, the experiment is worth 0.416. Assuming a linear utility and constant cost for the observations, this implies that the decision maker would be interested in the experiment as long as observations had unit price lower than 0.046.

13.3   Lindley information

13.3.1   Definition

In this section, we consider in more detail the decision problem of Chapter 10 and example 3 that is reporting a distribution regarding an unknown quantity θ with values in Θ. This can be thought of as a Bayesian way of modeling a situation in which data are not gathered to solve a specific decision problem but rather to learn about the world or to provide several decision makers with information that can be useful in multiple decision problems. This information is embodied in the distribution that is reported.

As seen in our examples in Sections 13.1.1 and 13.1.3, if a denotes the reported distribution and the utility function is $u(a(\theta)) = \log a(\theta)$ the Bayes decision is $a^* = \pi$, where π is the current distribution on θ. The value associated with the decision problem is

$$V(\pi) = \int_\Theta \log(\pi(\theta))\pi(\theta)d\theta. \qquad (13.24)$$

Given the outcome of experiment $\mathcal{E}$, consisting of observing x from the distribution $f(x|\theta)$, we can use the Bayes rule to determine π_x. By the argument above, the value associated with the decision problem after observing x is

$$V(\pi_x) = \int_\Theta \log(\pi_x(\theta))\pi_x(\theta)d\theta, \qquad (13.25)$$

and the observed information is

$$\begin{aligned}
\mathcal{V}_x(\mathcal{E}) &= V(\pi_x) - \mathcal{U}_{\pi_x}(a^*) \\
&= \int_\Theta \log(\pi_x(\theta))\pi_x(\theta)d\theta - \int_\Theta \log(\pi(\theta))\pi_x(\theta)d\theta \\
&= \int_\Theta \pi_x(\theta)\log\left(\frac{\pi_x(\theta)}{\pi(\theta)}\right)d\theta \equiv \mathcal{I}_x(\mathcal{E}).
\end{aligned} \qquad (13.26)$$

Prior to experimentation, the expected information from $\mathcal{E}$ is, using (13.12),

$$\begin{aligned}
\mathcal{V}(\mathcal{E}) &\equiv E_x[V(\pi_x)] - V(\pi) \\
&= \int_\mathcal{X}\int_\Theta \log\left(\frac{\pi_x(\theta)}{\pi(\theta)}\right)\pi_x(\theta)m(x)d\theta dx \qquad (13.27) \\
&= \int_\Theta\int_\mathcal{X} \log\left(\frac{f(x|\theta)}{m(x)}\right)f(x|\theta)\pi(\theta)dxd\theta \qquad (13.28) \\
&= E_x\left[E_{\theta|x}\left[\log\left(\frac{\pi_x(\theta)}{\pi(\theta)}\right)\right]\right] \qquad (13.29) \\
&= E_x\left[E_{\theta|x}\left[\log\left(\frac{f(x|\theta)}{m(x)}\right)\right]\right] \equiv \mathcal{I}(\mathcal{E}). \qquad (13.30)
\end{aligned}$$

This is called *Lindley information*, and was introduced in Lindley (1956). We use the notation $\mathcal{I}(\mathcal{E})$ for consistency with his paper. Lindley information was originally derived from information-theoretic, rather than decision-theoretic, arguments, and coincides with the expected value with respect to x of the Kullback–Leibler (KL) divergence between π_x and π, which is given by expression (13.26).

In general, if P and Q are two probability measures defined on the same space, with p and q their densities with respect to the common measure ν, the KL divergence between P and Q is defined as

$$KL(P, Q) = \int_\mathcal{X} \log\frac{p(x)}{q(x)}p(x)d\nu(x)$$

(see Kullback and Leibler 1951, Kullback 1959). The KL divergence is nonnegative but it is not symmetric. In decision-theoretic terms, the reason is that it quantifies the loss of representing uncertainty by Q when the truth is P, and that may differ from the loss of representing uncertainty by P when the truth is Q. For further details on properties of the KL divergence see Cover and Thomas (1991) or Schervish (1995).

13.3.2   Properties

Suppose that the experiment $\mathcal{E}_{12}$ consists of observing x_1 and x_2, and that the partial experiments of observing just x_1 and just x_2 are denoted by $\mathcal{E}_1$ and $\mathcal{E}_2$, respectively. Observing x_2 after x_1 has previously been observed has expected information $\mathcal{I}_{x_1 x_2}(\mathcal{E}_2 | \mathcal{E}_1)$. As discussed in Section 13.1.3, the expected information of $\mathcal{E}_2$ after $\mathcal{E}_1$ has been performed is the average of $\mathcal{I}_{x_1 x_2}(\mathcal{E}_2 | \mathcal{E}_1)$ with respect to x_1 and x_2 and it is denoted by $\mathcal{I}(\mathcal{E}_2 | \mathcal{E}_1)$. We learned from Theorem 13.3 that the overall information in $\mathcal{E}_{12}$ can be decomposed into the sum of the information contained in the first experiment and the additional information gained by the second one given the first. A special case using Lindley information is given by the next theorem. It is interesting to revisit it because here the proof clarifies how this decomposition can be related directly to the decomposition of a joint probability into a marginal and a conditional.

Theorem 13.4

$$\mathcal{I}(\mathcal{E}_{12}) = \mathcal{I}(\mathcal{E}_1) + \mathcal{I}(\mathcal{E}_2 | \mathcal{E}_1). \qquad (13.31)$$

Proof: From the definition of Lindley information we have

$$\mathcal{I}(\mathcal{E}_1) = E_{x_1} E_{\theta | x_1} \left[\log \left(\frac{f(x_1 | \theta)}{m(x_1)} \right) \right]$$

$$= E_{x_1} E_{x_2 | x_1} E_{\theta | x_1, x_2} \left[\log \left(\frac{f(x_1 | \theta)}{m(x_1)} \right) \right]. \qquad (13.32)$$

Also, from the definition of the information provided by $\mathcal{E}_2$ after $\mathcal{E}_1$ has been performed, we have

$$\mathcal{I}(\mathcal{E}_2 | \mathcal{E}_1) = E_{x_1 x_2} \left[\mathcal{I}_{x_1 x_2}(\mathcal{E}_1 | \mathcal{E}_2) \right]$$

$$= E_{x_1} E_{x_2 | x_1} E_{\theta | x_1, x_2} \left[\log \left(\frac{f(x_2 | \theta, x_1)}{m(x_2 | x_1)} \right) \right]. \qquad (13.33)$$

Adding expressions (13.32) and (13.33) we obtain

$$\mathcal{I}(\mathcal{E}_1) + \mathcal{I}(\mathcal{E}_2 | \mathcal{E}_1) = E_{x_1} E_{x_2 | x_1} E_{\theta | x_1, x_2} \left[\log \left(\frac{f(x_1 | \theta) f(x_2 | \theta, x_1)}{m(x_1) m(x_2 | x_1)} \right) \right]$$

$$= E_{x_1} E_{x_2 | x_1} E_{\theta | x_1, x_2} \left[\log \left(\frac{f(x_1, x_2 | \theta)}{m(x_1, x_2)} \right) \right]$$

$$= \mathcal{I}(\mathcal{E}_{12})$$

which is the desired result.   □

The results that follow are given without proof. Some of the proofs are good exercises. The first result guarantees that the information measure for any experiment is the same as the information for the corresponding sufficient statistic.

Theorem 13.5   Let $\mathcal{E}_1$ be the experiment consisting of the observation of x, and let $\mathcal{E}_2$ be the experiment consisting of the observation of t(X), where t is a sufficient statistic. Then $\mathcal{I}(\mathcal{E}_1) = \mathcal{I}(\mathcal{E}_2)$.

The next two results consider the case in which x_1 and x_2 are independent, conditional on θ, that is $f(x_1, x_2|\theta) = f(x_1|\theta)f(x_2|\theta)$.

Theorem 13.6   If x_1 and x_2 are conditionally independent given θ, then

$$\mathcal{I}(\mathcal{E}_2|\mathcal{E}_1) \leq \mathcal{I}(\mathcal{E}_2), \tag{13.34}$$

with equality if, and only if, x_1 and x_2 are unconditionally independent.

This proposition says that the information provided by the second of two conditionally independent observations is, on average, smaller than that provided by the first. This naturally reflects the diminishing returns of scientific experimentation.

Theorem 13.7   If x_1 and x_2 are independent, conditional on θ, then

$$\mathcal{I}(\mathcal{E}_1) + \mathcal{I}(\mathcal{E}_2) \geq \mathcal{I}(\mathcal{E}_{12}) \tag{13.35}$$

with equality if, and only if, x_1 and x_2 are unconditionally independent.

In the case of identical experiments being repeated sequentially, we have a more general result along the lines of the last proposition. Let $\mathcal{E}_1$ be any experiment and let $\mathcal{E}_2, \mathcal{E}_3, \ldots$ be conditionally independent and identical repetitions of $\mathcal{E}_1$. Let also

$$\mathcal{E}^{(n)} = (\mathcal{E}_1, \ldots, \mathcal{E}_n). \tag{13.36}$$

Theorem 13.8 $\mathcal{I}(\mathcal{E}^{(n)})$ is a concave, increasing function of n.

Proof: We need to prove that

$$0 \leq \mathcal{I}(\mathcal{E}^{(n+1)}) - \mathcal{I}(\mathcal{E}^{(n)}) \leq \mathcal{I}(\mathcal{E}^{(n)}) - \mathcal{I}(\mathcal{E}^{(n-1)}). \tag{13.37}$$

By applying equation (13.31) we have $\mathcal{I}(\mathcal{E}^{(n+1)}) = \mathcal{I}(\mathcal{E}_{n+1}|\mathcal{E}^{(n)}) + \mathcal{I}(\mathcal{E}^{(n)})$. Next, using the fact that Lindley information is nonnegative, we obtain $\mathcal{I}(\mathcal{E}^{(n+1)}) - \mathcal{I}(\mathcal{E}^{(n)}) \geq 0$, which proves the left side inequality.

Using again equation (13.31) on both $\mathcal{I}(\mathcal{E}^{(n+1)})$ and $\mathcal{I}(\mathcal{E}^{(n)})$ we can observe that the right side inequality is equivalent to $\mathcal{I}(\mathcal{E}_{n+1}|\mathcal{E}^{(n)}) \leq \mathcal{I}(\mathcal{E}_n|\mathcal{E}^{(n-1)})$. As $\mathcal{E}_{n+1} \equiv \mathcal{E}_n$, the inequality becomes $\mathcal{I}(\mathcal{E}_{n+1}|\mathcal{E}^{(n)}) \leq \mathcal{I}(\mathcal{E}_{n+1}|\mathcal{E}^{(n-1)})$. Also, since $\mathcal{E}^{(n)} = (\mathcal{E}^{(n-1)}, \mathcal{E}_n)$ we can rewrite the inequality as $\mathcal{I}(\mathcal{E}_{n+1}|\mathcal{E}^{(n-1)}, \mathcal{E}_n) \leq \mathcal{I}(\mathcal{E}_{n+1}|\mathcal{E}^{(n-1)})$. The proof of this inequality follows from an argument similar to that of (13.31). □

As the value of V is convex in the prior, increased uncertainty means decreased value of stopping before experimenting. We would then expect that increased uncertainty means increased value of information. This is in fact the case.

Theorem 13.9   *For fixed $\mathcal{E}$, $\mathcal{I}(\mathcal{E})$ is a concave function of $\pi(\theta)$.*

This means that if π_1 and π_2 are two prior distributions for θ, and $0 \leq \alpha \leq 1$, then the information provided by an experiment $\mathcal{E}$ with a mixture prior $\pi_0(\theta) = \alpha\pi_1(\theta) + (1-\alpha)\pi_2(\theta)$ is at least as large as the linear combination of the information provided by the experiment under priors π_1 and π_2. In symbols, $\mathcal{I}_0(\mathcal{E}) \geq \alpha\mathcal{I}_1(\mathcal{E}) + (1-\alpha)\mathcal{I}_2(\mathcal{E})$ where $\mathcal{I}_i(\mathcal{E})$ is the expected information provided by $\mathcal{E}$ under prior π_i for $i = 0, 1, 2$.

13.3.3   Computing

In realistic applications closed form expressions for $\mathcal{I}$ are hard to derive. Numerical evaluation of expected information requires Monte Carlo integration (see, for instance, (Gelman *et al.* (1995)). The idea is as follows. First, simulate points θ_i from $\pi(\theta)$ and, for each, x_i from $f(x|\theta_i)$, with $i = 1, \ldots, I$. Using the simulated sample obtain the Monte Carlo estimate for $E_\theta E_{x|\theta}\left[\log[f(x|\theta)]\right]$, that is

$$E_\theta E_{x|\theta}\left[\log(f(x|\theta))\right] \approx \frac{1}{I}\sum_{i=1}^{I}\log(f(x_i|\theta_i)). \qquad (13.38)$$

Next, evaluate $E_x\left[\log(m(x))\right]$, again using Monte Carlo integration, by calculating

$$E_x\left[\log(m(x))\right] \approx \frac{1}{I}\sum_{i=1}^{I}\log(m(x_i)). \qquad (13.39)$$

When the marginal is not available in closed form, this requires another Monte Carlo integration to evaluate $m(x_i)$ for each i as

$$m(x_i) \approx \frac{1}{J}\sum_{j=1}^{J}f(x_i|\theta_j),$$

where the θ_j are drawn from $\pi(\theta)$. Using these quantities, Lindley information is estimated by taking the difference between (13.38) and (13.39). Carlin and Polson (1991) use an implementation of this type. Müller and Parmigiani (1993) and Müller and Parmigiani (1995) discuss computational issues in the use of Markov-chain Monte Carlo methods to evaluate information-theoretic measures such as entropy, KL divergence, and Lindley information, and introduce fast algorithms for optimization problems that arise in Bayesian design. Bielza *et al.* Insua (1999) extend those algorithms to multistage decision trees.

13.3.4   Optimal design

An important reason for measuring the expected information provided by a prospective experiment is to optimize the experiment with respect to some design variables (Chaloner and Verdinelli 1995). Consider a family of experiments $\mathcal{E}_D$ indexed by the design variable D. The likelihood function for the experiment $\mathcal{E}_D$ is $f_D(x|\theta)$, $x \in \mathcal{X}$. Let $\pi_D(\theta|x)$ denote the posterior distribution resulting from the observation of x and let $m_D(x)$ denote the marginal distribution of x. If D is the sample size of the experiment, we know that $\mathcal{I}(\mathcal{E}_D)$ is monotone, and choosing a sample size requires trading off information against other factors like cost.

In many cases, however, there will be a natural trade-off built into the choice of D and $\mathcal{I}(\mathcal{E}_D)$ will have a maximum. Consider this simple example. Let the random variable Y have conditional density $f(y|\theta)$. We are interested in learning about θ, but cannot observe y exactly. We can only observe whether or not it is greater than some cutpoint D. This problem arises in applications in which we are interested in estimating the rate of occurrence of an event, but it is impractical to monitor the phenomenon studied continuously. What we observe is then a Bernoulli random variable x that has success probability $F(D|\theta)$. Choices of D that are very low will give mostly zeros and be uninformative, while choices that are very high will give mostly ones and also turn out uninformative. The best D is somewhere in the center of the distribution of y, but where exactly?

To solve this problem we first derive general optimality conditions. Let us assume that D is one dimensional, continuous, or closely approximated by a continuous variable, and that $f_D(\theta|x)$ is differentiable with respect to D. Lindley information is

$$\mathcal{I}(\mathcal{E}_D) = \int_{\mathcal{X}} \int_{\Theta} \log\left(\frac{\pi_D(\theta|x)}{\pi(\theta)}\right) \pi_D(\theta|x) m_D(x) d\theta dx.$$

Define

$$f_D'(x|\theta) \equiv \frac{\partial f_D(x|\theta)}{\partial D} \qquad\qquad \pi_D'(\theta|x) \equiv \frac{\partial \pi_D(\theta|x)}{\partial D}.$$

The derivative of $\mathcal{I}(\mathcal{E}_D)$ with respect to D turns out to be

$$\frac{d\mathcal{I}(\mathcal{E}_D)}{dD} = \int_{\Theta} \int_{\mathcal{X}} \log\left(\frac{\pi_D(\theta|x)}{\pi(\theta)}\right) f_D'(x|\theta) \pi(\theta) dx d\theta. \qquad (13.40)$$

To see that (13.40) holds, differentiate the integrand of $\mathcal{I}(\mathcal{E}_D)$ to get

$$\left[\log\left(\frac{\pi_D(\theta|x)}{\pi(\theta)}\right) f_D'(x|\theta) + f_D(x|\theta)\frac{\pi_D'(\theta|x)}{\pi_D(\theta|x)}\right] \pi(\theta).$$

It follows from differentiating Bayes' rule that

$$\pi_D' = \pi \frac{f_D' m_D - f_D m_D'}{m_D^2}.$$

Thus

$$\frac{f_D \pi_D'}{\pi_D} = \frac{m_D \pi_D'}{\pi} = \frac{f_D' m_D - f_D m_D'}{m_D} = f_D' - f_D \frac{m_D'}{m_D}.$$

It can be proved that the following area conditions hold:

$$\int_\Theta \int_\mathcal{X} f_D(x|\theta) \pi(\theta) \frac{m_D'(x)}{m_D(x)} dx d\theta = \int_\mathcal{X} m_D'(x) dx = 0,$$

$$\int_\Theta \int_\mathcal{X} f_D'(x|\theta) \pi(\theta) dx d\theta = 0.$$

Equation (13.40) follows from this.

With simple manipulations, (13.40) can be rewritten as

$$\frac{d\mathcal{I}(\mathcal{E}_D)}{dD} = \int_\Theta \int_\mathcal{X} \log (f_D(x|\theta)) f_D'(x|\theta) \pi(\theta) dx\, d\theta - \int_\mathcal{X} \log (m_D(x))\, m_D'(x)\, dx. \quad (13.41)$$

This expression parallels the alternative expression for (13.29), given by Lindley (1956):

$$\mathcal{I}(\mathcal{E}_D) = \int_\Theta \int_\mathcal{X} \log (f_D(x|\theta)) f_D(x|\theta) \pi(\theta) dx d\theta - \int_\mathcal{X} \log (m_D(x))\, m_D(x) dx. \quad (13.42)$$

Equations (13.40) and (13.41) can be used in calculating the optimal D whenever Lindley information enters as a part of the design criterion. For example, if D is sample size, and the cost of observation c is fixed, then the real-valued approximation to the optimal sample size satisfies

$$\frac{d\mathcal{I}(\mathcal{E}_D)}{dD} - c = 0. \quad (13.43)$$

Returning to the optimal cutoff problem, in which $c = 0$, we have

$$f_D(x|\theta) = \begin{cases} F(D|\theta), & \text{if } x = 1 \\ 1 - F(D|\theta), & \text{if } x = 0, \end{cases} \quad (13.44)$$

which gives

$$f_D'(x|\theta) = \begin{cases} f(D|\theta), & \text{if } x = 1 \\ -f(D|\theta), & \text{if } x = 0. \end{cases} \quad (13.45)$$

Similarly, if $F(x)$ is the marginal cdf of x,

$$m_D(x) = \begin{cases} \int_\Theta F(D|\theta) \pi(\theta) d\theta = F(D), & \text{if } x = 1 \\ 1 - \int_\Theta F(D|\theta) \pi(\theta) d\theta = 1 - F(D), & \text{if } x = 0 \end{cases} \quad (13.46)$$

with

$$m'_D(x) = \begin{cases} \int_\Theta f(D|\theta)\pi(\theta)d\theta = f(D), & \text{if } x = 1 \\ -\int_\Theta f(D|\theta)\pi(\theta)d\theta = -f(D), & \text{if } x = 0 \end{cases} \qquad (13.47)$$

By inserting (13.45) through (13.47) into (13.41), we obtain

$$\int_\Theta \log\left(\frac{F(D|\theta)}{1 - F(D|\theta)}\right)f(D|\theta)\pi(\theta)d\theta = \log\left(\frac{F(D)}{1 - F(D)}\right)f(D). \qquad (13.48)$$

Now, by dividing both sides by $f(D)$ we obtain that a necessary condition for D to be optimal is

$$\int_\Theta \log\left(\frac{F(D|\theta)}{1 - F(D|\theta)}\right)\pi(\theta|x = D)d\theta = \log\left(\frac{F(D)}{1 - F(D)}\right); \qquad (13.49)$$

in words, we must choose the cutoff so that the expected conditional logit is equal to the marginal logit. Expectation is taken with respect to the posterior distribution of θ given that x is at the cutoff point.

13.4   Minimax and the value of information

Theorem 13.2 requires that actions should be chosen according to the expected utility principle. What happens if an agent is minimax and a is chosen accordingly? In that case, the minimax principle can lead to deciding without experimenting, in instances where it is very hard to accept the conclusion that the new information is worthless. The point was first raised by Savage, as part of his argument for the superiority of the regret (or loss) form over the negative utility form of the minimax principle:

> Reading between the lines, it appears that Wald means to work with loss and not negative income. For example, on p.124 he says that if a certain experiment is to be done, and the only thing to decide is what to do after seeing its outcome, then the cost of the experiment (which may well depend on the outcome) is irrelevant to the decision; this statement is right for loss but wrong for negative income. (Savage 1951, p. 65)

While Savage considered this objection not to be relevant for the regret case, one can show that the same criticism applies to both forms. One can construct examples where the minimax regret principle attributes the same value V to an ancillary statistic and to a consistent estimator of the state of nature for arbitrary n. In other instances (Hodges and Lehmann 1950) the value is highest for observing an inconsistent estimator of the unknown quantity, even if a consistent estimator is available.

A simple example will illustrate the difficulty with minimax (Parmigiani 1992). This is constructed so that minimax is well behaved in the negative utility version of the problem, and not so well behaved in the regret (or loss) version. Examples can be constructed where the reverse is true. A box contains two marbles in one of three possible configurations: both marbles are red, the first is blue and the second is red, or both are blue. You can choose between two actions, whose consequences are summarized in the table below, whose entries are negative payoffs (a 4 means you pay $4):

	RR	BR	BB
a_1	2	0	4
a_2	4	4	0

A calculation left as an exercise would show that the minimax action for this problem is a mixed strategy that puts weight 2/3 on a_1 and 1/3 on a_2, and has risk $2 + 2/3$ in states RR and BB and $1 + 1/3$ in state BR.

Suppose now that you have the option to see the first marble at a cost of $0.1. To determine whether the observation is worthwhile, you look at its potential use. It would not be very smart to choose act a_2 after seeing that the first marble is red, so the undominated decision functions differ only in what they recommend to do after having seen that the first marble is blue. Call $a_1(B)$ the function choosing a_1, and $a_2(B)$ that choosing a_2. Then the available unmixed acts and their consequences are summarized in the table

	RR	BR	BB
a_1	2	0	4
a_2	4	4	0
$a_1(B)$	2.1	0.1	4.1
$a_2(B)$	2.1	4.1	0.1

As you can see, $a_1(B)$ is not an admissible decision rule, because it is dominated by a_1, but $a_2(B)$ is admissible. The equalizer strategy yields an even mixture of a_1 and $a_2(B)$ which has average loss of $2.05 in every state. So with positive probability, minimax applied to the absolute losses prescribes to make the observation.

Let us now consider the same regret form of the loss table. We get that by shifting each column so that the minimum is 0, which gives

	RR	BR	BB
a_1	0	0	4
a_2	2	4	0

The minimax strategy for this table is an even mixture of a_1 and a_2 and has risk 1 in state RR and 2 in the others. If you have the option of experimenting, the unmixed table is

	RR	BR	BB
a_1	0	0	4
a_2	2	4	0
$a_1(B)$	0.1	0.1	4.1
$a_2(B)$	0.1	4.1	0.1

In minimax decision making, you act as though an intelligent opponent decided which state occurs. In this case, state RR does not play any role in determining the minimax procedure, being dominated by state BR from the point of view of your opponent. You have no incentive to consider strategies $a_1(B)$ and $a_2(B)$: any mixture of one of these with the two acts that do not include observation will increase the average loss over the value of 2 in either column.

So in the end, if you follow the minimax regret principle you never perform the experiment. The same conclusion is reached no matter how little the experiment costs, provided that the cost is positive. The assumption that you face an opponent whose gains are your losses makes you disregard the fact that, in some cases, observing the first marble gives you the opportunity to choose an act with null loss, and only makes you worry that the marble may be blue, in which case no further cue is obtained. This concern can lead you to "ignore extensive evidence," to put it in Savage's terms. What makes the negative utility version work the opposite way is the "incentive" given to the opponent to choose the state where the experiment is valuable to you.

Chernoff (1954) noticed that an agent minimizing the maximum regret could in some cases choose action a_1 over action a_2 when these are the only available options, but choose a_2 when some other option a_3 is made available. This annoying feature originates from the fact that turning a payoff table into a regret table requires subtracting constants to each column, and these constants can be different when new rows appear. In our example, the added rows arise from the option to experiment: the losses of a rule based on an experiment are linear combinations of entries already present in the same column as payoffs for the terminal actions, and cannot exceed the maximum loss when no experimentation is allowed. This is a different mechanism whereby minimax decision making to "ignore extensive evidence" compared to the earlier example from Chernoff, because the constants for the standardization to regret are the same after one adds the two rows corresponding to experimental options.

13.5   Exercises

Problem 13.1   Prove equation (13.7).

Problem 13.2   (Savage 1954) In the example of Section 13.2.1 suppose the decision maker could directly observe the quality of the grapes. Show that the decision maker's best action would then yield 2 utilities, and show that it could not possibly lead the decision maker to buy 2 pounds of grapes.

Problem 13.3   Using simulation, approximate the distribution of $V_x(\mathcal{E})$ in the example in Section 13.1.4. Suppose each observation costs $230. Compute the marginal cost-effectiveness ratio for performing the experiment in the example (versus the status quo of no experimentation). Use the negative of the loss as the measure of effectiveness.

Problem 13.4   Derive the four minimax strategies advertised in Section 13.4.

Problem 13.5   Consider an experiment consisting of a single Bernoulli observation, from a population with success probability θ. Consider a simple versus simple hypothesis testing situation in which $\mathcal{A} = \{a_0, a_1\}$, $\Theta = \{0.50, 0.75\}$, and with utilities as shown in Table 13.7. Compute $\mathcal{V}(\mathcal{E}) = E_x[V(\pi_x)] - V(\pi)$.

Problem 13.6   Show that $\mathcal{I}$ is invariant to one-to-one transformations of the parameter θ.

Problem 13.7   Let $f(x|\theta) = pf_1(x|\theta) + (1-p)f_2(x|\theta)$. Let $\mathcal{E}, \mathcal{E}_1, \mathcal{E}_2$ be the experiments of observing a sample from $f(x|\theta), f_1(x|\theta)$ and $f_2(x|\theta)$ respectively. Show that

$$\mathcal{I}(\mathcal{E}) \leq p\mathcal{I}(\mathcal{E}_1) + (1-p)\mathcal{I}(\mathcal{E}_2). \tag{13.50}$$

Problem 13.8   Consider a population of individuals cross-classified according to binary variables x_1 and x_2. Suppose that the overall proportions of individuals with characteristics x_1 and x_2 are known to be p_1 and p_2 respectively. Say $p_1 < p_2 < 1 - p_2 < 1 - p_1$. However, the proportion ω of individuals with both characteristics is unknown. We are considering four experiments:

1. Random sample of individuals with $x_1 = 1$

2. Random sample of individuals with $x_1 = 0$

3. Random sample of individuals with $x_2 = 1$

4. Random sample of individuals with $x_2 = 0$.

Table 13.7   Utilities for Problem 13.5.

Actions	States of nature	
	0.50	0.75
a_0	0	-1
a_1	-1	0

Using the result stated in Problem 13.7 show that experiment 1 has the highest information on ω irrespective of the prior.

Hint: Work with the information on $\theta = \omega/p_1$ and then translate the result in terms of ω.

Problem 13.9   Prove Theorem 13.5.

Hint: Use the factorization theorem about sufficient statistics.

Problem 13.10   Prove Theorem 13.6. Use its result to prove Theorem 13.7.

Problem 13.11   Consider an experiment $\mathcal{E}$ that consists of observing n conditionally independent random variables $x_1, \ldots, x_n$, with $x_i \sim N(\theta, \sigma^2)$, with σ known. Suppose also that a priori $\theta \sim N(\mu_0, \tau_0^2)$. Show that

$$\mathcal{I}(\mathcal{E}) = \frac{1}{2} \log \left(1 + n \frac{\tau_0^2}{\sigma^2} \right).$$

You can use facts about conjugate priors from Bernardo and Smith (2000) or Berger (1985). However, please rederive $\mathcal{I}$.

Problem 13.12   Prove Theorem 13.9.

Hint: As in the proof of Theorem 13.1, consider any pair of prior distributions π_1 and π_2 of θ and $0 < \alpha < 1$. Then, calculate Lindley information for any experiment $\mathcal{E}$ with prior distribution $\alpha \pi_1 + (1 - \alpha) \pi_2$ for the unknown θ.

14

Sample size

In this chapter we discuss one of the most common decisions in statistical practice: the choice of the sample size for a study. We initially focus on the case in which the decision maker selects a fixed sample size n, collects a sample $x_1, \ldots, x_n$, and makes a terminal decision based on the observed sample. Decision-theoretic approaches to sample size determination formally model the view that the value of an experiment depends on the use that is planned for the results, and in particular on the decision that the results must help address. This approach, outlined in greater generality in Chapter 13, provides conceptual and practical guidance for determining an optimal sample size in a very broad variety of experimental situations. Here we begin with an overview of decision-theoretic concepts in sample size. We next move to a general simulation-based algorithm to solve optimal sample size problems in complex practical applications. Finally, we illustrate both theoretical and computational concepts with examples.

The main reading for this chapter is Raiffa and Schlaifer (1961, chapter 5), who are generally credited for providing the first complete formalization of Bayesian decision-theoretic approaches to sample size determination. The general ideas are implicit in earlier decision-theoretic approaches. For example Blackwell and Girshick (1954, page 170) briefly defined the framework for Bayesian optimal fixed sample size determination.

Featured book (Chapter 5):

Raiffa, H. and Schlaifer, R. (1961). *Applied Statistical Decision Theory*, Harvard University Press, Boston, MA.

Textbook references on Bayesian optimal sample sizes include DeGroot (1970) and Berger (1985). For a review of Bayesian approaches to sample size determination (including less formal decision-theoretic approaches) see Pham-Gia and

---

Turkkan (1992), Adcock (1997), and Lindley (1997); for practical applications see, for instance, Yao *et al.* (1996) and Tan and Smith (1998).

14.1   Decision-theoretic approaches to sample size

14.1.1   Sample size and power

From a historical perspective, a useful starting point for understanding decision-theoretic sample size determination is the Neyman–Pearson theory of testing a simple hypothesis against testing a simple alternative (Neyman and Pearson 1933). A standard approach in this context is to seek the smallest sample size that is sufficient to achieve a desired power at a specified significance level. From the standpoint of our discussion, there are three key features in this procedure:

1. After the data are observed, an optimal decision rule—in this case the use of a uniformly most powerful test—is used.

2. The sample size is chosen to be the smallest that achieves a desired performance level.

3. The same optimality criteria—in this case significance and power—are used for selecting both a terminal decision rule and the sample size.

As in the value of information analysis of Chapter 13, the sample size depends on the benefit expected from the data after it will be put to use, and the quantification of this benefit is user specified. This structure foreshadows the decision-theoretic approach to sample size that was to be developed in later years in the influential work by Wald (1947b), described below.

Generally, in choosing the sample size for a study, one tries to weigh the trade-off between carrying out a small study and improving the final decision with respect to some criterion. This can be achieved in two ways: either by finding the smallest sample size that guarantees a desired level of the criterion, or by explicitly modeling the trade-off between the utilities and costs of experimentation. Both of these approaches are developed in more detail in the remainder of this section.

14.1.2   Sample size as a decision problem

In previous chapters we dealt with problems where a decision maker has a utility $u(a(\theta))$ for decision a when the state of nature is θ. In the statistical decision-theoretic material we also cast the same problem in terms of a loss $L(\theta, a)$ for making decision a when the state of nature is θ, and the decision is based on a sample $x^n = (x_1, x_2, \ldots, x_n)$ of size n drawn from $f(x|\theta)$. We now extend this to also choosing optimally the sample size n of the experiment. We first specify a more general detailed function $u(\theta, a, n)$ or, equivalently, a loss function $L(\theta, a, n)$ for observing a sample of size n and making decision a when the state of nature is θ. The sample size n is really an action here, so the concepts of utility and loss are not altered in substance. We commonly use the form

$$u(\theta, a, n) = u(a(\theta)) - C(n) \qquad (14.1)$$

$$L(\theta, a, n) = L(\theta, a) + C(n). \qquad (14.2)$$

The function $L(\theta, a)$ in (14.2) refers, as in previous chapters, to the choice of a after the sample is observed and is referred to as the *terminal* loss function. The function $C(n)$ represents the cost of collecting a sample of size n. This formulation can be traced to Wald (1947a) and Wald (1949).

The practical implications of accounting for the costs of experimentation in (14.2) is explained in an early paper by Grundy *et al.*:

> One of the results of scientific research is the development of new processes for use in technology and agriculture. Usually these new processes will have been worked out on a small scale in the laboratory, and experience shows that there is a considerable risk in extrapolating laboratory results to factory or farm scale where conditions are less thoroughly controlled. It will usually pay, therefore, to carry out a programme of full-scale experiments before allowing a new process to replace an old one that is known to give reasonably satisfactory results. However, full-scale experimentation is expensive and its costs have to be set against any increase in output which may ensue if the new process fulfills expectations. The question then arises of how large an experimental programme can be considered economically justifiable. If the programme is extensive, we are not likely to reject a new process that would in fact have been profitable to install, but the cost of the experiments may eat up most of the profits that result from a correct decision; if we economize on the experiments, we may fail to adopt worthwhile improvements in technique, the results of which have been masked by experimental errors. (Grundy *et al.* 1956, p. 32)

Similar considerations apply to a broad range of other scientific contexts. Raiffa and Schlaifer further comment on the assumption of additivity of the terminal utility $u(a(\theta))$ and experimental costs $C(n)$:

> this assumption of additivity by no means restrict us to problems in which all consequences are monetary. . . . In general, sampling and terminal utilities will be additive whenever consequences can be measured or scaled in terms of *any* common *numéraire* the utility of which is linear over a suitably wide range; and we point out that number of patients cured or number of hours spent on research may well serve as such a *numéraire* in problems where money plays no role at all. (Raiffa and Schlaifer 1961, p. xiii)

14.1.3   Bayes and minimax optimal sample size

A Bayesian optimal sample size can be determined for any problem by specifying four components: (a) a likelihood function for the experiment; (b) a prior distribution on the unknown parameters; (c) a loss (or utility) function for the terminal decision problem; and (d) a cost of experimentation. Advantages include explicit consideration of costs and benefits, formal quantification of parameter uncertainty, and consistent treatment of this uncertainty in the design and inference stages.

Specifically, say π represents a priori knowledge about the unknown parameters, and δ_n^* is the Bayes rule with respect to that prior (see Raiffa and Schlaifer 1961, or DeGroot 1970). As a quick reminder, the Bayes risk is defined, as in equation (7.9), by

$$r(\pi, \delta_n) = \int_\Theta R(\theta, \delta_n)\pi(\theta)d\theta$$

and the Bayes rule δ_n^* minimizes the Bayes risk $r(\pi, \delta_n)$ among all the possible δ_n when the sample size is n. Thus the optimal Bayesian sample size n^* is the one that minimizes

$$r(\pi, n) = r(\pi, \delta_n^*) + C(n). \tag{14.3}$$

By expanding the definition of the risk function and reversing the order of integration, we see that

$$r(\pi, n) = \int_{\mathcal{X}} \int_\Theta L(\theta, \delta_n^*(x^n), n) \, \pi(\theta|x^n)d\theta \, m(x^n)dx^n. \tag{14.4}$$

Similarly, the Bayesian formulation can be reexpressed in terms of utility as the maximization of the function

$$\mathcal{U}_\pi(n) = \int_{\mathcal{X}} \int_\Theta u(\theta, \delta_n^*(x^n), n) \, \pi(\theta|x^n)d\theta \, m(x^n)dx^n. \tag{14.5}$$

Both formulations correspond to the solution of a two-stage decision tree similar to those of Sections 12.3.1 and 13.1.3, in which the first stage is the selection of the sample size, and the second stage is the solution of the statistical decision problem. In this sense Bayesian sample size determination is an example of preposterior analysis. Experimental data are conditioned upon in the inference stage and averaged over in the design stage.

By contrast, in a minimax analysis, we would analyze the data using the minimax decision function δ_n^M chosen so that

$$\inf_{\delta_n \in \mathcal{D}} \sup_{\theta \in \Theta} R(\theta, \delta_n) = \sup_{\theta \in \Theta} R(\theta, \delta_n^M).$$

Thus the best that one can expect to achieve after a sample of size n is observed is $\sup_{\theta \in \Theta} R(\theta, \delta_n^M)$, and the minimax sample size n^M minimizes

$$\sup_\theta R(\theta, \delta_n^M) + C(n). \tag{14.6}$$

Exploiting the formal similarity between the Bayesian and minimax schemes, Blackwell and Girshick (1954, page 170) suggest a general formalism for the optimal sample size problem as the minimization of a function of the form

$$\rho_n(\pi, \delta_n) + C(n), \tag{14.7}$$

where δ_n is a terminal decision rule based on n observations. This formulation encompasses both frequentist and Bayesian decision-theoretic approaches. In the Bayesian case δ_n is replaced by δ_n^*. For the minimax case, δ_n is the minimax rule, and π the prior that makes the minimax rule a Bayes rule (the so-called least favorable prior), if it exists. This formulation allows us in principle to specify criteria that do not make use of the optimal decision rule in evaluating ρ, a feature that may be useful when seeking approximations.

14.1.4   A minimax paradox

With some notable exceptions (Chernoff and Moses 1959), explicit decision-theoretic approaches are rare among frequentist sample size work, but more common within Bayesian work (Lindley 1997, Tan and Smith 1998, Müller 1999). Within the frequentist framework it is possible to devise sensible ad hoc optimal sample size criteria for specific families of problems, but it is more difficult to formulate a general rationality principle that can reliably handle a broad set of design situations. To illustrate this point consider this example, highlighting a limitation of the minimax approach.

Suppose we are interested in testing $H_0 : \theta \leq \theta_0$ versus $H_1 : \theta > \theta_0$ based on observations sampled from $N(\theta, \sigma^2)$, the normal distribution with unknown mean θ and known variance σ^2. We can focus on the class of tests that reject the null hypothesis if the sample mean is larger than a cutoff d, $\delta_d(\bar{x}_n) = I_{\{\bar{x}_n \geq d\}}$, where I_E is the indicator function of the event E. The sample mean is a sufficient statistic and this class is admissible (see Berger 1985 for details).

Let $F(.|\theta)$ be the cumulative probability function of $\bar{x}_n$ given θ. Under the 0–1–k loss function of Table 14.1, the associated risk is given by

$$R(\theta, \delta_d) = \begin{cases} k(1 - F(d|\theta)) & \text{if } \theta \leq \theta_0 \\ F(d|\theta) & \text{if } \theta > \theta_0. \end{cases}$$

Table 14.1   The 0–1–k loss function. No loss is incurred if H_0 holds and action a_0 is chosen, or if H_1 holds and action a_1 is chosen. If H_1 holds and decision a_0 is made, the loss is 1; if H_0 holds and decision a_1 is made, the loss is k.

	H_0	H_1
a_0: accept H_0	0	1
a_1: reject H_0	k	0

Let $\mathcal{P}(\theta) = E_{\bar{x}_n | \theta}[\delta_d(\bar{x}_n)] = 1 - F(d|\theta)$ denote the power function (Lehmann 1983). In the frequentist approach to hypothesis testing, the power function evaluated at θ_0 gives the type I error probability. When evaluated at any given $\theta = \theta_1$ chosen within the alternative hypothesis, the power function gives the power of the test against the alternative θ_1.

Since $\mathcal{P}$ is increasing in θ, it follows that

$$\sup_\theta R(\theta, \delta_d) = \max \{k\mathcal{P}(\theta_0), 1 - \mathcal{P}(\theta_0)\},$$

which is minimized by choosing d such that $k\mathcal{P}(\theta_0) = 1 - \mathcal{P}(\theta_0)$, that is $\mathcal{P}(\theta_0) = 1/(1 + k)$. When one specifies a type I error probability of α, the implicit k can be determined by solving $1/(1 + k) = \alpha$, which implies $k = (1 - \alpha)/\alpha$.

Solving the equation $\mathcal{P}(\theta_0) = 1/(1 + k)$ with respect to the cutoff point d, we find a familiar solution:

$$d^M = \theta_0 + z_{\frac{k}{1+k}} \frac{\sigma}{\sqrt{n}}.$$

Since $\inf_{\delta \in \mathcal{D}} \sup_\theta R(\theta, \delta) = \sup_\theta R(\theta, \delta_{d^M})$, $\delta^M = \delta_{d^M}$ is the minimax rule.

It follows that $\sup_\theta R(\theta, \delta_{d^M}) + C(n) = k/(k + 1) + C(n)$ is an increasing function in n, as long as C is itself increasing, which is always to be expected. Thus, the optimal sample size is zero! The reason is that, with the 0–1–k loss function, the minimax risk is constant in θ. The minimax strategy is rational if an intelligent opponent whose gains are our losses is choosing θ. In this problem such an opponent will always choose $\theta = \theta_0$ and make it as hard as possible for us to distinguish between the two hypotheses. This is another example of the "ultra-pessimism" discussed in Section 13.4.

Let us now study the problem from a Bayesian viewpoint. Using the 0–1–k loss function, the posterior expected losses of accepting and rejecting the null hypothesis are $\pi(H_1|x^n)$ and $k\pi(H_0|x^n)$, respectively. The Bayes decision minimizes the posterior expected loss. Thus, the Bayes rule rejects the null hypothesis whenever $\pi(H_1|x^n) > k/(1 + k)$ (see Section 7.5.1). Suppose we have a $N(\mu_0, \tau_0^2)$ prior on θ. Solving the previous inequality in $\bar{x}_n$, we find that the Bayes decision rule is $\delta^*(x^n) = I_{\{\bar{x}_n > d^B\}}$ with the cutoff given by

$$d^B = \theta_0 + z_{\frac{k}{k+1}} \frac{\sigma}{\sqrt{n}} \sqrt{1 + \frac{\sigma^2}{n\tau_0^2}}.$$

With a little bit of work, we can show that the Bayes risk r is given by

$$r(\pi, n) = (1 + k) \int_0^\infty \Phi \left(\frac{-u\sqrt{n}}{\sigma} + z_{\frac{k}{1+k}} \sqrt{1 + \frac{\sigma^2}{n\tau_0^2}} \right) \frac{1}{\sqrt{2\pi}\tau_0} e^{-\frac{u^2}{2\tau_0^2}} du$$

$$- k\Phi \left(z_{\frac{k}{1+k}} \frac{\sigma}{\sqrt{n}\tau_0} \right) + \frac{k}{2} + C(n). \tag{14.8}$$

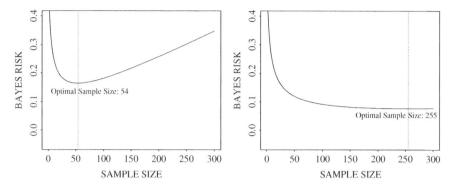

Figure 14.1   Bayes risk $r(\pi, n)$ as a function of the sample size n with the 0–1–k loss function. In the left panel the cost per observation is $c = 0.001$, in the right panel the cost per observation is $c = 0.0001$. While the optimal sample size is unique in both cases, in the case on the right a broad range of choices of sample size achieve a similar risk.

The above expression can easily be evaluated numerically and optimized with respect to n, as we did in Figure 14.1. While it is possible that the optimal n may be zero if costs are high enough, a range of results can emerge depending on the specific combination of cost, loss, and hyperparameters. For example, if $C(n) = cn$, with $c = 0.001$, and if $k = 19, \sigma = 1$, and $\tau_0 = 1$, then the optimal sample size is $n_B = 54$. When the cost per observation c is 0.0001, the optimal sample size is $n_B = 255$.

14.1.5   Goal sampling

A different Bayesian view of sample size determination can be traced back to Lindley's seminal paper on measuring the value of information, discussed in Chapter 13. There, he suggests that a:

> consequence of the view that one purpose of statistical experimentation is to gain information will be that the statistician will stop experimentation when he has enough information. (Lindley 1956, p. 1001)

A broad range of frequentist and Bayesian methods for sample size determination can be described as choosing the smallest sample that is sufficient to achieve, in expectation, some set goals of experimentation. An example of the former is seeking the smallest sample size that is sufficient to achieve a desired power at a specified significance level. An example of the latter is seeking the smallest sample size necessary to obtain, in expectation, a fixed width posterior probability interval for a parameter of interest. We refer to this as goal sampling. It can be implemented sequentially, by collecting data until a goal is met, or nonsequentially, by planning a sample that is sufficient to reach the goal in expectation.

Choosing the smallest sample size to achieve a set of goals can be cast in decision-theoretic terms, from both the Bayesian and frequentist viewpoints. This approach can be described as a constrained version of that described in Section 14.1.3, where one sets a desired value of ρ and minimizes (14.7) subject to that constraint. Because both cost and performance are generally monotone in n, an equivalent formulation is to choose the smallest n necessary to achieve a desired ρ. This formulation is especially attractive when cost of experimentation is hard to quantify or when the units of ρ are difficult to compare to the monetary unit c.

In hypothesis testing, we can also use this idea to draw correspondences between Bayesian and classical approaches to sample size determination based on controlling probabilities of classification error. We illustrate this approach in the case of testing a simple hypothesis against a simple alternative with normal data. Additional examples are in Inoue $et\ al.$ (2005).

Suppose that $x_1, \ldots, x_n$ is a sample from a $N(\theta, 1)$ distribution. It is desired to test $H_0 : \theta = \theta_0$ versus $H_1 : \theta = \theta_1$, where $\theta_1 > \theta_0$, with the 0–1 loss function. This is a special case of the 0–1–k loss of Table 14.1 with $k = 1$ and assigns a loss of 0 to both correct decisions and a loss of 1 to both incorrect decisions. Assume a priori that $\pi(H_0) = 1 - \pi(H_1) = \pi = 1/2$. The Bayes rule is to accept H_0 whenever $\pi(H_0|x^n) > 1/2$ or, equivalently, $\bar{x}_n < (\theta_0 + \theta_1)/2$. Let $F(.|\theta)$ denote the cumulative probability function of $\bar{x}_n$ given θ. Under the 0–1 loss function, the Bayes risk is

$$r(\pi, \delta_n^*) = \frac{1}{2}\left(1 - F\left(\frac{\theta_0 + \theta_1}{2}|\theta_0\right)\right) + \frac{1}{2}F\left(\frac{\theta_0 + \theta_1}{2}|\theta_1\right) \tag{14.9}$$

$$= 1 - \Phi\left(\frac{(\theta_1 - \theta_0)}{2}\sqrt{n}\right),$$

where $\Phi(z)$ is the cdf of a standard normal density. The Bayes risk in (14.9) is monotone and approaches 0 as the sample size approaches infinity.

The goal sampling approach is to choose the minimum sample size needed to achieve a desired value of the expected Bayes risk, say r_0. In symbols, we find n_{r_0} that solves $r(\pi, \delta_n^*) = r_0$. Applying this to equation (14.9) we obtain

$$n_{r_0} = \left(\frac{2z_{1-r_0}}{\theta_1 - \theta_0}\right)^2, \tag{14.10}$$

where z_{1-r_0} is such that $\Phi(z_{1-r_0}) = 1 - r_0$.

A standard frequentist approach to sample size determination corresponds to determining the sample size under specified constraints in terms of probabilities of type I and type II error (see, for example, Desu and Raghavarao 1990, Shuster 1993, Lachin 1998). In our example, the frequentist optimal sample size is given by

$$n_F = \left(\frac{z_\alpha + z_\beta}{\theta_1 - \theta_0}\right)^2, \tag{14.11}$$

where α and β correspond to the error probabilities of type I and II, respectively.

The frequentist approach described above does not make use of an explicit loss function. Lindley notes that:

> Statisticians today commonly do not deny the existence of either utilities or prior probabilities, they usually say that they are difficult to specify. However, they often use them (in an unspecified way) in order to determine significance levels and power. (Lindley 1962, p. 44)

Although based on different concepts, we can see in this example that classical and Bayesian optimal sample sizes are the same whenever

$$|z_{1-r_0}| = |z_\alpha + z_\beta|/2. \qquad (14.12)$$

Inoue *et al.* (2005) discuss a general framework for investigating the relationship between the two approaches, based on identifying mappings that connect the Bayesian and frequentist inputs necessary to obtain the same sample size, as done in (14.12) for the mapping between the frequentist inputs α and β and the Bayesian input r_0. Their examples illustrate that one can often find correspondences even if the underlying loss or utility functions are different.

Adcock (1997) formalizes goal-based Bayesian methods as those based on setting

$$E_{x^n}[T(x^n)] = r_0 \qquad (14.13)$$

and solving for n. Here $T(x^n)$ is a test statistic derived from the posterior distribution on the unknowns of interest. It could be, for instance, the posterior probability of a given interval, the length of the 95% highest posterior interval, and so forth. $T(x^n)$ plays the role of the posterior expected loss, though direct specification of T bypasses both the formalization of a decision-theoretic model and the associated computing. For specific test statistics, software is available (Pham-Gia and Turkkan 1992, Joseph *et al.* 1995, and Joseph and Wolfson 1997) to solve sample size problems of the form (14.13).

Lee and Zelen (2000) observe that even in frequentist settings it can be useful to set α and β so that they result in desirable posterior probabilities of the hypothesis being true conditional on the study's result. Their argument relies on a frequentist analysis of the data, but stresses the importance of (a) using quantities that are conditional on evidence when making terminal decisions, and (b) choosing α and β in a way that reflects the context to which the results are going to be applied. In the specifics of clinical trials, they offer the following considerations:

> we believe the two fundamental issues are (a) if the trial is positive "What is the probability that the therapy is truly beneficial?" and (b) if the trial is negative, "What is the probability that the therapies are comparable?" The frequentist view ignores these fundamental considerations and can result in positive harm because of the use of inappropriate error rates. The positive harm arises because an excessive number of false positive therapies may be introduced into practice. Many positive trials may be

unethical to duplicate and, even if replicated, could require many years to complete. Hence a false positive trial outcome may generate many years of patients' receiving non beneficial therapies. (Lee and Zelen 2000, p. 96)

14.2 Computing

In practice, finding a Bayesian optimal sample size is often too complex for analytic solutions. Here we describe two general computational approaches for Bayesian sample size determination and illustrate them in the context of a relatively simple example. Let x denote the number of successes in a binomial experiment with success probability θ. We want to optimally choose the sample size n of this experiment assuming a priori a mixture of two experts' opinions as represented by

$$\pi(\theta) = 0.5 \, Beta(\theta|3, 1) + 0.5 \, Beta(\theta|3, 3).$$

For the terminal decision problem of estimating θ assume that the loss function is $L(\theta, a) = |a - \theta|$. Let $\delta_n^*(x)$ denote the Bayes rule, in this case the posterior median (Problem 7.7). Let the sampling cost be 0.0008 per observation, that is $C(n) = 0.0008n$. The Bayes risk is

$$r(\pi, n) = \int_0^1 \sum_{x=0}^n [|\delta_n^*(x) - \theta| + 0.0008n] \binom{n}{x} \theta^x (1 - \theta)^{n-x} \pi(\theta) d\theta. \qquad (14.14)$$

This decision problem does not yield an analytical solution. We explore an alternative solution that is based on Monte Carlo simulation (Robert and Casella 1999). A straightforward implementation of Monte Carlo integration to evaluate (14.14) is as follows. Select a grid of plausible values of n, and a simulation size I.

1. For each n, draw (θ_i, x_i) for $i = 1, \ldots, I$. This can be done by generating θ_i from the prior and then, conditionally on θ_i, generating x_i from the likelihood.

2. Compute $l_i = L(\theta_i, \delta^*(x_i)) + C(n)$.

3. Compute $\tilde{r}(\pi, n) = (1/I) \sum_{i=1}^I l_i$.

4. Numerically minimize $\tilde{r}$ with respect to n.

This is predicated on the availability of a closed form for δ_n^*. When that is not the case things become more complicated, but other Monte Carlo schemes are available.

In practice the true r can be slow varying and smooth in n, while $\tilde{r}$ can be subject to residual noise from step 3, unless I is very large. Müller and Parmigiani (1995) discuss an alternative computational strategy that exploits the smoothness of r by considering a Monte Carlo sample over the whole design space, and then fit a curve for the Bayes risk $r(\pi, n)$ as a function of n. The optimization problem can then be solved in a deterministic way by minimization of the fitted curve. In the context of sample size determination, the procedure can be implemented as follows:

1. Select i points n_i, $i = 1, \ldots, I$, possibly including duplications.

2. Draw (θ_i, x_i) for $i = 1, \ldots, I$. This can be done by generating θ_i from the prior and then x_i from the likelihood.

3. Compute $l_i = L(\theta_i, \delta^*(x_i)) + C(n_i)$.

4. Fit a smooth curve $\hat{r}(\pi, n)$ to the resulting points (n_i, l_i).

5. Evaluate deterministically the minimum of $\hat{r}(\pi, n)$.

This effectively eliminates the loop over n implied by step 1 of the previous method. This strategy can also be applied when L is a function of n and when C is a function of x and θ.

To illustrate we choose the simulation points n_i to be integers chosen uniformly in the range (0, 120). The Monte Carlo sample size is $I = 200$. Figure 14.2(a) shows the simulated pairs (n_i, l_i). We fitted the generated data points by a nonlinear regression of the form

$$\hat{r}(\pi, n) = 0.0008n + 0.1929(1 + bn)^{-a} \tag{14.15}$$

where the values of a and b are estimated based on the Monte Carlo sample. This curve incorporates information regarding (i) the sampling cost; (ii) the value of the expected payoff when no observations are taken, which can be computed easily and is 0.1929; and (iii) the rate of convergence of the expected absolute error to zero, assumed here to be polynomial. The results are illustrated in Figure 14.2(a). Observe that even with a small Monte Carlo sample there is enough information to make a satisfactory sample size choice.

The regression model (14.15) is simple and has the advantage that it yields a closed form solution for the approximate optimal sample size, given by

$$\hat{n}^* = \frac{1}{\hat{b}} \left[\left(\frac{0.1929\hat{a}\hat{b}}{0.0008} \right)^{\frac{1}{\hat{a}+1}} - 1 \right]. \tag{14.16}$$

The sample shown gives $\hat{n}^* = 36$. More flexible model specifications may be preferable in other applications.

In this example, we can also rewrite the risk as

$$r(\pi, n) = \sum_{x=0}^{n} \int_0^1 [|\delta_n^*(x) - \theta| + 0.0008n]\pi_x(\theta)d\theta \, m(x) \tag{14.17}$$

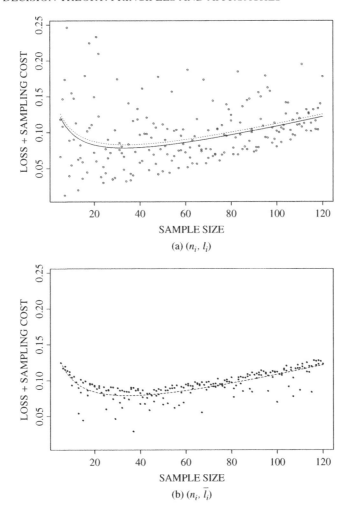

*Figure 14.2   Monte Carlo estimation of the Bayes risk curve. The points in panel
(a) have coordinates (n_i, l_i). The lines show $\hat{r}(\pi, n)$ using 200 replicates (dotted) and
2000 replicates (solid). The points in panel (b) have coordinates $(n_i, \bar{l}_i)$. The dashed
line shows $\bar{r}(\pi, n)$.*

where m is the marginal distribution of the data and π_x is the posterior distribution of
θ, that is a mixture of beta densities. Using the incomplete beta function, the inner
integral can be evaluated directly. We denote the inner integral as $\bar{l}_i$ when (n_i, x_i) is
selected. We can now interpolate directly the points $(n_i, \bar{l}_i)$, with substantial reduction
of the variability around the fitted curve. This is clearly recommendable whenever
possible. The resulting estimated curve $\bar{r}(\pi, n)$ is shown in Figure 14.2(b). The new
estimate is $\bar{n}^* = 35$, not far from the previous approach, which performs quite well
in this case.

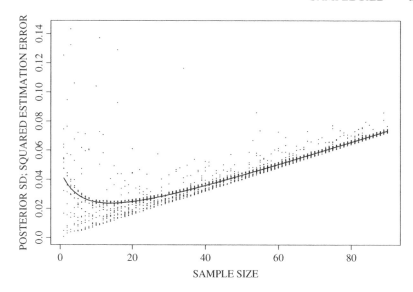

Figure 14.3   Estimation of the Bayes risk curve under quadratic error loss. Small points are the sampled values of $(\delta^*(x) - \theta)^2 + cn$, where $\delta^*(x)$ denotes the posterior mean. Large points are the posterior variances. The solid line is the fit to the posterior variances.

We end this section by considering two alternative decision problems. The first is an estimation problem with squared error loss $L(\theta, a) = (\theta - a)^2$. Here the optimal terminal decision rule is the posterior mean and the posterior expected loss is the posterior variance (see Section 7.6.1). Both of these can be evaluated easily, which allows us to validate the Monte Carlo approach. Figure 14.3 illustrates the results. The regression model is that of equation (14.15).

The last problem is the estimation of the expected information $\mathcal{V}(\mathcal{E}^{(n)})$. We can proceed as before by defining

$$v_i = \log\left(\frac{\pi(\theta_i|x_i)}{\pi(\theta_i)}\right) \tag{14.18}$$

and using curve fitting to estimate the expectation of v_i as a function of n. In this example we choose to use a loess (Cleveland and Grosse 1991) fit for the simulated points. Figure 14.4 illustrates various aspects of the simulation. The top panel nicely illustrates the decreasing marginal returns of experimentation implied by Theorem 13.8. The middle panel shows the change in utility as a function of the prior, and illustrates how the most useful experiments are those generated by parameter values that have low a priori probability, as these can radically change one's state of knowledge.

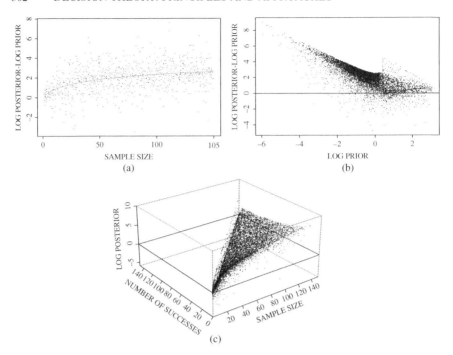

Figure 14.4   Estimation of the expected information curve. Panel (a) shows the difference between log posterior and log prior for a sample of θ, x, n. Panel (b) shows the difference between log posterior and log prior versus the log prior. For parameters corresponding to higher values of the log prior, the experiment is less informative, as expected. The points are a low-dimensional projection of what is close to a three-dimensional surface. Panel (c) shows the log of the posterior versus the sample size and number of successes x. The only source of variability is the value of θ.

14.3   Examples

In this section we discuss in detail five different examples. The first four have analytic solutions while the last one makes use of the computational approach of Section 14.2.

14.3.1   Point estimation with quadratic loss

We begin with a simple estimation problem in which preposterior calculations are easy, the solution is in closed form, and it provides a somewhat surprising insight.

Assume that $x_1, \ldots x_n$ are conditionally independent $N(\theta, \sigma^2)$, with σ^2 known. The terminal decision is to estimate θ with quadratic loss function $L(\theta, a, n) = (\theta - a)^2 + C(n)$. The prior distribution on θ is $N(\mu_0, \tau_0^2)$, with μ_0 and τ_0^2 known.

Since $\bar{x}_n \sim N(\theta, \sigma^2/n)$, the posterior distribution of θ given the observed sample mean $\bar{x}_n$ is $N(\mu_x, \tau_x^2)$ where, as in Appendix A.4,

$$\mu_x = \frac{\sigma^2}{\sigma^2 + n\tau_0^2} \mu_0 + \frac{n\tau_0^2}{\sigma^2 + n\tau_0^2} \bar{x}_n \quad \text{and} \quad \tau_x^2 = \frac{\sigma^2 \tau_0^2}{\sigma^2 + n\tau_0^2}.$$

The posterior expected loss, not including the cost of observation, is τ_x^2. Because this depends on the data only via the sample size n but is independent of $\bar{x}_n$, we have

$$r(\pi, n) = \frac{\sigma^2 \tau_0^2}{\sigma^2 + n\tau_0^2} + C(n). \tag{14.19}$$

If each additional observation costs a fixed amount c, that is $C(n) = c\,n$, replacing n with a continuous variable and taking derivatives we get that the approximate optimal sample size is

$$n^* \approx \frac{\sigma}{\sqrt{c}} - \frac{\sigma^2}{\tau_0^2}. \tag{14.20}$$

Considering the two integer values next to n^* and choosing the one with the smallest Bayes risk gives the optimal solution, because this function is convex. Figure 14.5 shows this optimization problem when $\tau_0^2 \to \infty$, in which case the optimal sample size is $n^* \approx \sigma/\sqrt{c}$.

In equation (14.20), the optimal sample size is a function of two ratios: the ratio of experimental standard deviation to cost that characterize the cost-effectiveness of a single sample point; and the ratio of experimental variance to prior variance that characterizes the incremental value of each observation in terms of knowledge. As a result, the optimal sample size increases with the prior variance, because less is

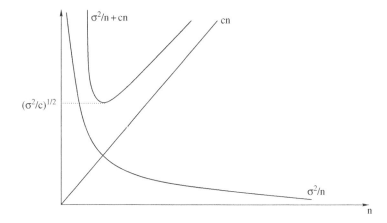

Figure 14.5   Risk functions and components in the optimal sample size problem of Section 14.3.1, when τ_0^2 is large compared to σ^2.

known a priori, and decreases with cost. However, n^* is not a monotone function of the sampling variance σ^2. When σ^2 is very small few observations suffice; when it is large, observations lose cost-effectiveness and a smaller overall size is optimal. The largest sample sizes obtain for intermediate values of σ^2. In particular n^*, as a function of σ, is maximized at $\sigma = \tau_0^2/(2\sqrt{c})$.

14.3.2 Composite hypothesis testing

Assume again that $x_1, \ldots, x_n$ is a sample from a $N(\theta, \sigma^2)$ population with σ^2 known, and that we are now interested in testing $H_0 : \theta \leq 0$ versus $H_1 : \theta > 0$. As usual, let a_i denote the decision of accepting hypothesis H_i, $i = 0, 1$. While we generally study loss function such as that of Table 14.1, a useful alternative for this decision problem is

$$L(\theta, a_0) = \begin{cases} 0 & \theta \leq 0 \\ \theta & \theta > 0 \end{cases} \quad \text{and} \quad L(\theta, a_1) = \begin{cases} -\theta & \theta \leq 0 \\ 0 & \theta > 0. \end{cases}$$

This loss function reflects both whether the decision is correct and whether the actual θ is far from 0 if the decision is incorrect. The loss specification is completed by setting $L(\theta, a, n) = L(\theta, a) + C(n)$. Assume that the prior density of θ is $N(\mu_0, \tau_0^2)$ with $\mu_0 = 0$. To determine the optimal sample size we first determine the optimal $\delta_n^*(x^n)$ by minimizing the posterior expected loss $\mathcal{L}$. Then we perform the preposterior analysis by evaluating $\mathcal{L}$ at the optimal $\delta_n^*(x^n)$ and averaging the result with respect to the marginal distribution of the data.

The sample mean $\bar{x}_n$ is sufficient for x^n, so $\pi(\theta|x^n) = \pi(\theta|\bar{x}_n)$ and we can handle all calculations by simply considering $\bar{x}_n$. The posterior expected losses for a_0 and a_1 are, respectively,

$$\mathcal{L}_{\pi_{\bar{x}_n}}(a_0) = \int_0^\infty \theta \pi(\theta|\bar{x}_n) d\theta$$

$$\mathcal{L}_{\pi_{\bar{x}_n}}(a_1) = -\int_{-\infty}^0 \theta \pi(\theta|\bar{x}_n) d\theta,$$

so that the Bayes rule is a_0 whenever

$$\int_0^\infty \theta \pi(\theta|\bar{x}_n) d\theta < -\int_{-\infty}^0 \theta \pi(\theta|\bar{x}_n) d\theta$$

and a_1, otherwise. This inequality can be rewritten in terms of the posterior mean: we choose a_0 whenever

$$\int_0^\infty \theta \pi(\theta|\bar{x}_n) d\theta + \int_{-\infty}^0 \theta \pi(\theta|\bar{x}_n) d\theta \equiv E[\theta|\bar{x}_n] < 0.$$

In this example

$$E[\theta|\bar{x}_n] = \bar{x}_n \frac{n\tau_0^2}{\sigma^2 + n\tau_0^2}$$

because $\mu_0 = 0$, so we decide in favor of a_0 (a_1) if $\bar{x}_n < (>)0$. In other words, the Bayes rule is

$$\delta_n^*(\bar{x}) = \begin{cases} a_0 & \bar{x}_n \leq 0 \\ a_1 & \bar{x}_n > 0. \end{cases}$$

This defines the terminal decision rule, that is the optimal strategy after the data have been observed, as a function of the sample size and sufficient statistic. Now we need to address the decision concerning the sample size. The posterior expected loss associated with the terminal decision is

$$L(\delta_n^*(\bar{x}_n)) = \begin{cases} \int_0^\infty \theta\pi(\theta|\bar{x}_n)d\theta & \bar{x}_n \leq 0 \\ -\int_{-\infty}^0 \theta\pi(\theta|\bar{x}_n)d\theta & \bar{x}_n > 0. \end{cases}$$

The Bayes risk for the sample size decision is the cost of experimentation plus the expectation of the expression above with respect to the marginal distribution of $\bar{x}_n$. Specifically,

$$r(\pi, n) = \int_{-\infty}^0 \int_0^\infty \theta\pi(\theta|\bar{x}_n)d\theta\, m(\bar{x}_n)d\bar{x}_n$$

$$- \int_0^\infty \int_{-\infty}^0 \theta\pi(\theta|\bar{x}_n)d\theta\, m(\bar{x}_n)d\bar{x}_n + C(n).$$

By adding and subtracting $\int_{-\infty}^0 \int_{-\infty}^0 \theta\pi(\theta|\bar{x}_n)m(\bar{x}_n)d\theta d\bar{x}_n$ and changing the order of integration in the first term we get

$$r(\pi, n) = \int_{-\infty}^0 m(\bar{x}_n)E[\theta|\bar{x}_n]d\bar{x}_n - \int_{-\infty}^0 \theta\pi(\theta)d\theta + C(n)$$

$$= \frac{n\tau_0^2}{\sigma^2 + n\tau_0^2} \int_{-\infty}^0 \bar{x}_n m(\bar{x}_n)d\bar{x}_n - \int_{-\infty}^0 \theta\pi(\theta)d\theta + C(n).$$

Because $\theta \sim N(0, \tau_0^2)$ then $\int_{-\infty}^0 \theta\pi(\theta)d\theta = -\tau_0/\sqrt{2\pi}$. Also, the marginal distribution of the sample mean $\bar{x}_n$ is $N(0, (\tau_0^2 n + \sigma^2)/n)$. Thus,

$$\int_{-\infty}^0 \bar{x}_n m(\bar{x}_n)d\bar{x}_n = -\sqrt{\frac{\tau_0^2 n + \sigma^2}{2\pi n}}$$

and

$$r(\pi, n) = \frac{\tau_0}{\sqrt{2\pi}} \left(1 - \tau_0\sqrt{\frac{n}{\tau_0^2 n + \sigma^2}}\right) + C(n).$$

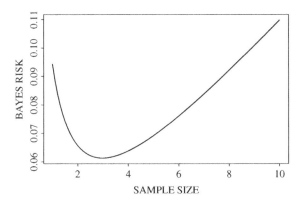

Figure 14.6   The Bayes risk for the composite hypothesis testing problem.

In particular, if $\sigma^2 = 1, \tau_0^2 = 4$, and the cost of a fixed sample of size n is $C(n) = 0.01\,n$ then

$$r(\pi, n) = \frac{2}{\sqrt{2\pi}}\left(1 - 2\sqrt{\frac{n}{4n+1}}\right) + 0.01n.$$

Figure 14.6 shows $r(\pi, n)$ as a function of n taken to vary continuously. Among integer values of n the optimal sample size turns out to be 3.

14.3.3   A two-action problem with linear utility

This example is based on Raiffa and Schlaifer (1961). Consider a health care agency, say the WHO, facing the choice between two actions a_0 and a_1, representing, for example, two alternative prevention programs. Both programs have a setup cost for implementation $K_i, i = 0, 1$, and will benefit a number k_i of individuals. Each individual in either program can expect a benefit of x additional units of utility, for example years of life translated into monetary terms. In the population, the distribution of x is normal with unknown mean θ and known variance σ^2. The utility to the WHO of implementing each of the two programs is

$$u(a_i(\theta)) = K_i + k_i\theta, \text{ for } i = 0, 1 \text{ with } k_0 < k_1. \tag{14.21}$$

We will assume that the WHO has the possibility of gathering a sample x^n from the population, to assist in the choice of a program, and we are going to study the optimal sample size for this problem. We are going to take a longer route than strictly necessary and first work out in detail both the expected value of perfect information and the expected value of experimental information for varying n. We are then going to optimize the latter assuming that the expected cost sampling is $C(n) = cn$.

Throughout, we will assume that θ has a priori normal distribution with mean μ_0 and variance $\tau_0^2 = \sigma^2/n_0$. Expressing τ_0^2 in terms of the population variance will simplify analytical expressions.

Let θ_b be the "break-even value," that is the value of θ under which the two programs have the same utility. Solving $u(a_0(\theta_b)) = u(a_1(\theta_b))$ leads to

$$\theta_b = \frac{K_0 - K_1}{k_1 - k_0}.$$

For a given value of θ, the optimal action a_θ is such that

$$a_\theta = \begin{cases} a_0 & \text{if } \theta \leq \theta_b \\ a_1 & \text{if } \theta \geq \theta_b. \end{cases} \tag{14.22}$$

As θ is unknown, a_θ is not implementable.

Before experimentation, the expected utility of action a_i is $E[u(a_i(\theta))] = K_i + k_i \mu_0$. Thus, the optimal decision is

$$a^* = \begin{cases} a_0 & \text{if } \mu_0 \leq \theta_b \\ a_1 & \text{if } \mu_0 \geq \theta_b. \end{cases} \tag{14.23}$$

As seen in Chapter 13, the conditional value of perfect information for a given value θ is

$$V_\theta(\mathcal{E}^\infty) = u(a_\theta(\theta)) - u(a^*(\theta)). \tag{14.24}$$

Using the linear form of the utility function and the fact that

$$(K_1 - K_0) = -(k_1 - k_0)\theta_b, \tag{14.25}$$

we obtain

$$V_\theta(\mathcal{E}^\infty) = \begin{cases} u(a_0(\theta)) - u(a_1(\theta)) = (k_1 - k_0)(\theta_b - \theta), & \text{if } \theta \leq \theta_b, \ \mu_0 \geq \theta_b \\ u(a_1(\theta)) - u(a_0(\theta)) = (k_1 - k_0)(\theta - \theta_b), & \text{if } \theta \geq \theta_b, \ \mu_0 \leq \theta_b \\ 0, & \text{otherwise.} \end{cases} \tag{14.26}$$

In Chapter 13, we saw that the expected value of perfect information is

$$V(\mathcal{E}^\infty) = E_\theta[V_\theta(\mathcal{E}^\infty)]. \tag{14.27}$$

It follows from (14.26) that

$$V(\mathcal{E}^\infty) = \begin{cases} (k_1 - k_0) \int_{\theta_b}^\infty (\theta - \theta_b)\pi(\theta)d\theta, & \text{if } \mu_0 \leq \theta_b \\ (k_1 - k_0) \int_{-\infty}^{\theta_b} (\theta_b - \theta)\pi(\theta)d\theta, & \text{if } \mu_0 \geq \theta_b. \end{cases}$$

We define the following integrals:

$$A(\theta_b) = \int_{\theta_b}^\infty (\theta - \theta_b)\pi(\theta)d\theta \tag{14.28}$$

$$B(\theta_b) = \int_{-\infty}^{\theta_b} (\theta_b - \theta)\pi(\theta)d\theta. \tag{14.29}$$

Using the results of Problem 14.5 we obtain

$$\mathcal{V}(\mathcal{E}^\infty) = (k_1 - k_0)\zeta(z)\sigma/\sqrt{n_0}, \tag{14.30}$$

where $z = \sqrt{n_0}/\sigma|\theta_b - \mu_0|$, $\zeta(z) = \phi(z) - z(1 - \Phi(z))$, where ϕ and Φ denote, respectively, the density and the cdf of the standard normal distribution.

We now turn our attention to the situation in which the decision maker performs an experiment $\mathcal{E}^{(n)}$, observes a sample $x^n = (x_1, \ldots, x_n)$, and then chooses between a_0 and a_1. The posterior distribution of θ given x^n is normal with posterior mean $\mu_x = (n_0\mu_0 + n\bar{x}_n)/(n_0 + n)$ and posterior variance $\tau_x^2 = \sigma^2/(n_0 + n)$. Thus, the optimal "a posteriori" decision is

$$\delta_n^*(x^n) = \begin{cases} a_0 & \text{if } \mu_x \leq \theta_b \\ a_1 & \text{if } \mu_x \geq \theta_b. \end{cases} \tag{14.31}$$

Recalling Chapter 13, the value of performing the experiment $\mathcal{E}^{(n)}$ and observing x^n as compared to no experimentation is given by the conditional value of the sample information, or observed information, $\mathcal{V}_{x^n}(\mathcal{E}^{(n)})$. The expectation of it taken with respect to the marginal distribution of x^n is the expected value of sample information, or expected information.

In this problem, because the terminal utility function is linear in θ, the observed information is functionally similar to the conditional value of perfect information; that is, the observed information is

$$\mathcal{V}_{x^n}(\mathcal{E}^{(n)}) = V(\pi_x) - \mathcal{U}_{\pi_x}(a^*). \tag{14.32}$$

One can check that the expected information is $\mathcal{V}(\mathcal{E}^{(n)}) = E_{\mu_x}(\mathcal{V}_{x^n}(\mathcal{E}^{(n)}))$. This expectation is taken with respect to the distribution of μ_x, that is a normal with mean μ_0 and variance σ^2/n_x where

$$\frac{1}{n_x} = \frac{1}{n_0} - \frac{1}{n_0 + n}.$$

Following steps similar to those used in the calculation of (14.30), it can be shown that the expected information is

$$\mathcal{V}(\mathcal{E}^{(n)}) = \begin{cases} (k_1 - k_0)A_{\mu_x}(\theta_b) & \text{if } \mu_0 \leq \theta_b \\ (k_1 - k_0)B_{\mu_x}(\theta_b) & \text{if } \mu_0 \geq \theta_b \end{cases} \tag{14.33}$$

$$= (k_1 - k_0)\zeta(z)\sigma/\sqrt{n_x},$$

where $z = |\theta_b - \mu_0|\sqrt{n_x}/\sigma$.

The expected net gain of an experiment that consists of observing a sample of size n is therefore

$$\mathcal{U}_\pi(n) = \mathcal{V}(\mathcal{E}^{(n)}) - cn = (k_1 - k_0)\zeta(z)\sigma/\sqrt{n_x} - cn. \tag{14.34}$$

Let $d = (k_1 - k_0)\sigma/(c\sqrt{n_0{}^3})$, $q = n/n_0$, and $z_0 = |\theta_b - \mu_0|\sqrt{n_0}/\sigma$. The "dimensionless net gain" is defined by Raiffa and Schlaifer (1961) as

$$g(d, q, z_0) = \frac{\mathcal{U}_\pi(n)}{cn_0} \tag{14.35}$$

$$= d\left(\frac{q}{q+1}\right)^{1/2} \zeta\left(z_0\left(\frac{q+1}{q}\right)^{1/2}\right) - q.$$

To solve the sample size problem we can equivalently maximize g with respect to q and then translate the result in terms of n. Assuming q to be a continuous variable and taking the first- and second-order derivatives of g with respect to q gives

$$g'(d, q, z_0) = \frac{1}{2}dq^{-1/2}(q+1)^{-3/2}\phi\left(z_0\left(\frac{q+1}{q}\right)^{1/2}\right) - 1$$

$$g''(d, q, z_0) = \frac{1}{4}d[q(q+1)]^{-5/2}\phi\left(z_0\left(\frac{q+1}{q}\right)^{1/2}\right)[z_0^2 + (z_0^2 - 1)q - 4q^2]. \tag{14.36}$$

The first-order condition for a local maximum is that $g'(d, q, z_0) = 0$. The function g always admits a local maximum when $z_0 = 0$. However, when $z_0 > 0$ a local maximum may or may not exist (see Problem 14.6).

By solving the equation $g'(d, q, z_0) = 0$ we obtain

$$q^{1/2}(q+1)^{3/2}e^{z_0^2/2q} = \frac{d}{2}\phi(z_0)$$

$$\Longleftrightarrow q^2\underbrace{\left[\left(\frac{q+1}{q}\right)^{3/2}e^{z_0^2/2q}\right]}_{\text{approaches 1 for large } q} = \frac{d}{2}\phi(z_0).$$

The large-q approximate solution is then $q^* = \sqrt{\phi(z_0)d/2}$. The implied sample size is

$$n^* = n_0\sqrt{\phi(z_0)d/2} = \sqrt{\phi(z_0)(k_1 - k_0)c^{-1}\sigma n_0^{1/2}/2}. \tag{14.37}$$

14.3.4   Lindley information for exponential data

As seen in Chapter 13, if the data to be collected are exchangeable, then the expected information $\mathcal{I}(\mathcal{E}^{(n)})$ is increasing and convex in n. If, in addition to this, the cost of observation is also an increasing function in n, and information and cost are additive, then there are at most two (adjacent) solutions. We illustrate this using the following example taken from Parmigiani and Berry (1994).

Consider exponential survival data with unknown failure rate θ and conjugate prior $Gamma(\alpha_0, \beta_0)$. Here α_0 can be interpreted as the number of events in a hypothetical prior study, while β_0 can be interpreted as the total time at risk in that study.

Let $t = \sum_i^n x_i$ be the total survival time. The expected information in a sample of size n is

$$\mathcal{I}(\mathcal{E}^{(n)}) = \int_0^\infty \int_0^\infty \log \left[\frac{\Gamma(\alpha_0)(\beta_0 + t)^{\alpha_0 + n}}{\Gamma(\alpha_0 + n)\beta_0^{\alpha_0}} \theta^n e^{-\theta t} \right]$$

$$\times \frac{\beta_0^{\alpha_0}}{\Gamma(\alpha_0)\Gamma(n)} \theta^{\alpha_0 + n - 1} t^{n+1} e^{-\theta(\beta_0 + t)} dt \, d\theta. \qquad (14.38)$$

This integration is a bit tedious but can be worked out in terms of the gamma function Γ and the digamma function ψ as

$$\mathcal{I}(\mathcal{E}^{(n)}) = \log \frac{\Gamma(\alpha_0)}{\Gamma(\alpha_0 + n)} + (\alpha_0 + n)\psi(\alpha_0 + n) - \alpha_0\psi(\alpha_0) - n. \qquad (14.39)$$

The result depends on the number of prior events α_0 but not on the prior total time at risk β_0.

From (14.38), the expected information for one observation is

$$\mathcal{I}(\mathcal{E}^{(1)}) = \frac{1}{\alpha_0} + \psi(\alpha_0) - \log \alpha_0.$$

In particular, if $\alpha_0 = 1$, then $\mathcal{I}(\mathcal{E}^{(1)}) = 1 - C = 0.4288$, where C is Euler's constant. Using a first-order Stirling approximation we get

$$\log \Gamma(n) \approx \frac{1}{2} \log 2\pi - n + \left(n - \frac{1}{2} \right) \log n$$

and

$$\psi(n) \approx \log n - \frac{1}{2n}.$$

Applying these to the expression of the expected information gives

$$\mathcal{I}(\mathcal{E}^{(n)}) \approx K(\alpha_0) + \frac{1}{2} \log(\alpha_0 + n)$$

where $K(\alpha_0) = \log \Gamma(\alpha_0) - \alpha_0(\psi(\alpha_0) - 1) - \frac{1}{2} \log 2\pi - \frac{1}{2}$.

Using this approximation and taking n to be continuous, when the sampling cost per observation is c (in information units), the optimal fixed sample size is

$$n^* = \frac{1}{2c} - \alpha_0.$$

The optimal solution depends inversely on the cost and linearly on the prior sample size.

14.3.5   Multicenter clinical trials

In a multicenter clinical trial, the same treatment is carried out on different populations of patients at different institutions or locations (centers). The goals of multicenter trials are to accrue patients faster than would be possible at a single center, and to collect evidence that is more generalizable, because it is less prone to center-specific factors that may affect the conclusions. Multicenter clinical trials are usually more expensive and difficult to perform than single-center trials. From the standpoint of sample size determination, two related questions arise: the appropriate number of centers and the sample size in each center. In this section we illustrate a decision-theoretic approach to the joint determination of these two sample sizes using Bayesian hierarchical modeling. Our discussion follows Parmigiani and Berry (1994).

Consider a simple situation in which each individual in the trial receives an experimental treatment for which there is no current alternative, and we record a binary outcome, representing successful recovery. Within each center i, we will assume that patients are exchangeable with an unknown success probability θ_i that is allowed to vary from center to center as a result of differences in the local populations, in the modalities of implementation of care, and in other factors. In a hierarchical model (Stangl 1995, Gelman *et al.* 1995), the variability of the center-specific parameters θ_i is described by a further probability distribution, which we assume here to be a $Beta(\alpha, \beta)$ where α and β are now unknown. The θ_i are conditionally independent draws from this distribution, implying that we have no a priori reasons to suppose that any of the centers may have a higher propensity to success. The problem specification is completed by a prior distribution $\pi(\alpha, \beta)$ on (α, β).

To proceed, we assume that the number of patients in each center is the same, that is n. Let k denote the number of centers in the study. Because the treatment would eventually be applied to the population at large, our objective will be to learn about the general population of centers rather than about any specific center. This learning is captured by the posterior distribution of α and β. Thus we will optimally design the pair (n, k) to maximize the expected information $\mathcal{I}$ on α and β. Alternatively, one could include explicit consideration of one or more specific centers. See for example, Mukhopadhyay and Stangl (1993).

Let x be a k-dimensional vector representing the number of successes in each center. The marginal distribution of the data is

$$m(x) = \int_0^\infty \int_0^\infty \left[B^{-k}(\alpha, \beta) \prod_{i=1}^k \binom{n}{x_i} B(\alpha + x_i, \beta + n - x_i) \right] \pi(\alpha, \beta) d\alpha d\beta$$

and

$$\pi(\alpha, \beta | x) = \frac{\pi(\alpha, \beta) \prod_{i=1}^k B(\alpha + x_i, \beta + n - x_i) \binom{n}{x_i}}{m(x) B^k(\alpha, \beta)},$$

where B is the beta function.

The following results consider a prior distribution on α and β given by independent and identically distributed gamma distributions with shape 6 and rate 2, corresponding to relatively little initial knowledge about the population of centers. This prior distribution is graphed in Figure 14.7(a). The implied prior predictive distribution on the probability of success in a hypothetical new center $k + 1$, randomly drawn from the population of centers, is shown in Figure 14.7(b).

Figure 14.8 shows the posterior distributions obtained by three different allocations of 16 patients: all in one center $(n = 16, k = 1)$, four in each of four centers $(n = 4, k = 4)$, and all in different centers $(n = 1, k = 16)$. All three posteriors are based on a total of 12 observed successes. As centers are exchangeable, the posterior distribution does not depend on the order of the elements of x. So for $k = 1$ and $k = 16$ the posterior graphed is the only one arising from 12 successes. For $k = 4$ the posterior shown is one of six possible.

Higher values of the number of centers k correspond to more concentrated posterior distributions; that is, to a sharper knowledge of the population of centers. A different way of illustrating the same point is to graph the marginal predictive distribution of the success probability θ_{k+1} in a hypothetical future center, for the three scenarios considered. This is shown in Figure 14.9.

Using the computational approach described in Section 14.2 and approximating k and n by real variables, we estimated the surface $\mathcal{I}(\mathcal{E}^{(n,k)})$. Figure 14.10 gives the contours of the surface. Choices that have the same total number kn of patients are represented by dotted lines. The expected information increases if one increases k and decreases n so that kn remains fixed.

If each patient, in information units, costs a fixed amount c, and if there is a fixed "startup" cost s in each center, the expected utility function is

$$\mathcal{U}_\pi(n, k) = \mathcal{I}(\mathcal{E}^{(n,k)}) - cnk - sk.$$

For example, assume $s = 0.045$ and $c = 0.015$. A contour plot of the utility surface is shown in Figure 14.11. This can be used to identify the optimum, in this case the pair $(n = 11, k = 6)$.

More generally, the entire utility surface provides useful information for planning the study. In practice, planning complex studies will require informal consideration of a variety of constraints and factors that are difficult to quantify in a decision-theoretic fashion. Examining the utility surface, we can divide the decision space into regions with positive and negative utility. In this example, some choices of k and n have negative expected utility and are therefore worse than no sampling. In particular, all designs with only one center (the horizontal line at $k = 1$) are worse than no experimentation. We can also identify regions that are, say, within 10% of the utility of the maximum. This parallels the notion of credible intervals and provides information on robustness of optimal choices.

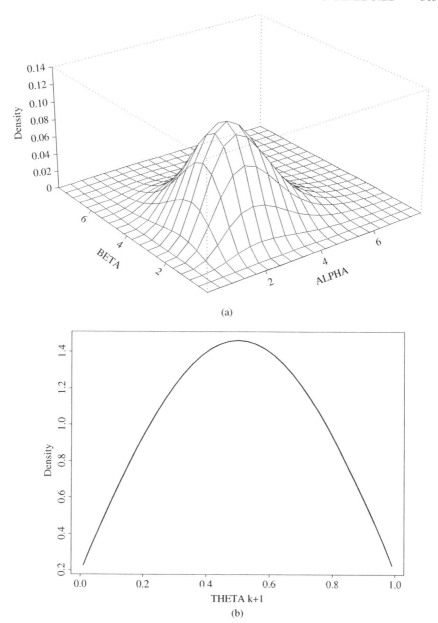

Figure 14.7   Prior distributions. Panel (a), prior distribution on α and β. Panel (b), prior predictive distribution of the success probability in center k + 1. Figure from Parmigiani and Berry (1994).

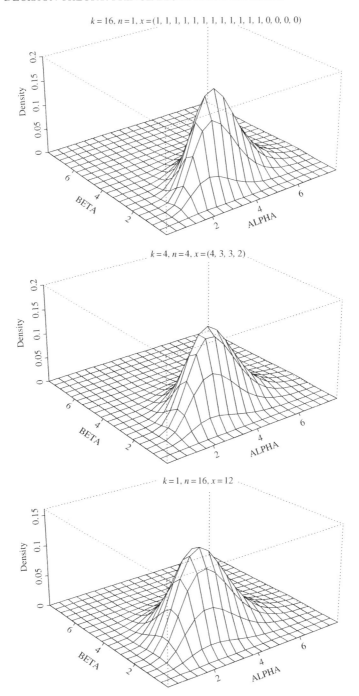

Figure 14.8   Posterior distributions on α and β under alternative designs; all are based on 12 successes in 10 patients. Figure from Parmigiani and Berry (1994).

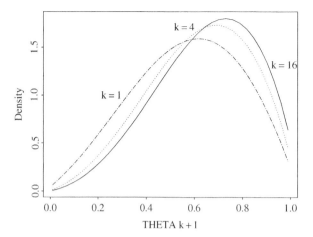

Figure 14.9   Marginal distributions of the success probability θ_{k+1} under alternative designs; all are based on 12 successes in 16 patients. Outcomes are as in Figure 14.8. Figure from Parmigiani and Berry (1994).

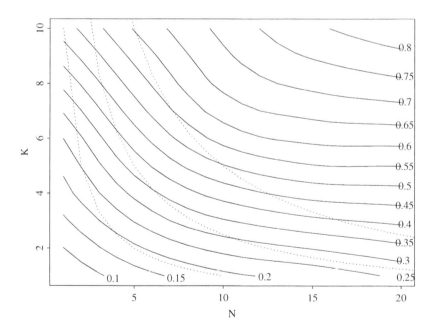

Figure 14.10   Contour plot of the estimated information surface as a function of the number of centers k and the number of patients per center n; dotted lines identify designs with the same number of patients, respectively 10, 25, and 50. Figure from Parmigiani and Berry (1994).

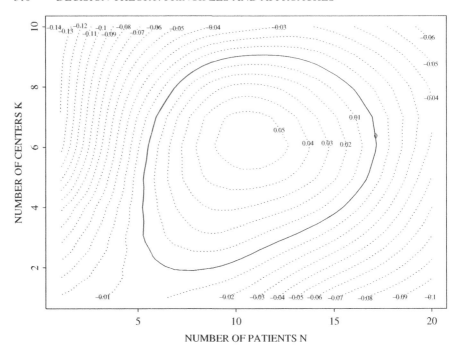

Figure 14.11   Contour plot of the estimated utility surface as a function of the number of centers k and the number of patients per center n. Pairs inside the 0-level curve are better than no experimentation. Points outside are worse. Figure from Parmigiani and Berry (1994).

14.4   Exercises

Problem 14.1   A medical laboratory must test N samples of blood to see which have traces of a rare disease. The probability that any individual from the relevant population has the disease is a known θ. Given θ, individuals are independent. Because θ is small it is suggested that the laboratory combine the blood of a individuals into equal-sized pools of $n = N/a$ samples, where a is a divisor of N. Each pool of samples would then be tested to see if it exhibited a trace of the infection. If no trace were found then the individual samples comprising the group would be known to be uninfected. If on the other hand a trace were found in the pooled sample it is then proposed that each of the a samples be tested individually. Discuss finding the optimal a when $N = 4$, $\theta = 0.02$, and then when $N = 300$. What can you say about the case in which the disease frequency is an unknown θ?

We work out the known θ case and leave the unknown θ for you. First we assume that $N = 4$, $\theta = 0.02$, and that it costs $1 to test a sample of blood, whether pooled or unpooled. The possible actions are $a = 1$, $a = 2$, and $a = 4$. The possible experimental outcomes for the pooled samples are:

(a) If $a = 4$ is chosen

$x_4 = 1$ Positive test result.
$x_4 = 0$ Negative test result.

(b) If $a = 2$ is chosen

$x_2 = 2$ Positive test result in both pooled samples.
$x_2 = 1$ Positive test result in exactly one of the pooled samples.
$x_2 = 0$ Negative test result in both pooled samples.

Recalling that $N = 4$, $\theta = 0.02$, and that nonoverlapping groups of individuals are also independent, we get

$$f(x_4 = 1|\theta) = 1 - (1 - 0.02)^4 = 0.0776$$
$$f(x_4 = 0|\theta) = (1 - 0.02)^4 = 0.9224$$
$$f(x_2 = 2|\theta) = [1 - (1 - 0.02)^2]^2 = 0.0016$$
$$f(x_2 = 1|\theta) = (1 - 0.02)^4 = 0.9224$$
$$f(x_2 = 0|\theta) = 1 - f(x_2 = 1|\theta) - f(x_2 = 0|\theta) = 0.076.$$

Figure 14.12 presents a simple one-stage decision tree for this problem, solved by using the probabilities just obtained. The best decision is to take one pooled sample of size 4.

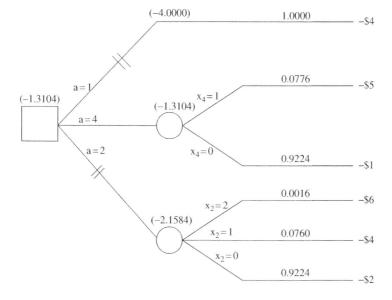

Figure 14.12   Decision tree for the pooling example.

Now, let us assume that $N = 300$. We are going to graph the optimal a for θ as $0.0, 0.1, 0.2, 0.3, 0.4$, or 0.5. The cost is still \$1. For all $a > 1$ divisors of N let us define $n_a = N/a$ and x_a the number of subgroups that exhibited a trace of infection. For $a = 1$ the cost will be N dollars for sure. Therefore, the possible values of x_a are $\{0, 1, \ldots, n_a\}$. Let us also define θ_a as the probability that a particular subgroup exhibits a trace of infection, which is the same as the probability that at least one of the a individuals in that subgroup exhibits a trace of infection. Then

$$\theta_a = 1 - (1 - \theta)^a \tag{14.40}$$

where, again, θ is the probability that a particular individual presents a trace of infection. Finally,

$$x_a \sim Binomial(n_a, \theta_a). \tag{14.41}$$

If $x_a = 0$ then only one test is needed for each group and the cost is $C_a = n_a$. If $x_a = 1$, then we need one test for each group, plus a tests for each individual in the specific group that presented a trace of infection. Thus, defining C_a as the cost associated with decision a we have

$$C_a = n_a + ax_a. \tag{14.42}$$

Also, $C_a = n_a + an_a$ if $x_a = n_a$. Hence,

$$E[C_a|\theta] = n_a + aE[x_a|\theta]$$
$$= n_a + an_a\theta_a$$
$$= N/a + N[1 - (1 - \theta)^a]$$

or, factoring terms,

$$E[C_a|\theta] = N[1 + 1/a - (1 - \theta)^a] \qquad a > 1. \tag{14.43}$$

Table 14.2 gives the values of $E[C_a|\theta]$ for all the divisors of $N = 300$ and for a range of values of θ. Figure 14.13 shows a versus $E[C_a|\theta]$ for the six different values of θ. If $\theta = 0.0$ nobody is infected, so we do not need to do anything. If $\theta = 0.5$ it is better to apply the test for every individual instead of pooling them in subgroups. The reason is that θ is high enough that many of the groups will be retested with high probability. Table 14.3 shows the optimal a for each value of θ.

Problem 14.2   Suppose that $x_1, x_2, \ldots$ are conditionally independent Bernoulli experiments, all with success probability θ. You are interested in estimating θ and your terminal loss function is

$$L(\theta, a) = \frac{(\theta - a)^2}{\theta(1 - \theta)}.$$

Table 14.2   Computation of $E[C_a|\theta]$ for different values of θ. The cost associated with optimal solutions is in bold. If $\theta = 0$ no observation is necessary.

			θ			
a	0.0	0.1	0.2	0.3	0.4	0.5
1	300	300	300	300	**300**	**300**
2	150	207	258	303	342	375
3	100	181	**246**	**297**	335	362
4	75	**178**	252	302	336	356
5	60	182	261	309	336	350
6	50	190	271	314	336	345
10	30	225	297	321	328	329
12	25	240	304	320	324	324
15	20	258	309	318	319	319
20	15	278	311	314	314	314
25	12	290	310	311	311	312
30	10	297	309	309	309	310
50	6	304	305	306	306	306
60	5	304	304	305	305	305
75	4	303	304	304	304	304
100	3	302	303	303	303	303
150	2	302	302	302	302	302
300	**1**	301	301	301	301	301

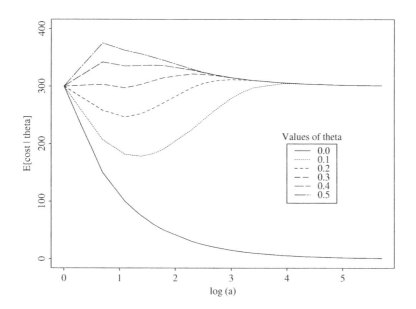

Figure 14.13   Computation of $E[C_a|\theta]$ for different values of θ.

Table 14.3   Optimal decisions for different values of θ.

θ	0.1	0.2	0.3	0.4	0.5
Optimal a	4	3	3	1	1

This is the traditional squared error loss, standardized using the variance of a single observation. With this loss function, errors of the same absolute magnitude are considered worse if they are made near the extremes of the interval $(0, 1)$. You have a $Beta(1.5, 1.5)$ prior on θ. Each observation costs you c. What is the optimal fixed sample size? You may start using computers at any stage in the solution of this problem.

Problem 14.3   You are the Mayor of Solvantville and you are interested in estimating the difference in the concentration of a certain contaminant in the water supply of homes in two different sections of town, each with its own well. If you found that there is a difference you would further investigate whether there may be contamination due to industrial waste. How big a sample of homes do you need for your purposes? Specify a terminal loss function, a cost of observation, a likelihood, and a prior. Keep it simple, but justify your assumptions and point out limitations.

Problem 14.4   In the context of Section 14.3.2 determine the optimal n by minimizing $r(\pi, n)$.

Problem 14.5   Suppose x has a normal distribution with mean θ and variance σ^2. Let $f(.)$ denote its density function.

(a) Prove that $\int_{-\infty}^{u} x f(x)dx = \theta \Phi(z) - \sigma \phi(z)$, where $z = (u - \theta)/\sigma$, and ϕ and Φ denote, respectively, the density and the cumulative distribution function of the standard normal distribution.

(b) Define $\zeta(y) = \phi(y) - y(1 - \Phi(y))$. Prove that $\zeta(-y) = y + \zeta(y)$.

(c) Let $A(u) = \int_{u}^{\infty}(x - u)f(x)dx$ and $B(u) = \int_{-\infty}^{u}(u - x)f(x)dx$ be, respectively, the right and left hand side linear loss integrals. Let $z = (u - \theta)/\sigma$. Prove that

   (i) $A(u) = \sigma \zeta(z)$.

   (ii) $B(u) = \sigma \zeta(-z)$.

Problem 14.6   In the context of the example of Section 14.3.3:

(a) Derive the function g in equation (14.35) and prove that its first- and second-order derivatives with respect to r are as given in (14.36).

(b) Prove that when $z_0 > 0$, a local maximum may or may not exist (that is, in the latter case, the optimal sample size is zero).

Problem 14.7   Can you find the optimal sample size in Example 14.3.1 when σ is not known? You can use a conjugate prior and standard results such as those of Sec. 3.3 in Gelman *et al.* (1995).

15

Stopping

In Chapter 14 we studied how to determine an optimal fixed sample size. In that discussion, the decision maker selects a sample size before seeing any of the data. Once the data are observed, the decision maker then makes a terminal decision, typically an inference. Here, we will consider the case in which the decision maker can make observations one at a time. At each time, the decision maker can either stop sampling (and choose a terminal decision), or continue sampling. The general technique for solving this sequential problem is dynamic programming, presented in general terms in Chapter 12.

In Section 15.1, we start with a little history of the origin of sequential sampling methods. Sequential designs were originally investigated because they can potentially reduce the expected sample size (Wald 1945). Our revisitation of Wald's sequential stopping theory is almost entirely Bayesian and can be traced back to Arrow *et al.* (1949). In their paper, they address some limitations of Wald's proof of the optimality of the sequential probability ratio test, derive a Bayesian solution, and also introduce a novel backward induction method that was the basis for dynamic programming. We illustrate the gains that can be achieved with sequential sampling using a simple example in Section 15.2. Next, we formalize the sequential framework for the optimal choice of the sample size in Sections 15.3.1, 15.3.2, and 15.3.3. In Section 15.4, we present an example of optimal stopping using the dynamic programming technique.

In Section 15.5, we discuss sequential sampling rules that do not require specifying the costs of experimentation, but rather continue experimenting until enough information about the parameters of interest is gained. An example is to sample until the posterior probability of a hypothesis of interest reaches a prespecified level. We examine the question of whether such a procedure could be used to design studies that always reach a foregone conclusion, and conclude that from a

Bayesian standpoint, that is not the case. Finally, in Section 15.6 we discuss the role of stopping rules in the terminal decision, typically inferences, and connect our decision-theoretic approach to the Likelihood Principle.

Featured articles:

Wald, A. (1945). Sequential tests of statistical hypotheses, *Annals of Mathematical Statistics* **16**: 117–186.

Arrow, K., Blackwell, D. and Girshick, M. (1949). Bayes and minimax solutions of sequential decision problems, *Econometrica* **17**: 213–244.

   Useful reference texts for this chapter are DeGroot (1970), Berger (1985), and Bernardo and Smith (2000).

15.1   Historical note

The earliest example of a sequential method in the statistical literature is, as far as we know, provided by Dodge and Romig (1929) who proposed a two-stage procedure in which the decision of whether or not to draw a second sample was based on the observations of the first sample. It was only years later that the idea of sequential analysis would be more broadly discussed. In 1943, US Navy Captain G. L. Schuyler approached the Statistical Research Group at Columbia University with the problem of determining the sample size needed for comparing two proportions. The sample size required was very large. In a letter addressed to Warren Weaver, W. Allen Wallis described Schuyler's impressions:

> When I presented this result to Schuyler, he was impressed by the largeness of the samples required for the degree of precision and certainty that seemed to him desirable in ordnance testing. Some of these samples ran to many thousands of rounds. He said that when such a test program is set up at Dahlgren it may prove wasteful. If a wise and seasoned ordnance expert like Schuyler were on the premises, he would see after the first few thousand or even few hundred [rounds] that the experiment need not be completed, either because the new method is obviously inferior or because it is obviously superior beyond what was hoped for. He said that you cannot give any leeway to Dahlgren personnel, whom he seemed to think often lack judgement and experience, but he thought it would be nice if there were some mechanical rule which could be specified in advance stating the conditions under which the experiment might be terminated earlier than planned. (Wallis 1980, p. 325)

   W. Allen Wallis and Milton Friedman explored this idea and came up with the following conjecture:

> Suppose that N is the planned number of trials and W_N is a most powerful critical region based on N observations. If it happens that on the basis of

the first n trials ($n < N$) it is already certain that the completed set of N must lead to a rejection of the null hypothesis, we can terminate the experiment at the n-trial and thus save some observations. For instance, if W_N is defined by the inequality $x_1^2 + \ldots + x_N^2 \geq c$, and if for some $n < N$ we find that $x_1^2 + \ldots + x_n^2 \geq c$, we can terminate the process at this stage. Realization of this naturally led Friedman and Wallis to the conjecture that modifications of current tests may exist which take advantage of sequential procedure and effect substantial improvements. More specifically, Friedman and Wallis conjectured that a sequential test may exist that controls the errors of the first and second kinds to exactly the same extent as the current most powerful test, and at the same time requires an expected number of observations substantially smaller than the number of observations required by the current most powerful test. (Wald 1945, pp. 120–121)

They first approached Wolfowitz with this idea. Wolfowitz, however, did not show interest and was doubtful about the existence of a sequential procedure that would improve over the most powerful test. Next, Wallis and Friedman brought this problem to the attention of A. Wald who studied it and, in April of 1943, proposed the *sequential probability ratio test*. In the problem of testing a simple null versus a simple alternative Wald claimed that:

The sequential probability ratio test frequently results in a saving of about 50% in the number of observations as compared with the current most powerful test. (Wald 1945, p. 119)

Wald's finding was of immediate interest, as he explained:

Because of the substantial savings in the expected number of observations effected by the sequential probability ratio test, and because of the simplicity of this test procedure in practical applications, the National Defense Research Committee considered these developments sufficiently useful for the war effort to make it desirable to keep the results out of the reach of the enemy, at least for a certain period of time. The author was, therefore, requested to submit his findings in a restricted report. (Wald 1945, p. 121)

It was only in 1945, after the reports of the Statistical Research Group were no longer classified, that Wald's research was published. Wald followed his paper with a book published in 1947 (Wald 1947b).

In Chapter 7 we discussed Wald's contribution to statistical decision theory. In our discussion, however, we only considered the situation in which the decision maker has a fixed number of observations. Extensions of the theory to sequential decision making are in Wald (1947a) and Wald (1950). Wald explores minimax and Bayes rules in the sequential setting. The paper by Arrow *et al.* (1949) is, however, the first example we know of utilizing *backwards induction* to derive the formal Bayesian optimal sequential procedure.

15.2   A motivating example

This example, based on DeGroot (1970), illustrates the insight of Wallis and Friedman, and shows how a sequential procedure improves over the fixed sample size design.

Suppose it is desired to test the null hypothesis $H_0 : \theta = \theta_0$ versus the alternative $H_1 : \theta = \theta_1$. Let a_i denote the decision of accepting hypothesis H_i, $i = 0, 1$. Assume that the utility function is

$$u(a_i(\theta)) = -I_{\{a_i \neq \theta_i\}}. \tag{15.1}$$

The decision maker can take observations $x_i, i = 1, 2, \ldots$. Each observation costs c units and it has a probability α of providing an uninformative outcome, while in the remainder of the cases it provides an answer that reveals the value of θ without ambiguity. Formally x_i has the following probability distribution:

$$f(x|\theta) = \begin{cases} (1 - \alpha), & \text{if } \theta = \theta_0 \text{ and } x = 1, \text{ or } \theta = \theta_1 \text{ and } x = 2 \\ 0, & \text{if } \theta = \theta_0 \text{ and } x = 2, \text{ or } \theta = \theta_1 \text{ and } x = 1 \\ \alpha, & \text{if } x = 3. \end{cases} \tag{15.2}$$

Let $\pi = \pi(\theta = \theta_0)$ denote the prior probability of the hypothesis H_0, and assume that $\pi \leq 1/2$. If no observations are made, the Bayes decision is a_1 and the associated expected utility is $-\pi$. Now, let us study the case in which observations are available. Let y_x count the number of observations with value x, with $x = 1, 2, 3$. Given a sequence of observations $x^n = (x_1, \ldots, x_n)$, the posterior distribution is

$$\pi_{x^n} = \pi(\theta = \theta_0|x^n) = \frac{\pi f(x^n|\theta_0)}{m(x^n)} = \begin{cases} 1, & \text{if } y_2 = 0 \text{ and } y_3 < n \\ 0, & \text{if } y_2 > 0 \\ \pi, & \text{if } y_3 = n. \end{cases} \tag{15.3}$$

There is no learning about θ if all the observed x are equal to 3, the uninformative case. However, in any other case, the value of θ is known with certainty. It is futile to continue experimenting after $x = 1$ or $x = 2$ is observed. For sequences x^n that reveal the value of θ with certainty, the expected posterior utility, not including sampling costs, is 0. When $\pi_{x^n} = \pi$, the expected posterior utility is $-\pi$. Thus, the expected utility of taking n observations, accounting for sampling costs, is

$$\mathcal{U}_\pi(n) = -\pi \alpha^n - cn. \tag{15.4}$$

Suppose that the values of c and α are such that $\mathcal{U}_\pi(1) > \mathcal{U}_\pi(0)$, that is it is worthwhile to take at least one observation, and let n^* denote the positive integer that maximizes equation (15.4). An approximate value is obtained by assuming continuity in n. Differentiating with respect to n, and setting the result to 0, gives

$$n^* = \left[\log\left(\frac{1}{\alpha}\right)\right]^{-1}\left[\log\left(\frac{\pi\log(1/\alpha)}{c}\right)\right]. \tag{15.5}$$

The decision maker can reduce costs of experimentation by taking the n^* observations sequentially with the possibility of stopping as soon as he or she observes a value different from 3. Using this procedure, the sample size N is a random variable. Its expected value is

$$E[N|\theta_0] = \sum_{j=1}^{n^*} jP(N = j|\theta = \theta_0)$$

$$= \sum_{j=1}^{n^*-1} j\alpha^{j-1}(1-\alpha) + n^*\alpha^{n^*-1} = \frac{1-\alpha^{n^*}}{1-\alpha}. \tag{15.6}$$

This is independent of θ, so $E[N|\theta_1] = E[N|\theta_0]$. Thus, $E[N] = \left(1-\alpha^{n^*}\right)/(1-\alpha)$.

This sequential sampling scheme imposes an upper bound of n^* observations on the expected value of N. To illustrate the reduction on the number of observations when using the sequential sampling, let $\alpha = 1/4, c = 0.000001, \pi = 1/4$. We obtain $n^* = 10$ and $E[N] = 1.33$; that is, on average, we have a great reduction in the sample size needed.

Figure 15.1 illustrates this further using a simulation. We chose values of θ from its prior distribution with $\pi = 1/2$. Given θ, we sequentially simulated observations using (15.2) with $\alpha = 1/4$ until we obtained a value different from 3, or reached the maximum sample size n^* and recorded the required sample size. We repeated

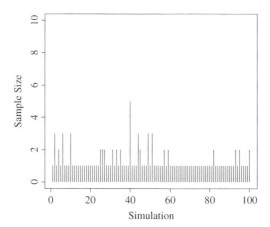

Figure 15.1   Sample size for 100 simulated sequential experiments in which $\pi = 1/2$ and $\alpha = 1/4$.

this procedure 100 times. In the figure we show the resulting number of observations with the sequential procedure.

15.3   Bayesian optimal stopping

15.3.1   Notation

As in Chapter 14, suppose that observations $x_1, x_2, \ldots$ become available sequentially one at a time. After taking n observations, that is observing $x^n = (x_1, \ldots, x_n)$, the joint probability density function (or joint probability mass function depending on the application) is $f(x^n|\theta)$. As usual, assuming that π is the prior distribution for θ, after n observations, the posterior density is $\pi(\theta|x^n)$ and the marginal distribution of x^n is

$$m(x^n) = \int_\Theta f(x^n|\theta)\pi(\theta)d\theta.$$

The terminal utility function associated with a decision a when the state of nature is θ is $u(a(\theta))$. The experimental cost of taking n observations is denoted by $C(n)$. The overall utility is, therefore,

$$u(\theta, a, n) = u(a(\theta)) - C(n).$$

The terminal decision rule after n observations (that is, after observing x^n) is denoted by $\delta_n(x^n)$. As in the nonsequential case, for every possible experimental outcome x^n, $\delta_n(x^n)$ tell us which action to choose if we stop sampling after x^n. In the sequential case, though, a decision rule requires the full sequence $\delta = \{\delta_1(x^1), \delta_2(x^2), \ldots\}$ of terminal decision rules. Specifying δ tells us what to do if we stop sampling after $1, 2, \ldots$ observations have been made. But when should we stop sampling?

Let $\zeta_n(x^n)$ be the decision rule describing whether or not we stop sampling after n observations, that is

$$\zeta_n(x^n) = \begin{cases} 1, \text{ stop after observing } x^n \\ 0, \text{ continue sampling.} \end{cases}$$

If $\zeta_0 \equiv 1$ there is no experimentation. We can now define stopping rule, stopping time, and sequential decision rule.

Definition 15.1   *The sequence $\zeta = \{\zeta_0, \zeta_1(x^1), \zeta_2(x^2), \ldots\}$ is called a* stopping rule.

Definition 15.2   *The number N of observations after which the decision maker decides to stop sampling and make the final decision is called* stopping time.

The stopping time is a random variable whose distribution depends on ζ. Specifically,

$$N = \min\{n \geq 0, \quad \text{such that} \quad \zeta_n(x^n) = 1\}.$$

We will also use the notation $\{N = n\}$ for the set of observations $(x_1, x_2, \ldots)$ such that the sampling stops at n, that is for which

$$\zeta_0 = 0, \ \zeta_1(x^1) = 0, \ \ldots, \ \zeta_{n-1}(x^{n-1}) = 0, \ \zeta_n(x^n) = 1.$$

Definition 15.3 *A sequential decision rule d consists of a stopping rule and a decision rule, that is $d = (\zeta, \delta)$. If $P_\theta(N < \infty) = 1$ for all values of θ, then we say that the sequential stopping rule ζ is proper.*

From now on, in our discussion we will consider just the class of proper stopping rules.

15.3.2 Bayes sequential procedure

In this section we extend the definition of Bayes rule to the sequential framework. To do so, we first note that

$$P_\theta(N < \infty) = \sum_{n=0}^{\infty} P_\theta(N = n)$$

$$= P_\theta(N = 0) + \sum_{n=1}^{\infty} \int_{\{N=n\}} f(x^n|\theta)dx^n.$$

The counterpart of the frequentist risk R in this context is the long-term average utility of the sequential procedure d for a given θ, that is

$$U(\theta, d) = u(\theta, \delta_0, 0)P_\theta(N = 0) + \sum_{n=1}^{\infty} \left\{ \int_{\{N=n\}} u(\theta, \delta(x^n), n)f(x^n|\theta)dx^n \right\}$$

$$= u(\delta_0(\theta))P_\theta(N = 0) + \sum_{n=1}^{\infty} U(\theta, \delta_n) - \sum_{n=1}^{\infty} C(n)P_\theta(N = n),$$

where

$$U(\theta, \delta_n) = \int_{\{N=n\}} u(\delta_n(x^n)(\theta))f(x^n|\theta)dx^n,$$

with $\delta_n(x^n)(\theta)$ denoting that for each x^n the decision rule $\delta_n(x^n)$ produces an action, which in turn is a function from states to outcomes.

The counterpart of the Bayes risk is the expected utility of the sequential decision procedure d, given by

$$\mathcal{U}_\pi(d) = \int_\Theta U(\theta, d)\pi(\theta)d\theta.$$

Definition 15.4 *The Bayes sequential procedure d^π is the procedure which maximizes the expected utility $\mathcal{U}_\pi(d)$.*

Theorem 15.1   *If $\delta_n^\pi(x^n)$ is a Bayes rule for the fixed sample size problem with observations $x_1, \ldots, x_n$ and utility $u(\theta, a, n)$, then $\delta^\pi = (\delta_1^\pi, \delta_2^\pi, \ldots)$ is a Bayes sequential decision rule.*

Proof: Without loss of generality, we assume that $P_\theta(N = 0) = 0$:

$$\mathcal{U}_\pi(d) = \int_\Theta U(\theta, d)\pi(\theta)d\theta$$

$$= \int_\Theta \sum_{n=1}^\infty \int_{\{N=n\}} u(\theta, \delta_n(x^n), n)f(x^n|\theta)\pi(\theta)dx^n d\theta.$$

If we interchange the integrals on the right hand side of the above expression we obtain

$$\mathcal{U}_\pi(d) = \sum_{n=1}^\infty \int_{\{N=n\}} \int_\Theta u(\theta, \delta_n(x^n), n)f(x^n|\theta)\pi(\theta)d\theta dx^n.$$

Note that $f(x^n|\theta)\pi(\theta) = \pi(\theta|x^n)m(x^n)$. Using this fact, let

$$\mathcal{U}_{\pi_{x^n}, N=n}(\delta_n) = \int_\Theta u(\theta, \delta_n(x^n), n)\pi(\theta|x^n)d\theta$$

denote the posterior expected utility at stage n. Thus,

$$\mathcal{U}_\pi(d) = \sum_{n=1}^\infty \int_{\{N=n\}} \mathcal{U}_{\pi_{x^n}, N=n}(\delta_n(x^n))m(x^n)dx^n.$$

The maximum of $\mathcal{U}_\pi(d)$ is attained if, for each x^n, $\delta_n(x^n)$ is chosen to maximize the posterior expected utility $\mathcal{U}_{\pi_{x^n}, N=n}(\delta_n(x^n))$ which is done in the fixed sample size problem, proving the theorem.   □

This theorem tells us that, regardless of the stopping rule used, once the experiment is stopped, the optimal decision is the formal Bayes rule conditional on the observations collected. This result echoes closely our discussion of normal and extensive forms in Section 12.3.1. We will return to the implications of this result in Section 15.6.

15.3.3   Bayes truncated procedure

Next we discuss how to determine the optimal stopping time. We will consider *bounded* sequential decision procedures d (DeGroot 1970) for which there is a positive number k such that $P(N \leq k) = 1$. The technique for solving the optimal stopping rule is backwards induction as we previously discussed in Chapter 12.

Say we carried out n steps and observed x^n. If we continue we face a new sequential decision problem with the same terminal decision as before, the possibility of observing $x_{n+1}, x_{n+2}, \ldots$, and utility function $u(a(\theta)) - C(j)$, where j is the number of additional observations. Now, though, our updated probability distribution on θ, that is π_{x^n}, will serve as the prior. Let $\mathcal{D}^n$ be the class of all proper sequential decision rules for this problem. The expected utility of a rule d in $\mathcal{D}^n$ is $\mathcal{U}_{\pi_{x^n}}(d)$, while the maximum achievable expected utility if we continue is

$$W(\pi_{x^n}, n) = \sup_{d \in \mathcal{D}^n} \mathcal{U}_{\pi_{x^n}}(d).$$

To decide whether or not to stop we compare the expected utility of the continuation problem to the expected utility of stopping immediately. At time n, the posterior expected utility of collecting an additional 0 observations is

$$W_0(\pi_{x^n}, n) = \sup_a \int_\Theta [u(a(\theta)) - C(n)] \, \pi_{x^n}(\theta) d\theta.$$

Therefore, we would continue sampling if $W(\pi_{x^n}, n) > W_0(\pi_{x^n}, n)$.

A catch here is that computing $W(\pi_{x^n}, n)$ may not be easy. For example, dynamic programming tells us we should begin at the end, but here there is no end. An easy fix is to truncate the problem. Unless you plan to be immortal, you can do this with little loss of generality. So consider now the situation in which the decision maker can take no more than k overall observations. Formally, we will consider a subset $\mathcal{D}^n_k$ of $\mathcal{D}^n$, consisting of rules that take a total of k observations. These rules are all such that $\zeta_k(x^k) = 1$. Procedures d^k with this feature are called *k-truncated procedures*. We can define the *k-truncated* value of the experiment starting at stage n as

$$W_{k-n}(\pi_{x^n}, n) = \sup_{d \in \mathcal{D}^n_k} \mathcal{U}_{\pi_{x^n}}(d).$$

Given that we made it to n and saw x^n, the best we could possible do if we continue for at most $k - n$ additional steps is $W_{k-n}(\pi_{x^n}, n)$. Because $\mathcal{D}^n_k$ is a subset of $\mathcal{D}^n$, the value of the k-truncated problem is non-decreasing in k.

The *Bayes k-truncated procedure* d^k is a procedure that lies in $\mathcal{D}^0_k$ and for which

$$\mathcal{U}_{\pi_{x^n}}(d^k) = W_k(\pi_{x^n}, 0).$$

Also, you can think of $W(\pi_{x^n}, n)$ as $W_\infty(\pi_{x^n}, n)$.

The relevant expected utilities for the stopping problem can be calculated by induction. We know the value is $W_0(\pi_{x^n}, n)$ if we stop immediately. Then we can recursively evaluate the value of stopping at any intermediate stage between n and k by

$$W_1(\pi_{x^n}, n) = \max \left\{ W_0(\pi_{x^n}, n), \; E_{x_{n+1}|x^n}[W_0(\pi_{x^{n+1}}, n + 1)] \right\}$$

$$W_2(\pi_{x^n}, n) = \max \left\{ W_0(\pi_{x^n}, n), \; E_{x_{n+1}|x^n}[W_1(\pi_{x^{n+1}}, n + 1)] \right\}$$

$$\vdots$$

$$W_{k-n}(\pi_{x^n}, n) = \max \left\{ W_0(\pi_{x^n}, n), \; E_{x_{n+1}|x^n}[W_{k-n-1}(\pi_{x^{n+1}}, n + 1)] \right\}.$$

Based on this recursion we can work out the following result.

Theorem 15.2   *Assume that the Bayes rule δ_n^* exists for all n, and $W_j(\pi_{x^n}, n)$ are finite for $j \leq k, n \leq (k-j)$. The Bayes k-truncated procedure is given by $d^k = (\zeta^k, \delta^*)$ where ζ^k is the stopping rule which stops sampling at the first n such that*

$$W_0(\pi_{x^n}, n) = W_{k-n}(\pi_{x^n}, n).$$

This theorem tell us that at the initial stage we compare $W_0(\pi, 0)$, associated with an immediate Bayes decision, to the overall k-truncated $W_k(\pi, 0)$. We continue sampling (observe x_1) only if $W_0(\pi, 0) < W_k(\pi, 0)$. Then, if this is the case, after x_1 has been observed, we compare $W_0(\pi_{x^1}, 1)$ of an immediate decision to the $(k-1)$-truncated problem $W_{k-1}(\pi_{x^1}, 1)$ at stage 1. As before, we continue sampling only if $W_0(\pi_{x^1}, 1) < W_{k-1}(\pi_{x^1}, 1)$. We proceed until $W_0(\pi_{x^n}, n) = W_{k-n}(\pi_{x^n}, n)$.

In particular, if the cost of each observation is constant and equal to c, we have $C(n) = nc$. Therefore,

$$W_0(\pi_{x^n}, n) = \sup_a \int_\Theta [u(a(\theta)) - nc]\pi_{x^n}(\theta)d\theta$$

$$= \sup_a \mathcal{U}_{\pi_{x^n}}(a) - nc$$

$$\equiv V_0(\pi_{x^n}) - nc.$$

It follows that

$$W_1(\pi_{x^n}, n) = \max\left\{V_0(\pi_{x^n}), E_{x_{n+1}|x^n}[V_0(\pi_{x^{n+1}})] - c\right\} - nc.$$

If we define, inductively,

$$V_j(\pi_{x^n}) = \max\left\{V_0(\pi_{x^n}), E_{x_{n+1}|x^n}[V_{j-1}(\pi_{x^{n+1}})] - c\right\}$$

then $W_j(\pi_{x^n}, n) = V_j(\pi_{x^n}) - nc$.

Corollary 2   *If the cost of each observation is constant and equal to c, the Bayes k-truncated stopping rule, ζ^k, is to stop sampling and make a decision for the first $n \leq k$ for which*

$$V_0(\pi_{x^n}) = V_{k-n}(\pi_{x^n}).$$

We move next to applications.

15.4   Examples

15.4.1   Hypotheses testing

This is an example of optimal stopping in a simple hypothesis testing setting. It is discussed in Berger (1985) and DeGroot (1970). Another example in the context of estimation is Problem 15.1. Assume that $x_1, x_2, \ldots$ is a sequential sample from a Bernoulli distribution with unknown probability of success θ. We wish to test H_0 :

Table 15.1   Utility function for the sequential hypothesis testing problem.

	True value of θ	
	$\theta = 1/3$	$\theta = 2/3$
a_0	0	-20
a_1	-20	0

$\theta = 1/3$ versus $H_1 : \theta = 2/3$. Let a_i denote accepting H_i. Assume $u(\theta, a, n) = u(a(\theta)) - nc$, with sampling cost $c = 1$ and with "0–20" terminal utility: that is, no loss for a correct decision, and a loss of 20 for an incorrect decision, as in Table 15.1. Finally, let π_0 denote the prior probability that H_0 is true, that is $Pr(\theta = 1/3) = \pi_0$. We want to find d^2, the optimal procedure among all those taking at most two observations.

We begin with some useful results. Let $y_n = \sum_{i=1}^{n} x_i$.

(i) The marginal distribution of x^n is

$$m(x^n) = f(x^n|\theta = 1/3)\pi_0 + f(x^n|\theta = 2/3)(1 - \pi_0)$$
$$= (1/3)^{y_n}(2/3)^{n-y_n}\pi_0 + (2/3)^{y_n}(1/3)^{n-y_n}(1 - \pi_0)$$
$$= (1/3)^n \left[2^{n-y_n}\pi_0 + 2^{y_n}(1 - \pi_0) \right].$$

(ii) The posterior distribution of θ given x is

$$\pi(\theta = 1/3|x^n) = \frac{\pi_0}{\pi_0 + 2^{2y_n - n}(1 - \pi_0)}$$
$$\pi(\theta = 2/3|x^n) = 1 - \pi(\theta = 1/3|x^n).$$

(iii) The predictive distribution of x_{n+1} given x^n is

$$m(x_{n+1}|x^n) = \frac{m(x^{n+1})}{m(x^n)}$$
$$= \begin{cases} \dfrac{1}{3} \dfrac{2^{n+1-y_n}\pi_0 + 2^{y_n}(1 - \pi_0)}{2^{n-y_n}\pi_0 + 2^{y_n}(1 - \pi_0)} & \text{if } x_{n+1} = 0 \\[3mm] \dfrac{1}{3} \dfrac{2^{n-y_n}\pi_0 + 2^{y_n+1}(1 - \pi_0)}{2^{n-y_n}\pi_0 + 2^{y_n}(1 - \pi_0)} & \text{if } x_{n+1} = 1. \end{cases}$$

Before making any observation, the expected utility of immediately making a decision is

$$W_0(\pi_0, 0) = V_0(\pi_0) = \max_a \left[\sum_\theta u(a(\theta))\pi(\theta) \right]$$

$$= \max\{-20(1 - \pi_0), -20\pi_0\}$$

$$= \begin{cases} -20\pi_0 & \text{if } 0 \le \pi_0 \le 1/2 \\ -20(1 - \pi_0) & \text{if } 1/2 \le \pi_0 \le 1. \end{cases}$$

Therefore, the Bayes decision is a_0 if $1/2 \le \pi_0 \le 1$ and a_1 if $0 \le \pi_0 \le 1/2$. Figure 15.2 shows the V-shaped utility function V_0 highlighting that we expect to be better off at the extremes than at the center of the range of π_0 in addressing to choice of hypothesis.

Next, we calculate $V_0(\pi_{x^1})$, the value of taking an observation and then stopping. Suppose that x_1 has been observed. If $x_1 = 0$, the posterior expected utility is

$$V_0(\pi(\theta|x_1 = 0)) = \begin{cases} \dfrac{-40\pi_0}{1 + \pi_0} & \text{if } 0 \le \pi_0 \le 1/3 \\[2ex] \dfrac{-20(1 - \pi_0)}{1 + \pi_0} & \text{if } 1/3 \le \pi_0 \le 1. \end{cases}$$

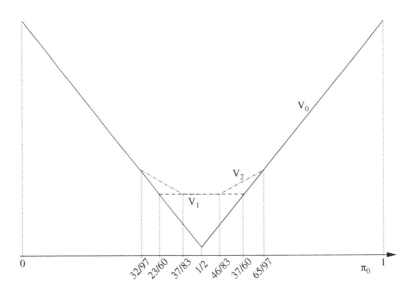

Figure 15.2   Expected utility functions V_0 (solid), V_1 (dashed), and V_2 (dot–dashed) as functions of the prior probability π_0. Adapted from DeGroot (1970).

If $x_1 = 1$ the posterior expected utility is

$$V_0(\pi(\theta|x_1 = 1)) = \begin{cases} \dfrac{-20\pi_0}{2 - \pi_0} & \text{if } 0 \leq \pi_0 \leq 2/3 \\[2mm] \dfrac{-40(1 - \pi_0)}{2 - \pi_0} & \text{if } 2/3 \leq \pi_0 \leq 1. \end{cases}$$

Thus, the expected posterior utility is

$$E_{x_1}[V_0(\pi_{x^1})] = V_0(\pi(\theta|x_1 = 0))m(0) + V_0(\pi(\theta|x_1 = 1))m(1)$$

$$= \begin{cases} -20\pi_0 & \text{if } 0 \leq \pi_0 \leq 1/3 \\ -20/3 & \text{if } 1/3 \leq \pi_0 \leq 2/3 \\ -20(1 - \pi_0) & \text{if } 2/3 \leq \pi_0 \leq 1. \end{cases}$$

Using the recursive relationship we derived in Section 15.3.3, we can derive the utility of the optimal procedure in which no more than one observation is taken. This is

$$V_1(\pi_0) = \max \left\{ V_0(\pi_0), E_{x_1}[V_0(\pi_{x^1})] - 1 \right\}$$

$$= \begin{cases} -20\pi_0 & \text{if } 0 \leq \pi_0 \leq 23/60 \\ -23/3 & \text{if } 23/60 \leq \pi_0 \leq 37/60 \\ -20(1 - \pi_0) & \text{if } 37/60 \leq \pi_0 \leq 1. \end{cases}$$

If the prior is in the range $23/60 \leq \pi_0 \leq 37/60$, the observation of x_1 will potentially move the posterior sufficiently to change the optimal decision, and to do so with sufficient confidence to offset the cost of observation.

The calculation of V_2 follows the same steps and works out so that

$$E_{x_1}[V_1(\pi_{x^1})] = \begin{cases} -20\pi_0 & \text{if } 0 \leq \pi_0 \leq 23/97 \\ -(83\pi_0 + 23)/9 & \text{if } 23/97 \leq \pi_0 \leq 37/83 \\ -20/3 & \text{if } 37/83 \leq \pi_0 \leq 46/83 \\ -(106 - 83\pi_0)/9 & \text{if } 46/83 \leq \pi_0 \leq 74/97 \\ -20(1 - \pi_0) & \text{if } 74/97 \leq \pi_0 \leq 1, \end{cases}$$

which leads to

$$V_2(\pi_0) = \max \left\{ V_0(\pi_0), E_{x_1}[V_1(\pi_{x^1})] - 1 \right\}$$

$$= \begin{cases} -20\pi_0 & \text{if } 0 \leq \pi_0 \leq 32/97 \\ -(83\pi_0 + 32)/9 & \text{if } 32/97 \leq \pi_0 \leq 37/83 \\ -23/3 & \text{if } 37/83 \leq \pi_0 \leq 46/83 \\ -(115 - 83\pi_0)/9 & \text{if } 46/83 \leq \pi_0 \leq 65/97 \\ -20(1 - \pi_0) & \text{if } 65/97 \leq \pi_0 \leq 1. \end{cases}$$

Because of the availability of the second observation, the range of priors for which it is optimal to start sampling is now broader, as shown in Figure 15.2.

To see how the solution works out in a specific case, suppose that $\pi_0 = 0.48$. With this prior, a decision with no data is a close call, so we expect the data to be useful. In fact, $V_0(\pi_0) = -9.60 < V_2(\pi_0) = -7.67$, so we should observe x_1. If $x_1 = 0$, we have $V_0(\pi_{x^1}) = V_1(\pi_{x^1}) = -7.03$. Therefore, it is optimal to stop and choose a_0. If $x_1 = 1$, $V_0(\pi_{x^1}) = V_1(\pi_{x^1}) = -6.32$ and, again, it is optimal to stop, but we choose a_1.

15.4.2   An example with equivalence between sequential and fixed sample size designs

This is another example from DeGroot (1970), and it shows that in some cases there is no gain from having the ability to accrue observations sequentially.

Suppose that $x_1, x_2, \ldots$ is a sequential sample from a normal distribution with unknown mean θ and specified variance $1/\omega$. The parameter ω is called precision and is a more convenient way to handle this problem as we will see. Assume that θ is estimated under utility $u(a(\theta)) = -(\theta - a)^2$ and that there is a fixed cost c per observation. Also, suppose that the prior distribution π of θ is a normal distribution with mean μ and precision h. Then the posterior distribution of θ given x^n is normal with posterior precision $h + n\omega$. Therefore, there is no uncertainty about future utilities, as

$$W_0(\pi, 0) = -\mathrm{Var}[\theta] = -\frac{1}{h}$$

$$W_0(\pi_{x^n}, n) = -\mathrm{Var}[\theta | x^n] - nc = -\frac{1}{h + n\omega} - nc$$

and so $V_0(\pi_{x^n}) = -1/(h + n\omega)$ for $n = 0, 1, 2, \ldots$. This result shows that the expected utility depends only on the number of observations taken and not on their observed values. This implies that the optimal sequential decision procedure is a procedure in which a fixed number of observations is taken, because since the beginning we can predict exactly what our state of information will be at any stage in the process. Try Problem 15.3 for a more general version of this result.

Exploring this example a bit more, we discover that in this simple case we can solve the infinite horizon version of the dynamic programming solution to the optimal stopping rule. From our familiar recursion equation

$$V_1(\pi_{x^n}) = \max\{V_0(\pi_{x^n}), E_{x_{n+1}|x^n}[V_0(\pi_{x^{n+1}})] - c\}$$

$$= \max\left\{-\frac{1}{h + n\omega}, -\frac{1}{h + (n+1)\omega} - c\right\}.$$

Now

$$\frac{1}{h + n\omega} < \frac{1}{h + (n+1)\omega} + c$$

whenever

$$\omega < c(h + n\omega)(h + (n + 1)\omega).$$

Using this fact, and an inductive argument (Problem 15.4), one can show that $W_j(\pi, n) = -1/(h + n\omega) - nc$ for $\beta_n < \omega \leq \beta_{n+1}$ where β_n is defined as $\beta_0 = 0, \beta_n = c(h + (n - 1)\omega)(h + n\omega)$ for $n = 1, \ldots, j$, and $\beta_{j+1} = \infty$. This implies that the optimal sequential decision is to take exactly n^* observations, where n^* is an integer such that $\beta_{n^*} < \omega \leq \beta_{n^*+1}$. If the decision maker is not allowed to take more than k observations, then if $n^* > k$, he or she should take k observations.

15.5 Sequential sampling to reduce uncertainty

In this section we discuss sampling rules that do not involve statements of costs of experimentation, following Lindley (1956) and DeGroot (1962). The decision maker's goal is to obtain enough information about the parameters of interest. Experimentation stops when enough information is reached. Specifically, let $V_x(\mathcal{E})$ denote the observed information provided by observing x in experiment $\mathcal{E}$ as defined in Chapter 13. Sampling would continue until $V_{x^n}(\mathcal{E}) \geq 1/\epsilon$, with ϵ chosen before experimentation starts. In estimation under squared error loss, for example, this would mean sampling until $\mathrm{Var}[\theta | x^n] < \epsilon$ for some ϵ.

Example 15.1   Suppose that $\Theta = \{\theta_0, \theta_1\}$ and let $\pi = \pi(\theta_0)$ and $\pi_{x^n} = \pi_{x^n}(\theta_0)$. Consider the following sequential procedure: sampling continues until the Shannon information of π_{x^n} is at least ϵ. In symbols, sampling continues until

$$V(\pi_{x^n}) = \pi_{x^n} \log(\pi_{x^n}) + (1 - \pi_{x^n}) \log(1 - \pi_{x^n}) \geq \epsilon. \tag{15.7}$$

For more details on Shannon information and entropy see Cover and Thomas (1991). One can show that Shannon's information is convex in π_{x^n}. Thus, inequality (15.7) is equivalent to sampling as long as

$$A' \leq \pi_{x^n} \leq B', \tag{15.8}$$

with A' and B' satisfying the equality in (15.7). If $\pi_{x^n} < A'$, then we stop and accept H_1; if $\pi_{x^n} > B'$ we stop and accept H_0.
Rewriting the posterior in terms of the prior and likelihood,

$$\pi_{x^n} = \frac{\pi f(x^n | \theta_0)}{\pi f(x^n | \theta_0) + (1 - \pi) f(x^n | \theta_1)} = \left[1 + \frac{1 - \pi}{\pi} \frac{f(x^n | \theta_1)}{f(x^n | \theta_0)}\right]^{-1}. \tag{15.9}$$

This implies that inequality (15.8) is equivalent to

$$A < \frac{f(x^n | \theta_1)}{f(x^n | \theta_0)} < B, \tag{15.10}$$

with

$$A = \left(\frac{\pi}{1-\pi}\right)\left(\frac{1-B'}{B'}\right) \quad \text{and} \quad B = \left(\frac{\pi}{1-\pi}\right)\left(\frac{1-A'}{A'}\right).$$

The procedure just derived is of the same form as Wald's *sequential probability ratio test* (Wald 1947b), though in Wald the boundaries for stopping are determined using frequentist operating characteristics of the algorithm.   ★

Example 15.2   Let $x_1, x_2, \ldots$ be sequential observations from a Bernoulli distribution with unknown probability of success θ. We are interested in estimating θ under the quadratic loss function. Suppose that θ is a priori $Beta(\alpha_0, \beta_0)$, and let α_n and β_n denote the parameters of the beta posterior distribution. The optimal estimate of θ is the posterior mean, that is $\delta^* = \alpha_n/(\alpha_n + \beta_n)$. The associated Bayes risk is the posterior variance $\mathrm{Var}[\theta|x^n] = \alpha_n\beta_n/[(\alpha_n + \beta_n)^2(\alpha_n + \beta_n + 1)]$.

Figure 15.3 shows the contour plot of the Bayes risk as a function of α_n and β_n. Say the sampling rule is to take observations as long as $\mathrm{Var}[\theta|x^n] \geq \epsilon = 0.02$. Starting with a uniform prior ($\alpha = \beta = 1$), the prior variance is approximately equal to 0.08. The segments in Figure 15.3 show the trajectory of the Bayes risk after observing a particular sample. The first observation is a success and thus $\alpha_n = 2, \beta_n = 1$, so our trajectory moves one step to the right. After $n = 9$ observations

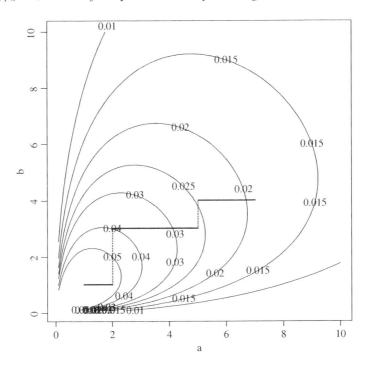

Figure 15.3   Contour plot of the Bayes risk as a function of the beta posterior parameters α_n (labeled a) and β_n (labeled b).

with $x = (1, 0, 0, 1, 1, 1, 0, 1, 1)$ the decision maker crosses the 0.02 curve, and stops experimentation.

For a more detailed discussion of this example, including the analysis of the problem using asymptotic approximations, see Lindley (1956) and Lindley (1957). ★

Example 15.3   Suppose that $x_1, x_2, \ldots$ are sequentially drawn from a normal distribution with unknown mean θ and known variance σ^2. Suppose that a priori θ has $N(0, 1)$ distribution. The posterior variance $\text{Var}[\theta | x^n] = \sigma^2/(n + \sigma^2)$ is independent of x^n. Let $0 < \epsilon < 1$. Sampling until $\text{Var}[\theta | x^n] < \epsilon$ is equivalent to taking a fixed sample size with $n > \sigma^2(1 - \epsilon)/\epsilon$. ★

15.6   The stopping rule principle

15.6.1   Stopping rules and the Likelihood Principle

In this section we return to our discussion of foundation and elaborate on Theorem 15.1, which states that the Bayesian optimal decision is not affected by the stopping rule. The reason for this result is a general factorization of the likelihood: for any stopping rule ζ for sampling from a sequence of observations $x_1, x_2, \ldots$ having fixed sample size parametric model $f(x^n | n, \theta) = f(x^n | \theta)$, the likelihood function is

$$f(n, x^n | \theta) = \zeta_n(x_n) \prod_{i=1}^{n-1} (1 - \zeta_i(x_i)) f(x^n | \theta) \propto f(x^n | \theta), \quad \theta \in \Theta, \qquad (15.11)$$

for all (n, x^n) such that $f(n, x^n | \theta) \neq 0$.

Thus the likelihood function is the same irrespective of the stopping rule. The stopping time provides no additional information about θ to that already contained in the likelihood function $f(x^n | \theta)$ or in the prior $\pi(\theta)$, a fact referred to as the *stopping rule principle* (Raiffa and Schlaifer 1961). If one uses any statistical procedure satisfying the Likelihood Principle the stopping rule should have no effect on the final reported evidence about θ (Berger and Wolpert 1988):

> The stopping rule principle *does not* say that one can ignore the stopping rule and then use any desired measure of evidence. It *does* say that reasonable measures of evidence should be such that they do not depend on the stopping rule. Since frequentist measures do depend on the stopping rule, they do not meet this criterion of reasonableness. (Berger and Berry 1988, p. 45)

Some of the difficulties faced by frequentist analysis will be illustrated with the next example (Berger and Wolpert 1988, pp. 74.1–74.2.). A scientist comes to the statistician's office with 100 observations that are assumed to be independent and identically distributed $N(\theta, 1)$. The scientist wants to test the hypothesis $H_0 : \theta = 0$ versus $H_1 : \theta \neq 0$. The sample mean is $\bar{x}_n = 0.2$, so that the test statistic is $z = \sqrt{n}|\bar{x}_n - 0| = 2$. At the level 0.05, a classical statistician could conclude that there is significant evidence against the null hypothesis.

With a more careful consultancy skill, suppose that the classical statistician asks the scientist the reason for stopping experimentation after 100 observations. If the scientist says that she decided to take only a batch of 100 observations, then the statistician could claim significance at the level 0.05. Is that right? From a classical perspective, another important question would be the scientist's attitude towards a non-significant result. Suppose that the scientist says that she would take another batch of 100 observations thus, considering a rule of the form:

(i) Take 100 observations.

(ii) If $\sqrt{100}|\bar{x}_{100}| \geq k$ then stop and reject the null hypothesis.

(iii) If $\sqrt{100}|\bar{x}_{100}| < k$ then take another 100 observations and reject the null hypothesis if $\sqrt{200}|\bar{x}_{200}| \geq k$.

This procedure would have level $\alpha = 0.05$, with $k = 2.18$. Since $\sqrt{100}|\bar{x}_{100}| = 2 < 2.18$, the scientist could not reject the null hypothesis and would have to take additional 100 observations.

This example shows that when using the frequentist approach, the interpretation of the results of an experiment depends not only on the data obtained and the way they were obtained, but also on the experimenter's intentions. Berger and Wolpert comment on the problems faced by frequentist sequential methods:

> Optional stopping poses a significant problem for classical statistics, even when the experimenters are extremely scrupulous. Honest frequentists face the problem of getting extremely convincing data too soon (i.e., before their stopping rule says to stop), and then facing the dilemma of honestly finishing the experiment, even though a waste of time or dangerous to subjects, or of stopping the experiment with the prematurely convincing evidence and then not being able to give frequency measures of evidence. (Berger and Wolpert 1988, pp. 77–78)

15.6.2   Sampling to a foregone conclusion

One of the reasons for the reluctance of frequentist statisticians to accept the stopping rule principle is the concern that when the stopping rule is ignored, investigators using frequentist measures of evidence could be allowed to reach any conclusion they like by sampling until their favorite hypothesis receives enough evidence. To see how that happens, suppose that $x_1, x_2, \ldots$ are sequentially drawn, normally distributed random variables with mean θ and unit variance. An investigator is interested in disproving the null hypothesis that $\theta = 0$. She collects the data sequentially, by performing a fixed sample size test, and stopping when the departure from the null hypothesis is significant at some prespecified level α. In symbols, this means continue sampling until $|\bar{x}_n| > k_\alpha/\sqrt{n}$, where $\bar{x}_n$ denotes the sample mean and k_α is chosen so that the level of the test is α. This stopping rule can be proved to be proper.

Thus, almost surely she will reject the null hypothesis. This may take a long time, but it is guaranteed to happen. We refer to this as "sampling to a foregone conclusion."

Kadane *et al.* (1996) examine the question of whether the expected utility theory, which leads to not using the stopping rule in reaching a terminal decision, is also prone to sampling to a foregone conclusion. They frame the question as follows: can we find, within the expected utility paradigm, a stopping rule such that, given the experimental data, the posterior probability of the hypothesis of interest is necessarily greater than its prior probability?

One thing they discover is that sampling to foregone conclusions is indeed possible when using improper priors that are finitely, but not countably, additive (Billingsley 1995). This means trouble for all the decision rules that, like minimax, are Bayes only when one takes limits of priors:

> Of course, there is a . . . perspective which also avoids foregone conclusions. This is to prescribe the use of merely finitely additive probabilities altogether. The cost here would be an inability to use improper priors. These have been found to be useful for various purposes, including reconstructing some basic "classical" inferences, affording "minimax" solutions in statistical decisions when the parameter space is infinite, approximating "ignorance" when the improper distribution is a limit of natural conjugate priors, and modeling what appear to be natural states of belief. (Kadane *et al.* 1996, p. 1235)

By contrast, though, if all distributions involved are countably additive, then, as you may remember from Chapter 10, the posterior distribution is a martingale, so a priori we expect it to be stable. Consider any real-valued function h. Assuming that the expectations involved exist, it follows from the law of total probability that

$$E_\theta[h(\theta)] = E_x[E_\theta[h(\theta)|x]].$$

So there can be no experiment designed to drive up or drive down for sure the conditional expectation of h, given x. Let $h = I_{\{\theta \in \Theta_0\}}$ denote the indicator that θ belongs to Θ_0, the subspace defined by some hypothesis H_0, and let $E_\theta[h(\theta)] = \pi(H_0) = \pi_0$. Suppose that $x_1, x_2, \ldots$ are sequentially observed. Consider a design with a minimum sample of size $k \geq 0$ and define the stopping time $N = \inf\{n \geq k : \pi(H_0|x^n) \geq q\}$, with $N = \infty$ if the set is empty. This means that sampling stops when the posterior probability is at least q. Assume that $q > \pi_0$. Then

$$\pi_0 = \pi(H_0) \geq \pi(H_0, N < \infty)$$

$$= \sum_{n=k}^{\infty} P(N = n)\pi(H_0|N = n)$$

$$= \sum_{n=k}^{\infty} P(N = n) \int_{x^n:N=n} \pi(H_0|x^n) dF(x^n|n)$$

$$\geq \sum_{n=k}^{\infty} P(N = n) \int_{x^n:N=n} q \, dF(x^n|n)$$

$$= qP(N < \infty),$$

where $F(x^n|n)$ is the cumulative distribution of x^n given $N = n$. From the inequality above, $P(N < \infty) \leq \pi_0/q < 1$. Then, with probability less than 1, a Bayesian stops the sequence of experiments and concludes that the posterior probability of hypothesis H_0 is at least q. Based on this result, a foregone conclusion cannot be reached for sure.

Now that we have reached the conclusion that our Bayesian journey could not be rigged to reach a foregone conclusion, we may stop.

15.7   Exercises

Problem 15.1 (Berger 1985)   This is an application of optimal stopping in an estimation problem. Assume that $x_1, x_2, \ldots$ is a sequential sample from a Bernoulli distribution with parameter θ. We want to estimate θ when the utility is $u(\theta, a, n) = -(\theta - a)^2 - nc$. Assume a priori that θ has a uniform distribution on $(0,1)$. The cost of sampling is $c = 0.01$. Find the Bayes three-truncated procedure d^3.

Solution
   First, we will state without proof some basic results:

   (i) The posterior distribution of θ given x^n is a $Beta(\sum_{i=1}^{n} x_i + 1, n - \sum_{i=1}^{n} x_i + 1)$.

   (ii) Let $y_n = \sum_{i=1}^{n} x_i$. The marginal distribution of x^n is given by

$$m(x^n) = \frac{\Gamma(y_n + 1)\Gamma(n - y_n + 1)}{\Gamma(n + 2)}.$$

   (iii) The predictive distribution of x_{n+1} given x^n is

$$m(x_{n+1}|x^n) = \frac{m(x^{n+1})}{m(x^n)} = \begin{cases} \dfrac{n + 1 - y_n}{n + 2} & \text{if } x_{n+1} = 0, \\ \dfrac{y_n + 1}{n + 2} & \text{if } x_{n+1} = 1. \end{cases}$$

Now, since the terminal utility is $u(a(\theta)) = -(\theta - a)^2$, the optimal action a^* is the posterior mean, and the posterior expected utility is minus the posterior variance. Therefore,

$$W_0(\pi_{x^n}, n) = \sup_a \left[\int_{\Theta} (u(a(\theta)) - c(n)) \, \pi_{x^n}(\theta)d\theta \right] = -\text{Var}[\theta|x^n] - 0.01n.$$

Using this result,

$$W_0(\pi, 0) = -\mathrm{Var}[\theta] = -1/12 \approx -0.0833,$$
$$W_0(\pi_{x^1}, 1) = -\mathrm{Var}[\theta|x^1] - 0.01 = -2/36 - 0.01 \approx -0.0655,$$
$$W_0(\pi_{x^2}, 2) = \begin{cases} -0.0700, & \text{if } y_2 = 1 \\ -0.0575, & \text{if } y_2 = 0 \text{ or } y_2 = 2, \end{cases}$$
$$W_0(\pi_{x^3}, 3) = \begin{cases} -0.0700, & \text{if } y_3 = 1 \text{ or } y_3 = 2 \\ -0.0567, & \text{if } y_3 = 0 \text{ or } y_3 = 3. \end{cases}$$

Observe that

$$W_j(\pi_{x^n}, n) = \max\left\{ W_0(\pi_{x^n}, n), E_{x_{n+1}|x^n}[W_{j-1}(\pi_{x^{n+1}}, n+1)] \right\}.$$

Note that

$$E_{x_1}[\mathrm{Var}[\theta|x^1]] = m(0)\mathrm{Var}[\theta|x_1 = 0] + m(1)\mathrm{Var}[\theta|x_1 = 1]$$
$$= (1/2)(2/36) + (1/2)(2/36) = 2/36 = 1/18$$
$$E_{x_2|x^1}[\mathrm{Var}[\theta|x^2]] = m(0|x^1)\mathrm{Var}[\theta|x^1, x_2 = 0] + m(1|x^1)\mathrm{Var}[\theta|x^1, x_2 = 1]$$
$$= \begin{cases} (2/3)(3/80) + (1/3)(4/80) = 1/24, & \text{if } y_1 = 0 \\ (1/3)(4/80) + (2/3)(3/80) = 1/24, & \text{if } y_1 = 1, \end{cases}$$

and, similarly,

$$E_{x_3|x^2}[\mathrm{Var}[\theta|x^3]] = \begin{cases} 18/600, & \text{if } y_2 = 0 \text{ or } y_2 = 2 \\ 6/150, & \text{if } y_2 = 1, \end{cases}$$

Thus,

$$W_1(\pi, 0) = \max\{W_0(\pi, 0), E_{x_1}[W_0(\pi_{x^1}, 1)]\}$$
$$= \max\{-\mathrm{Var}(\theta), -E_{x_1}[\mathrm{Var}[\theta|x^1]] - c\}$$
$$= \max\{-1/12, -(1/18 + 0.01)\} \approx -0.0655$$
$$W_1(\pi_{x^1}, 1) = \max\{W_0(\pi_{x^1}, 1), E_{x_2|x_1}[W_0(\pi_{x^2}, 2)]\}$$
$$= \max\{W_0(\pi_{x^1}, 1), -E_{x_2|x_1}[\mathrm{Var}[\theta|x^1, x_2]] - 2 \times c\}$$
$$= \max\{-0.0655, -0.0617\} = -0.0617$$
$$W_1(\pi_{x^2}, 2) = \max\{W_0(\pi_{x^2}, 2), E_{x_3|x^2}[W_0(\pi_{x^3}, 3)]\}$$
$$= \max\{W_0(\pi_{x^2}, 2), -E_{x_3|x^2}[\mathrm{Var}[\theta|x^2, x_3]] - 3 \times c\}$$
$$= \begin{cases} \max\{-0.0575, -0.0600\} = -0.0575, & \text{if } y_2 = 0 \text{ or } y_2 = 2 \\ \max\{-0.0700, -0.0700\} = -0.0700, & \text{if } y_2 = 1. \end{cases}$$

Proceeding similarly, we find,

$$W_2(\pi,0) = \max\{W_0(\pi,0), E_{x_1}[W_1(\pi_{x^1},1)]\} = -0.0617$$
$$W_2(\pi_{x^1},1) = \max\{W_0(\pi_{x^1},1), E_{x_2|x_1}[W_1(\pi_{x^2},2)]\} = -0.0617$$
$$W_3(\pi,0) = \max\{W_0(\pi,0), E_{x_1}[W_2(\pi_{x^1},1)]\} = -0.0617.$$

We can now describe the optimal three-truncated Bayes procedure. Since at stage $0, W_0(\pi,0) = -0.083 < W_3(\pi,0) = -0.0617, x_1$ should be observed. After observing x_1, as $W_0(\pi_{x^1},1) = -0.0655 < W_2(\pi_{x^1},1) = -0.0617$, we should observe x_2. After observing it, since we have $W_0(\pi_{x^2},2) = W_1(\pi_{x^2},2)$, it is optimal to stop. The Bayes action is

$$\delta_2^*(x^2) = \begin{cases} 1/4 \text{ if } y_2 = 0 \\ 1/2 \text{ if } y_2 = 1 \\ 3/4 \text{ if } y_2 = 2. \end{cases}$$

□

Problem 15.2   Sequential problems get very hard very quickly. This is one of the simplest you can try, but it will take some time. Try to address the same question as the worked example in Problem 15.1, except with Poisson observations and a *Gamma*$(1,1)$ prior. Set $c = 1/12$.

Problem 15.3   (From DeGroot 1970) Let π be the prior distribution of θ in a sequential decision problem. (a) Prove by using an inductive argument that, in a sequential decision problem with fixed cost per observation, if $V_0(\pi_{x^n})$ is constant for all observed values x^n, then every function V_j for $j = 1, 2, \ldots$ as well as V_0 has this property. (b) Conclude that the optimal sequential decision procedure is a procedure in which a fixed number of observations are taken.

Problem 15.4   (From DeGroot 1970) (a) In the context of Example 15.4.2, for $n = 0, 1, \ldots, j$, prove that

$$W_j(\pi,n) = -\frac{1}{h+n\omega} - nc \quad \text{for} \quad \beta_n < \omega \le \beta_{n+1}$$

where β_n is defined as $\beta_0 = 0, \beta_n = c(h + (n-1)\omega)(h + n\omega)$ for $n = 1, \ldots, j$ and $\beta_{j+1} = \infty$. (b) Conclude that if the decision maker is allowed to take no more than k observations, then the optimal procedure is to take $n^* \le k$ observations where n^* is such that $\beta_{n^*} < \omega \le \beta_{n^*+1}$.

Appendix

A.1 Notation

Notation	Explanation	Chapter
$\succ, \prec$	preference relations	2
$\sim$	equivalence relation	2
$\mathcal{Z}$	set of outcomes, rewards	3
z	outcome, generic element of $\mathcal{Z}$	3
z^0	best outcome	3
z_0	worst outcome	3
$z_{i_1 i_2 \ldots i_S 0k}$	outcome from actions $a_{i_1}^{(1)}, \ldots, a_{i_S}^{(s)}, a_0^{(s)}$ when the state of nature is $\theta_{0k}^{(s)}$	12
$z_{i_1 i_2 \ldots i_S k}$	outcome from actions $a_{i_1}^{(1)}, \ldots, a_{i_S}^{(S)}$ when the state of nature is $\theta_{i_S k}^{(s)}$	12
Θ	set of states of nature, parameter space	3
θ	event, state of nature, parameter, generic element of Θ, scalar	2
$\boldsymbol{\theta}$	state of nature, parameter, generic element of Θ, vector	7
$\theta_{0k}^{(s)}$	generic state of nature upon taking stopping action $a_0^{(s)}$ at stage s	12
$\theta_{i_S k}^{(S)}$	generic state of nature at last stage S upon taking action $a_{i_S}^{(S)}$	12
π	subjective probability	2
π_θ	price, relative to stake S, for a bet on θ	2
$\pi_{\theta_1 \mid \theta_2}$	price of a bet on θ_1 called off if θ_2 does not occur	2
π_θ^0	current probability assessment of θ	2

(continued overleaf)

Decision Theory: Principles and Approaches   G. Parmigiani, L. Y. T. Inoue
© 2009 John Wiley & Sons, Ltd

(Continued)

Notation	Explanation	Chapter
π_θ^T	future probability assessment of θ	2
π^M	least favorable prior	7
S_θ	stake associated with event θ	2
g_θ	net gains if event θ occurs	2
$\mathcal{A}$	action space	3
a	action, generic element of $\mathcal{A}$	3
$a(\theta)$	VNM action, function from states to outcomes	3
$\bar{z}$	expected value of the rewards for a lottery a	4
$z^*(a)$	certainty equivalent of lottery a	4
$\lambda(z)$	Arrow–Pratt (local) measure of risk aversion at z	4
P	set of probability functions on $\mathcal{Z}$	6
a	AA action, horse lottery, list of VNM lotteries $a(\theta)$ in P for $\theta \in \Theta$	6
$a(\theta, z)$	probability that lottery $a(\theta)$ assigns to outcome z	6
$a_{S,\theta}$	action of betting stake S on event θ	2
a_θ	action that would maximize decision maker's utility if θ was known	13
$\boldsymbol{a}$	action, vector	7
$a_{i_s}^{(s)}$	generic action, indexed by i_s, at stage s of a multistage decision problem	12
$a_0^{(s)}$	stopping action at stage s of a multistage decision problem	12
a^M	minimax action	7
a^*	Bayes action	3
$p(z)$	probability of outcome z	3
$\boldsymbol{p}$	lottery or gamble, probability distribution over $\mathcal{Z}$	3
χ_z	degenerate action with mass 1 at reward z	3
u	utility	3
$u(z)$	utility of outcome z	3
$u(a(\theta))$	utility of outcome $a(\theta)$	3
$u_\theta(z)$	state-dependent utility of outcome z	6
$\mathcal{S}(\boldsymbol{q})$	expected score of the forecast probability $\boldsymbol{q}$	10
$s(\theta, \boldsymbol{q})$	scoring rule for the distribution $\boldsymbol{q}$ and event θ	10
$RP(a)$	risk premium associated with action a	4
$\mathcal{X}$	sample space	7
x	random variable or observed outcome	7
x^n	random sample $(x_1, \ldots, x_n)$	14
$\boldsymbol{x}$	multivariate random sample	7
$x_i^{(s)}$	random variable with possible values $x_{i1}, \ldots, x_{iJ_s}$ observed at stage s, upon taking continuation action $a_{i_s}^{(s)}, i > 0$	12

(Continued)

Notation	Explanation	Chapter
$f(x\|\theta)$	probability density function of x or likelihood function	7
$m(x)$	marginal density function of x	7
$F(x\|\theta)$	distribution function of x	7
$\pi(\theta)$	prior probability of θ; may indicate a density if θ is continuous	7
$\pi(\theta\|x), \pi_x$	posterior density of θ given x	7
$E_x[g(x)]$	expectation of the function $g(x)$ with respect to $m(x)$	7
$E_{x\|\theta}[g(x)], E[g(x)\|\theta]$	expectation of the function $g(x)$ with respect to $f(x\|\theta)$	7
$\mathcal{U}_\pi(a)$	expected utility of action a, using prior π	3
$\mathcal{U}_\pi(d)$	expected utility of the sequential decision procedure d	15
δ	decision rule, function with domain $\mathcal{X}$ and range $\mathcal{A}$	7
δ	multivariate decision rule	7
δ^M	minimax decision rule	7
δ^*	Bayes decision rule	7
$\delta^R(x, \cdot)$	randomized decision rule for a given x	7
δ_n	terminal decision rule after n observations x^n	15
δ	sequence of decision rules $\delta_1(x^1), \delta_2(x^2), \ldots$	15
ζ_n	stopping rule after n observations x^n	15
ζ	sequence of stopping rules $\zeta_0, \zeta_1(x^1), \ldots$	15
d	sequential decision rule: pair (δ, ζ)	15
$L(\theta, a)$	loss function (in regret form)	7
$L_u(\theta, a)$	loss function as the negative of the utility function	7
$L(\theta, a, n)$	loss function for n observations	14
$u(\theta, a, n)$	utility function for n observations	14
$L(\theta, \delta^R(x))$	loss function for randomized decision rule δ^R	7
$\mathcal{L}_\pi(a)$	prior expected loss for action a	7
$\mathcal{L}_{\pi_x}(a)$	posterior expected loss for action a	7
$U(\theta, d)$	average utility of the sequential procedure d for given θ	15
$R(\theta, \delta)$	risk function of decision rule δ	7
$V(\pi)$	maximum expected utility with respect to prior π	13
$W_0(\pi_{x^n}, n)$	posterior expected utility, at time n, of colleting 0 additional observations	15
$W_{k-n}(\pi_{x^n}, n)$	posterior expected utility, at time n, of continuing for at most an additional $k - n$ steps	15

(continued overleaf)

(Continued)

Notation	Explanation	Chapter
$r(\pi, \delta)$	Bayes risk associated with prior π and decision δ	7
$r(\pi, n)$	Bayes risk adopting the optimal terminal decision with n observations	14
$\mathcal{D}$	class of decision rules	7
$\mathcal{D}^R$	class of all (including randomized) decision rules	7
$\mathcal{E}^\infty$	perfect experiment	13
$\mathcal{E}$	generic statistical experiment	13
$\mathcal{E}_{12}$	statistical experiment consisting of observing both variables x_1 and x_2	13
$\mathcal{E}^{(n)}$	statistical experiment consisting of experiments $\mathcal{E}_1, \mathcal{E}_2, \ldots, \mathcal{E}_n$ where $\mathcal{E}_2, \ldots, \mathcal{E}_n$ are conditionally i.i.d. repetitions of $\mathcal{E}_1$	13
$\mathcal{V}_\theta(\mathcal{E}^\infty)$	conditional value of perfect information for a given θ	13
$\mathcal{V}(\mathcal{E}^\infty)$	expected value of perfect information	13
$\mathcal{V}_x(\mathcal{E})$	observed value of information for a given x in experiment $\mathcal{E}$	13
$\mathcal{V}(\mathcal{E})$	expected value of information in the experiment $\mathcal{E}$	13
$\mathcal{V}(\mathcal{E}_{12})$	expected value of information in the experiment $\mathcal{E}_{12}$	13
$\mathcal{V}(\mathcal{E}_2|\mathcal{E}_1)$	expected information of x_2 conditional on observed x_1	13
$\mathcal{I}_x(\mathcal{E})$	observed (Lindley) information provided by observing x in experiment $\mathcal{E}$	13
$\mathcal{I}(\mathcal{E})$	expected (Lindley) information provided by the experiment $\mathcal{E}$	13
$\mathcal{I}(\mathcal{E}_{12})$	expected (Lindley) information provided by the experiment $\mathcal{E}_{12}$	13
$\mathcal{I}(\mathcal{E}_2|\mathcal{E}_1)$	expected (Lindley) information provided by E_2 conditional on E_1	13
c	cost per observation	14
$C(n)$	cost function	14
δ_n^*	Bayes rule based on n observations	14
$\mathcal{U}_\pi(n)$	expected utility of making n observations and adopting the optimal terminal decision	14
n^*	Bayesian optimal sample size	14
n^M	Minimax optimal sample size	14
η	nuisance parameter	7
$\Phi(.)$	cumulative distribution function of the $N(0,1)$	8
$\mathcal{M}$	class of parametric models in a decision problem	11
M	generic model within $\mathcal{M}$	11
$\mathcal{H}_s$	history set at stage s of a multistage decision problem	12

A.2   Relations

$$p(z) = \int_{\theta:a(\theta)=z} \pi(\theta)d\theta \tag{3.3}$$

$$\bar{z} = \sum_z z p(z) \tag{4.4}$$

$$z^*(a) = u^{-1}\left(\sum_z u(z)p(z)\right) \tag{4.9}$$

$$RP(a) = \bar{z} - z^*(a) \tag{4.10}$$

$$\lambda(z) = -u''(z)/u'(z) = -(d/dz)(\log u'(z)) \tag{4.13}$$

$$L_u(\theta,a) = -u(a(\theta)) \tag{7.1}$$

$$L(\theta,a) = L_u(\theta,a) - \inf_{a\in\mathcal{A}} L_u(\theta,a) \tag{7.2}$$

$$= \sup_{a'(\theta)} u(a'(\theta)) - u(a(\theta)) \tag{7.3}$$

$$\mathcal{L}_\pi(a) = \int_\Theta L(\theta,a)\pi(\theta)d\theta \tag{7.6}$$

$$\mathcal{L}_{\pi_x}(a) = \int_\Theta L(\theta,a)\pi(\theta|x)d\theta \tag{7.13}$$

$$\mathcal{U}_\pi(a) = \int_\Theta u(a(\theta))\pi(\theta)d\theta \tag{3.4}$$

$$= \sum_{z\in\mathcal{Z}} p(z)u(z) \tag{3.5}$$

$$\mathcal{S}(\boldsymbol{q}) = \sum_{j=1}^J s(\theta_j,\boldsymbol{q})\pi_j \tag{10.1}$$

$$V(\pi) = \sup_a \mathcal{U}_\pi(a) \tag{13.2}$$

$$a^M = \operatorname{argmin} \max_\theta L(\theta,a) \tag{7.4}$$

$$a^* = \operatorname{argmax}\, \mathcal{U}_\pi(a) \tag{3.6}$$

$$= \operatorname{argmin} \int_\Theta L(\theta,a)\pi(\theta)d\theta \tag{7.5}$$

$$a_\theta = \arg\sup_{a\in\mathcal{A}} u(a(\theta)) \tag{13.4}$$

$$m(x) = \int_\Theta \pi(\theta)f(x|\theta)d\theta \tag{7.12}$$

$$\pi(\theta|x) = \pi(\theta)f(x|\theta)/m(x) \tag{7.11}$$

$$L(\theta,\delta^R(x)) = E_{\delta^R(x,.)}L(\theta,a) = \int_{a\in\mathcal{A}} L(\theta,a)\delta^R(x,a)da \tag{7.16}$$

$$R(\theta,\delta) = \int_{\mathcal{X}} L(\theta,\delta)f(x|\theta)dx \tag{7.7}$$

$$r(\pi,\delta) = \int_\Theta R(\theta,\delta)\pi(\theta)d\theta \tag{7.9}$$

$$\delta^M \quad\text{s.t. } \sup_\theta R(\theta,\delta^M) = \inf_\delta \sup_\theta R(\theta,\delta) \tag{7.8}$$

$$\delta^* \quad\text{s.t. } r(\pi,\delta^*) = \inf_\delta r(\pi,\delta) \tag{7.10}$$

$$\mathcal{V}_\theta(\mathcal{E}^\infty) = u(a_\theta(\theta)) - u(a^*(\theta)) \tag{13.5}$$

$$\mathcal{V}(\mathcal{E}^\infty) = E_\theta[\mathcal{V}_\theta(\mathcal{E}^\infty)] \tag{13.6}$$

$$= E_\theta\left[\sup_a u(a(\theta))\right] - V(\pi) \tag{13.7}$$

$$\mathcal{V}_x(\mathcal{E}) = V(\pi_x) - \mathcal{U}_{\pi_x}(a^*) \tag{13.9}$$

$$\mathcal{V}(\mathcal{E}) = E_x[\mathcal{V}_x(\mathcal{E})] = E_x[V(\pi_x) - \mathcal{U}_{\pi_x}(a^*)] \tag{13.10}$$

$$= E[\mathcal{V}_x(\mathcal{E})] = E_x[V(\pi_x)] - V(\pi) \tag{13.12}$$

$$= E_x[V(\pi_x)] - V(E_x[\pi_x]) \tag{13.13}$$

$$\mathcal{V}(\mathcal{E}_2|\mathcal{E}_1) = E_{x_1 x_2}[V(\pi_{x_1 x_2})] - E_{x_1}[V(\pi_{x_1})] \tag{13.17}$$

$$\mathcal{V}(\mathcal{E}_{12}) = E_{x_1 x_2}[V(\pi_{x_1 x_2})] - V(\pi) \tag{13.16}$$

$$= \mathcal{V}(\mathcal{E}_1) + \mathcal{V}(\mathcal{E}_2|\mathcal{E}_1) \tag{13.18}$$

$$\mathcal{I}(\mathcal{E}) = \int_{\mathcal{X}} \int_\Theta \log(\pi_x(\theta)/\pi(\theta))\,\pi_x(\theta)m(x)d\theta dx \tag{13.27}$$

$$= \int_\Theta \int_{\mathcal{X}} \log(f(x|\theta)/m(x))f(x|\theta)\pi(\theta)dx d\theta \tag{13.28}$$

$$= E_x\left[E_{\theta|x}\left[\log(\pi_x(\theta)/\pi(\theta))\right]\right] \tag{13.29}$$

$$= E_x\left[E_{\theta|x}\left[\log(f(x|\theta)/m(x))\right]\right] \tag{13.30}$$

$$\mathcal{I}(\mathcal{E}_2|\mathcal{E}_1) = E_{x_1}E_{x_2|x_1}E_{\theta|x_1,x_2}\left[\log(f(x_2|\theta,x_1)/m(x_2|x_1))\right] \tag{13.33}$$

$$\mathcal{I}(\mathcal{E}_{12}) = \mathcal{I}(\mathcal{E}_1) + \mathcal{I}(\mathcal{E}_2|\mathcal{E}_1) \tag{13.31}$$

$$\mathcal{I}_x(\mathcal{E}) = \int_\Theta \pi_x(\theta)\log(\pi_x(\theta)/\pi(\theta))\,d\theta \tag{13.26}$$

$$u(a, \theta, n) = u(a(\theta)) - C(n) \tag{14.1}$$
$$L(\theta, a, n) = L(\theta, a) + C(n) \tag{14.2}$$
$$r(\pi, n) = r(\pi, \delta_n^*) + C(n) \tag{14.3}$$
$$= \int_\mathcal{X} \int_\Theta L(\theta, \delta_n^*(x^n), n) \, \pi(\theta|x^n) d\theta \; m(x^n) dx^n \tag{14.4}$$
$$\mathcal{U}_\pi(n) = \int_\mathcal{X} \int_\Theta u(\theta, \delta_n^*(x^n), n) \, \pi(\theta|x^n) d\theta \; m(x^n) dx^n \tag{14.5}$$

A.3 Probability (density) functions of some distributions

1. If $x \sim Bin(n, \theta)$ then

$$f(x|\theta) = \binom{n}{x} \theta^x (1 - \theta)^{n-x}.$$

2. If $x \sim N(\theta, \sigma^2)$ then

$$f(x|\theta, \sigma^2) = \frac{1}{\sqrt{2\pi}\sigma} \exp\left(-\frac{(x - \theta)^2}{2\sigma^2}\right).$$

3. If $x \sim Gamma(\alpha, \beta)$ then

$$f(x|\alpha, \beta) = \frac{\beta^\alpha}{\Gamma(\alpha)} x^{\alpha-1} e^{-\beta x}.$$

4. If $x \sim Exp(\theta)$ then

$$f(x|\theta) = \theta e^{-\theta x}.$$

5. If $x \sim Beta - Binomial(n, \alpha, \beta)$ then

$$m(x) = \frac{\Gamma(\alpha + \beta)\Gamma(n + 1)\Gamma(\alpha + x)\Gamma(\beta + n - x)}{\Gamma(\alpha)\Gamma(\beta)\Gamma(x + 1)\Gamma(n - x + 1)\Gamma(\alpha + \beta + n)}.$$

6. If $x \sim Beta(\alpha, \beta)$ then

$$m(x) = \frac{\Gamma(\alpha + \beta)}{\Gamma(\alpha)\Gamma(\beta)} x^\alpha (1 - x)^\beta.$$

A.4 Conjugate updating

In the following description, prior hyperparameters have subscript 0, while parameters of the posterior distributions have subscript x.

1. Sampling model: binomial
   (a) Data: $x|\theta \sim Bin(n, \theta)$.
   (b) Prior: $\theta \sim Beta(\alpha_0, \beta_0)$.

(c) Posterior: $\theta|x \sim Beta(\alpha_x, \beta_x)$ where

$$\alpha_x = \alpha_0 + x \ \text{ and } \ \beta_x = \beta_0 + n - x.$$

(d) Marginal: $x \sim BetaBinomial(n, \alpha_0, \beta_0)$.

2. Sampling model: normal

(a) Data: $x \sim N(\theta, \sigma^2)$ (σ^2 known).

(b) Prior: $\theta \sim N(\mu_0, \tau_0^2)$.

(c) Posterior: $\theta|x \sim N(\mu_x, \tau_x^2)$ where

$$\mu_x = \frac{\sigma^2}{\sigma^2 + \tau_0^2}\mu_0 + \frac{\tau_0^2}{\sigma^2 + \tau_0^2}x \ \text{ and } \ \tau_x^2 = \frac{\tau_0^2\sigma^2}{\sigma^2 + \tau_0^2}.$$

(d) Marginal: $x \sim N(\mu_0, \sigma^2 + \tau_0^2)$.

References

Adcock, C. J. (1997). Sample size determination: a review, *The Statistician* **46**: 261–283.

Akaike, H. (1973). Information theory and an extension of the maximum likelihood principle, *2nd International Symposium on Information Theory*, pp. 267–281.

Alam, K. (1979). Estimation of multinomial probabilities, *Annals of Statistics* **7**: 282–283.

Allais, M. (1953). Le comportement de l'homme rationnel devant le risque: critique des postulats et axiomes de l'école américaine, *Econometrica* **21**: 503–546.

Anscombe, F. J. and Aumann, R. J. (1963). A definition of subjective probability, *Annals of Mathematical Statistics* **34**: 199–205.

Arnold, S. F. (1981). *The theory of linear models and multivariate analysis*, John Wiley & Sons, Ltd, Chichester.

Arrow, K., Blackwell, D. and Girshick, M. (1949). Bayes and minimax solutions of sequential decision problems, *Econometrica* **17**: 213–244.

Bahadur, R. R. (1955). A characterization of sufficiency, *Annals of Mathematical Statistics* **26**(2): 286–293.
URL: *http://www.jstor.org/stable/2236883*

Baranchik, A. J. (1970). A family of minimax estimators of the mean of a multivariate normal distribution, *Annals of Mathematical Statistics* **41**: 642–645.

Baranchik, A. J. (1973). Inadmissibility of maximum likelihood estimators in some multiple regression problems with three or more independent variables, *Annals of Statistics* **1**: 312–321.

Barry, M. J., Cantor, R. and Clarke, J. R. (1986). *Basic Diagnostic Test Evaluation*, Society for Medical Decision Making, Washington, DC.

Basu, D. (1975). On the elimination of nuisance parameters, p. 38.

Bather, J. (2000). *Decision Theory: an introduction to dynamic programming and sequential decisions*, John Wiley & Sons, Ltd, Chichester.

Bellman, R. (1984). *Eye of the Hurricane*, World Scientific, Singapore.

Bellman, R. E. (1957). *Dynamic programming*, Princeton University Press.

Bellman, R. E. and Dreyfus, S. E. (1962). *Applied dynamic programming*, Princeton University Press.

Berger, J. O. and Berry, D. A. (1988). The relevance of stopping rules in statistical inference (with discussion), *in* J. Berger and S. Gupta (eds.), *Statistical Decision Theory and Related Topics IV*, Vol. 1, Springer-Verlag, New York, pp. 29–72.

Berger, J. O. (1985). *Statistical Decision Theory and Bayesian Analysis*, second ed., Springer-Verlag, New York.

Berger, J. O. and Delampady, M. (1987). Testing precise hypotheses, *Statistical Science* **2**: 317–335.

Berger, J. O. and Wolpert, R. L. (1988). *The Likelihood Principle*, second ed., Institute of Mathematical Statistics, Hayward, CA.

Bernardo, J. M. and Smith, A. F. M. (1994). *Bayesian Theory*, John Wiley & Sons, Ltd, Chichester.

Bernardo, J. M. and Smith, A. F. M. (2000). *Bayesian Theory*, John Wiley & Sons, Ltd, Chichester.

Bernoulli, D. (1738). Specimen theoriae novae de mensura sortis, *Commentarii academiae scientiarum imperialis Petropolitanae* **5**: 175–192.

Bernoulli, D. (1954). Exposition of a new theory on the measurement of risk, *Econometrica* **22**: 23–36.

Berry, D. A. (1987). Interim analysis in clinical trials: the role of the likelihood principle, *American Statistician* **41**: 117–122.

Berry, D. A. and Eick, S. (1995). Adaptive assignment versus balanced randomization in clinical trials: a decision analysis, *Statistics in Medicine* **14**: 231–246.

Berry, D. A., Müller, P., Grieve, A., Smith, M., Parke, T., Blazek, R., Mitchard, N. and Krams, M. (2001). Adaptive Bayesian designs for dose-ranging drug trials, *in* C. Gatsonis, B. Carlin and A. Carriquiry (eds.), *Case Studies in Bayesian Statistics – volume V*, Vol. 5, Springer-Verlag, New York, pp. 99–181.

Berry, D. A. and Fristedt, B. (1985). *Bandit problems: Sequential allocation of experiments*, Chapman & Hall.

Berry, D. A. and Hochberg, Y. (1999). Bayesian perspectives on multiple comparisons, *Journal of Statistical Planning and Inference* **82**: 215–227.

Berry, D. A. and Stangl, D. K. (eds.) (1996). *Bayesian Biostatistics*, Vol. 151 of *Statistics: Textbooks and Monographs*, Marcel Dekker, New York.

Berry, S. and Viele, K. (2008). A note on hypothesis testing with random sample sizes and its relationship to Bayes factors, *Journal of Data Science* **6**: 75–87.

Bielza, C., Müller, P. and Rios Insua, D. (1999). Decision analysis by augmented probability simulations, *Management Science* **45**: 995–1007.

Billingsley, P. (1995). *Probability and Measure,* third ed., John Wiley & Sons, Inc., New York.

Birnbaum, A. (1962). On the foundations of statistical inference (Com: P307-326), *Journal of the American Statistical Association* **57**: 269–306.

Blackwell, D. (1951). Comparison of experiments, *Proceedings of the Second Berkeley Symposium on Mathematical Statistics and Probability*, pp. 93–102.

Blackwell, D. (1953). Equivalent comparison of experiments, *Annals of Mathematical Statistics* **24**: 265–272.

Blackwell, D. and Girshick, M. (1954). *Theory of games and statistical decisions*, Chapman & Hall.

Blyth, C. R. (1951). On minimax statistical decision procedures and their admissibility, *Annals of Mathematical Statistics* **22**: 22–42.

Bonferroni, C. (1924). La media esponenziale in matematica finanziaria, *Annuario del Regio Istituto Superiore di Scienze Economiche e Commerciali di Bari*, Vol. 23–24, pp. 1–14.

Box, G. (1979). *Robustness in Statistics*, Academic Press.

Box, G. and Tiao, G. (1973). *Bayesian Inference in Statistical Analysis*, Addison-Wesley, Reading, MA.

Box, G. E. P. (1980). Sampling and Bayes' inference in scientific modelling and robustness, *Journal of the Royal Statistical Society. Series A (General)* **143**(4): 383–430.
URL: *http://www.jstor.org/stable/2982063*

Brandwein, A. C. and Strawderman, W. E. (1990). Stein estimation: the spherically symmetric case, *Statistical Science* **5**: 356–369.
URL: *http://www.jstor.org/stable/2245823*

Breiman, L. (2001). Statistical modeling: the two cultures, *Statistical Science* **16**: 199–231.

Brier, G. (1950). Verification of forecasts expressed in terms of probability, *Monthly Weather Review* **78**: 1–3.

Brockwell, A. and Kadane, J. B. (2003). A gridding method for Bayesian sequential decision problems, *Journal of Computational & Graphical Statistics* **12**: 566–584.
URL: *http://www. ingentaconnect.com/content/asa/jcgs/2003/00000012/00000003/ art00005*

Brown, L. D. (1976). Notes on statistical decision theory (unpublished lecture notes), *Technical report*, Ithaca, NY.

Brown, L. D. (1975). Estimation with incompletely specified loss functions (the case of several location parameters), *Journal of the American Statistical Association* **70**: 417–427.

Brown, L. D. (1981). A complete class theorem for statistical problems with finite sample spaces, *Annals of Statistics* **9**: 1289–1300.

Carlin, B. P. and Louis, T. (2008). *Bayesian Methods for Data Analysis,* third ed., CRC Press.

Carlin, B. P., Kadane, J. B. and Gelfand, A. E. (1998). Approaches for optimal sequential decision analysis in clinical trials, *Biometrics* **54**: 964–975.

Carlin, B. P. and Polson, N. G. (1991). An expected utility approach to influence diagnostics, *Journal of the American Statistical Association* **86**: 1013–1021.

Carota, C., Parmigiani, G. and Polson, N. G. (1996). Diagnostic measures for model criticism, *Journal of the American Statistical Association* **91**: 753–762.
URL: *http://www.jstor.org/pss/2291670*

Chaloner, K. and Verdinelli, I. (1995). Bayesian experimental design: a review, *Statistical Science* **10**: 273 304.

Chambers, J. M., Cleveland, W. S., Kleiner, B. and Tukey, P. A. (1983). *Graphical Methods for Data Analysis*, Wadsworth, Belmont, CA.

Chapman, G. and Elstein, A. (2000). Cognitive processes and biases in medical decision making, *in* G. Chapman and F. Sonnenberg (eds.), *Decision Making in Health Care: Theory, Psychology, and Applications*, Cambridge University Press, Boston, MA, pp. 183–210.

Chapman, G. B. and Sonnenberg, F. (eds.) (2000). *Decision Making in Health Care: Theory, Psychology, and Applications*, Cambridge University Press, Boston, MA.

Chernoff, H. (1954). Rational selection of a decision function, *Econometrica* **22**: 422–443.

Chernoff, H. and Moses, L. (1959). *Elementary Decision Theory*, John Wiley & Sons, Inc., New York.

Chisini, O. (1929). Sul concetto di media, *Periodico di Matematiche* **9**: 106–116.

Cifarelli, D. M. and Regazzini, E. (1996). De Finetti's contribution to probability and statistics, *Statistical Science* **11**: 253–282.

Clemen, R. T. and Reilly, T. (2001). *Making Hard Decisions*, Duxbury, Pacific Grove, CA.

Cleveland, W. S. and Grosse, E. (1991). Computational methods for local regression, *Statistics and Computing* **1**: 47–62.

Clyde, M. and George, E. (2004). Model uncertainty, *Statistical Science* **19**: 81–94.

Congdon, P. (2001). *Bayesian Statistical Modelling*, John Wiley & Sons, Ltd, Chichester.

Cover, T. M. and Thomas, J. A. (1991). *Elements of information theory*, John Wiley & Sons, Inc., New York.

Cox, D. R. (1958). Some problems connected with statistical inference, *Annals of Mathematical Statistics* **29**: 357–372.

Cox, D. R. (1962). Further results on tests of separate families of hypotheses, *Journal of the Royal Statistical Society. Series B (Methodological)* **24**(2): 406–424.
URL: *http://www.jstor.org/stable/2984232*

Critchfield, G. C. and Willard, K. E. (1986). Probabilistic analysis of decision trees using Monte Carlo simulation, *Medical Decision Making* **6**(2): 85–92.

Davies, K. R., Cox, D. D., Swartz, R. J., Cantor, S. B. and Follen, M. (2007). Inverse decision theory with applications to screening and diagnosis of cervical intraepithelial neoplasia, *Gynecologic Oncology* **107**(1 Suppl. 1): S187–S195.
URL: *http://dx.doi.org/10.1016/j.ygyno.2007.07.053*

Dawid, A. (1982). The well-calibrated Bayesian, *Journal of the American Statistical Association* **77**: 605–613.

de Finetti, B. (1931a). Sul concetto di media, *Giornale dell'Istituto Italiano degli Attuari* **2**: 369–396.

de Finetti, B. (1931b). Sul significato soggettivo della probabilità, *Fundamenta Mathematicae* **17**: 298–329.

de Finetti, B. (1937). Foresight: its logical laws, its subjective sources, *in* H. E. Kyburg and H. E. Smokler (eds.), *Studies in Subjective Probability*, Krieger, New York, pp. 55–118.

de Finetti, B. (1952). Sulla preferibilità, *Giornale degli Economisti e Annali di Economia* **11**: 685–709.

de Finetti, B. (1964a). Probabilità composte e teoria delle decisioni, *Rend Mat* **23**: 128–134.

de Finetti, B. (1964b). Teoria delle decisioni, *Lezioni di Metodologia Statistica per Ricercatori* **6**.

de Finetti, B. (1972). *Probability, Induction and Statistics*, John Wiley & Sons, Ltd, Chichester.

de Finetti, B. (1974). *Theory of Probability: A critical introductory treatment*, John Wiley & Sons, Ltd, Chichester.

Debreu, G. (1959). Topological method in cardinal utiity theory, *in* K. Arrow, S. Karlin and P. Suppes (eds.), *Mathematical Methods in the Social Sciences*, pp. 16–26.

DeGroot, M. H. (1962). Uncertainty, information, and sequential experiments, *Annals of Mathematical Statistics* **33**: 404–419.

DeGroot, M. H. (1970). *Optimal Statistical Decisions*, McGraw-Hill.

DeGroot, M. H. (1984). Changes in utility as information, *Theory and Decision* **17**: 287–303.

DeGroot, M. H. and Fienberg, S. E. (1982). Assessing probability assessors: calibration and refinement, *Statistical Decision Theory and Related Topics III, in two volumes, Volume 1*, pp. 291–314.

DeGroot, M. H. and Fienberg, S. E. (1983). The comparison and evaluation of forecasters, *The Statistician* **32**: 12–22.

Desu, M. and Raghavarao, D. (1990). *Sample Size Methodology*, Academic Press, New York.

Dodge, H. and Romig, H. (1929). A method of sampling inspection, *Bell System Technical Journal* **8**: 613–631.

Doubilet, P., Begg, C. B., Weinstein, M. C., Braun, P. and McNeil, B. J. (1985). Probabilistic sensitivity analysis using Monte Carlo simulation: a practical approach, *Medical Decision Making* **5(2)**: 157–177.

Draper, D. (1995). Assessment and propagation of model uncertainty (disc: P71-97), *Journal of the Royal Statistical Society, Series B, Methodological* **57**: 45–70.

Dreyfus, S. (2002). Richard Bellman on the birth of dynamic programming, *Operations Research* **50**(1): 48–51.
    URL: *http://www.eng.tau.ac.il/ ami/cd/or50/1526-5463-2002-50-01-0048.pdf*

Drèze, J. (1987). *Essays on economic decisions under uncertainty*, Cambridge University Press, New York.

Duda, R. O., Hart, P. E. and Stork, D. G. (2000). *Pattern Classification*, Wiley-Interscience.

Duncan, D. (1965). A Bayesian approach to multiple comparison, *Technometrics* pp. 171–122.

Edgeworth, F. Y. (1887). The method of measuring probability and utility, *Mind* **12** (47): 484–488.
    URL: *http://www.jstor.org/stable/2247259*

Efron, B. (1978). Controversies in the foundations of statistics, *American Mathematical Monthly* **85**: 231–246.

Efron, B. (1986). Double exponential families and their use in generalized linear regression, *Journal of the American Statistical Association* **82**: 709–721.

Efron, B. and Morris, C. (1973a). Combining possibly related estimation problems (with discussion), *Journal of the Royal Statistical Society, Series B, Methodological* **35**: 379–421.

Efron, B. and Morris, C. (1973b). Stein's estimation rule and its competitors – an empirical Bayes approach, *Journal of the American Statistical Association* **68**: 117–130.

Ellsberg, D. (1961). Risk, ambiguity and the savage axioms, *Quarterly Journal of Economics* **75**: 643–669.

Ewens, W. and Grant, G. (2001). *Statistical Methods in Bioinformatics*, Springer-Verlag, New York.

Ferguson, T. S. (1967). *Mathematical Statistics: A Decision–Theoretic Approach*, Academic Press, New York.

Ferguson, T. S. (1976). Development of the decision model, *On the History of Statistics and Probability*, pp. 333–346.

Ferreira, M. A. R. and Lee, H. K. H. (2007). *Multiscale Modeling: A Bayesian Perspective*, Springer-Verlag.

Fienberg, S. E. (1992). A brief history of statistics in three and one-half chapters: a review essay, *Statistical Science* **2**: 208–225.

Fishburn, P. C. (1986). The axioms of subjective probability, *Statistical Science*.

Fishburn, P. C. (1989). Retrospective on the utility theory of von Neumann and Morgenstern, *Journal of Risk and Uncertainty*, pp. 127–158.

Fishburn, P. C. (1970). *Utility theory for decision making*, John Wiley & Sons, Inc., New York.

Fishburn, P. C. (1981). Subjective expected utility: a review of normative theories, *Theory and Decision* **13**: 139–199.

Fishburn, P. C. (1982). *The Foundations of Expected Utility*, Reidel, Dordrecht.

Fishburn, P. C. (1983). Ellsberg revisited: a new look at comparative probability, *Annals of Statistics* **11**: 1047–1059.

Fisher, R. A. (1925). Theory of statistical estimation, *Proceedings of the Cambridge Philosophical Society* **22**: 700–725.

Fisher, R. A. (1965). Commentary and assessment, *in* J. H. Bennett (ed.), *Experiments in Plant Hybridisation*, Oliver and Boyd.

French, S. (1988). *Decision theory: an introduction to the mathematics of rationality*, Ellis Horwood, Chichester.

Gail, M. H. (2008). Discriminatory accuracy from single-nucleotide polymorphisms in models to predict breast cancer risk, *Journal of the National Cancer Institute* **100**(14): 1037–1041.
    URL: *http://dx.doi.org/10.1093/jnci/djn180*

Galton, F. (1888). Co-relations and their measurement, chiefly from anthropometric data, *Proceedings of the Royal Society of London* **45**: 135–145.
    URL: *http://www.jstor.org/stable/114860*

Gärdenfors, P. and Sahlin, N.-E. (1988). *Decision, Probability and Utility*, Cambridge University Press, Cambridge.

Garthwaite, P. H., Kadane, J. B. and O'Hagan, A. (2005). Statistical methods for eliciting probability distributions, *Journal of the American Statistical Association* **100**: 680–701.
    URL: *http://www.ingentaconnect.com/content/asa/jasa/2005/00000100/00000470/art00028*

Gauss, K. (1821). Theory of the combination of observations which leads to the smallest errors, *Gauss Werke*, pp. 1–93.

Gelman, A., Carlin, J., Stern, H. and Rubin, D. (1995). *Bayesian Data Analysis*, Chapman and Hall, London.

Gelman, A., Meng, X.-L. and Stern, H. (1996). Posterior predictive assessment of model fitness via realized discrepancies, *Statistica Sinica* **6**: 733–807.

Genest, C. and Zidek, J. V. (1986). Combining probability distributions: a critique and an annotated bibliography (C/R: P135-148), *Statistical Science* **1**: 114–135.

Genovese, C. and Wasserman, L. (2002). Operating characteristics and extensions of the false discovery rate procedure, *Journal of the Royal Statistical Society, Series B* **64**: 499–518.

Genovese, C. and Wasserman, L. (2003). Bayesian and frequentist multiple testing, *Bayesian Statistics 7*, Oxford University Press, pp. 145–161.

Glasziou, P. P., Simes, R. J. and Gelber, R. D. (1990). Quality adjusted survival analysis, *Statistics in Medicine* **9**(11): 1259–1276.

Goldstein, M. (1983). The prevision of a prevision, *Journal of the American Statistical Association* **78**: 817–819.

Goldstein, M. (1985). Temporal coherence, *in* J. Bernardo, M. DeGroot, D. Lindley and A. Smith (eds.), *Bayesian Statistics 2*, John Wiley & Sons, Ltd, Chichester, pp. 231–248.

Good, I. J. (1952). Rational decisions, *Journal of the Royal Statistical Society, Series B, Methodological* **14**: 107–114.

Good, I. J. (1967). On the principle of total evidence, *British Journal for the Philosophy of Science* **17**: 319–321.

Goodman, S. N. (1999). Towards evidence-based medical statistics. 2: The Bayes factor, *Annals of Internal Medicine* **130**(12): 1005–1013.

Grundy, P., Healy, M. and Rees, D. (1956). Economic choice of the amount of experimentation, *Journal of The Royal Statistical Society. Series B.* **18**: 32–55.

Hacking, I. (1976). *Logic of statistical inference*, Cambridge University Press.

Halmos, P. R. and Savage, L. J. (1949). Application of the Radon-Nikodym theorem to the theory of sufficient statistics, *Annals of Mathematical Statistics* **20**(2): 225–241.
URL: *http://www.jstor.org/stable/2236855*

Hastie, T., Tibshirani, R. and Friedman, J. (2003). *The Elements of Statistical Learning: Data Mining, Inference, and Prediction*, Springer-Verlag, New York.

Hirschhorn, J. N. and Daly, M. J. (2005). Genome-wide association studies for common diseases and complex traits, *Nature Reviews Genetics* **6**(2): 95–108.
URL: *http://dx.doi.org/10.1038/nrg1521*

Hochberg, Y. and Tamhane, A. C. (1987). *Multiple comparison procedures*, John Wiley & Sons, Inc., New York.

Hodges, J. L., J. and Lehmann, E. L. (1950). Some problems in minimax point estimation, *Annals of Mathematical Statistics* **21**: 182–197.

Howson, C. and Urbach, P. (1989). *Scientific reasoning: the Bayesian approach*, Open Court.

Huygens, C. (1657). De ratiociniis in ludo aleae, *Exercitationum Mathematicorum*.

Inoue, L. Y. T., Berry, D. A. and Parmigiani, G. (2005). Correspondences between Bayesian and frequentist sample size determination, *American Statistician* **59**: 79–87.

James, W. and Stein, C. (1961). Estimation with quadratic loss, *Proceedings of the Fourth Berkeley Symposium on Mathematical Statistics and Probability* **1**: 361–380.

Jeffreys, H. (1961). *Theory of Probability*, Oxford University Press, London.

Jensen, N. E. (1967). An introduction to Bernoullian utility theory: I Utility functions, *Swedish Journal of Economics* **69**: 163–183.

Johnson, B. M. (1971). On the admissible estimators for certain fixed sample binomial problems, *Annals of Mathematical Statistics* **42**: 1579–1587.

Jorland, G. (1987). The Saint Petersburg paradox 1713–1937, *The Probabilistic Revolution: Volume 1, Ideas in History*, MIT Press, pp. 157–190.

Joseph, L. and Wolfson, D. (1997). Interval-based versus decision-theoretic criteria for the choice of sample size, *The Statistician* **46**: 145–149.

Joseph, L., Wolfson, D. and Berger, R. (1995). Sample size calculations for binomial proportions via highest posterior density intervals, *The Statistician* **44**: 143–154.

Kadane, J. B. and Vlachos, P. (2002). Hybrid methods for calculating optimal few-stage sequential strategies: data monitoring for a clinical trial, *Statistics and Computing* **12**: 147–152.

Kadane, J. B., Schervish, M. J. and Seidenfeld, T. (1996). Reasoning to a foregone conclusion, *Journal of the American Statistical Association* **91**: 1228–1235.

Kadane, J. B. and Winkler, R. L. (1988). Separating probability elicitation from utilities, *Journal of the American Statistical Association* **83**: 357–363.

Kahneman, D., Slovic, P. and Tversky, A. (1982). *Judgment under Uncertainty: Heuristics and Biases*, Cambridge University Press, Cambridge.

Kahneman, D. and Tversky, A. (1979). Prospect theory: an analysis of decision under risk, *Econometrica* **47**: 263–291.

Kass, R. E. and Raftery, A. E. (1995). Bayes factors, *Journal of the American Statistical Association* **90**: 773–795.

Keeney, R. L., Raiffa, H. A. and Meyer, R. F. C. (1976). *Decisions with Multiple Objectives: Preferences and Value Tradeoffs*, John Wiley & Sons, Inc., New York.

Kolmogorov, A. N. (1930). Sur la notion de la moyenne, *Atti Accademia Nazionale dei Lincei* **9**: 388–391.

Kreps, D. M. (1988). *Notes on the theory of choice*, Westview Press, Boulder, CO.

Kullback, S. (1959). *Information Theory and Statistics*, John Wiley & Sons, Inc., New York.

Kullback, S. and Leibler, R. A. (1951). On information and sufficiency, *Annals of Mathematical Statistics* **22**: 79–86.

Lachin, J. (1998). Sample size determination, *in* P. Armitage and T. Colton (eds.), *Encyclopedia of Biostatistics*, John Wiley & Sons, Inc., New York.

Laplace, P.-S. (1812). *Théorie Analitique des Probabilités*, Courcier, Paris.

Lee, S. and Zelen, M. (2000). Clinical trials and sample size considerations: another perspective, *Statistical Science* **15**: 95–110.

Lehmann, E. L. (1983). *Theory of Point Estimation*, John Wiley & Sons, Inc., New York.

Lehmann, E. L. (1997). *Testing statistical hypotheses*, Springer-Verlag.

Levi, I. (1985). Imprecision and indeterminacy in probability judgment, *Philosophy of Science* **52**: 390–409.

Lewis, R. and Berry, D. (1994). Group sequential clinical trials: a classical evaluation of Bayesian decision-theoretic designs, *Journal of the American Statistical Association* **89**: 1528–1534.

Lindley, D. V. (1957). Binomial sampling schemes and the concept of information, *Biometrika* **44**: 179–186.

Lindley, D. V. (1961). Dynamic programming and decision theory, *Applied Statistics* **10**: 39–51.

Lindley, D. V. (1962). A treatment of a simple decision problem using prior probabilities, *The Statistician* **12**: 43–53.

Lindley, D. V. (1997). The choice of sample size, *The Statistician* **46**: 129–138.

Lindley, D. V. (1956). On a measure of the information provided by an experiment, *Annals of Mathematical Statistics* **27**: 986–1005.

Lindley, D. V. (1968a). The choice of variables in multiple regression (with discussion), *Journal of the Royal Statistical Society, Series B, Methodological* **30**: 31–66.

Lindley, D. V. (1968b). Decision making, *The Statistician* **18**: 313–326.

Lindley, D. V. (1980). L. J. Savage – his work in probability and statistics, *Annals of Statistics* **8**: 1–24.

Lindley, D. V. (1982a). The Bayesian approach to statistics, pp. 65–87.

Lindley, D. V. (1982b). Scoring rules and the inevitability of probability (with discussion), *International Statistical Review* **50**: 1–26.

Lindley, D. V. (1985). *Making Decisions,* second ed., John Wiley & Sons, Ltd, Chichester.

Lindley, D. V. (2000). The philosophy of statistics, *The Statistician* **49**(3): 293–337.
   URL: *http://www.jstor.org/stable/2681060*

Lusted, L. B. (1968). *Introduction to Medical Decision Making*, Thomas, Springfield, IL.

Lusted, L. B. (1971). Signal detectability and medical decision-making, *Science* **171**(977): 1217–1219.

Madigan, D. and Raftery, A. E. (1994). Model selection and accounting for model uncertainty in graphical models using Occam's window, *Journal of the American Statistical Association* **89**: 1535–1546.

Malenka, D. J., Baron, J. A., Johansen, S., Wahrenberger, J. W. and Ross, J. M. (1993). The framing effect of relative and absolute risk, *Journal of General Internal Medicine* **8**(10): 543–548.

McNeil, B., Weichselbaum, R. and Pauker, S. (1981). Speech and survival: tradeoffs between quality and quantity of life in laryngeal cancer, *New England Journal of Medicine* **305**: 982–987.

McNeil, B. J., Keeler, E. and Adelstein, S. J. (1975). Primer on certain elements of medical decision making, *New England Journal of Medicine* **293**: 211–215.

Mukhopadhyay, S. and Stangl, D. (1993). Balancing centers and observations in multicenter clinical trials, *Discussion paper*, ISDS, Duke University.

Muliere, P. and Parmigiani, G. (1993). Utility and means in the 1930s, *Statistical Science* **8**: 421–432.
   URL: *http://www.jstor.org/pss/2246176*

Müller, P. (1999). Simulation-based optimal design, *Bayesian Statistics 6 – Proceedings of the Sixth Valencia International Meeting*, pp. 459–474.

Müller, P., Berry, D. A., Grieve, A. P., Smith, M. and Krams, M. (2007a). Simulation-based sequential Bayesian design, *Journal of Statistical Planning and Inference* **137**(10): 3140–3150. Special Issue: Bayesian Inference for Stochastic Processes.
   URL: *http://www.sciencedirect.com/science/article/B6V0M-4N8M87Y-2/2/c917cdd 17116e8b1321faea40ee4ac7e*

Müller, P. and Parmigiani, G. (1993). Numerical evaluation of information theoretic measures, *in* D. A. Berry, K. M. Chaloner and J. K. Geweke (eds), *Bayesian Statistics and Econometrics: Essays in Honor of A. Zellner*, John Wiley & Sons, Inc., New York.

Müller, P. and Parmigiani, G. (1995). Optimal design via curve fitting of Monte Carlo experiments, *Journal of the American Statistical Association* **90**: 1322–1330.
   URL: *http://www.jstor.org/pss/2291522*

Müller, P. and Parmigiani, G. (1996). Numerical evaluation of information theoretic measures, *Bayesian Statistics and Econometrics: Essays in Honor of A. Zellner. D.A. Berry, K.M. Chaloner and J.F. Geweke*, John Wiley & Sons, Inc., New York, pp. 394–406.

Müller, P., Parmigiani, G. and Rice, K. (2007b). FDR and Bayesian multiple comparisons rules, *in* J. M. Bernardo, S. Bayarri, J. O. Berger, A. Dawid, D. Heckerman, A. F. M. Smith and M. West (eds.), *Bayesian Statistics 8*, Oxford University Press.

Müller, P., Parmigiani, G., Robert, C. and Rousseau, J. (2005). Optimal sample size for multiple testing: the case of gene expression microarrays, *Journal of the American Statistical Association* **99**: 990–1001.

Naglie, G., Krahn, M. D., Naimark, D., Redelmeier, D. A. and Detsky, A. S. (1997). Primer on medical decision analysis: Part 3 estimating probabilities and utilities, *Medical Decision Making* **17**(2): 136–141.

Nagumo, M. (1930). Uber eine Klasse der Mittelwerte, *Japanese Journal of Mathematics* **7**: 72–79.

Nakhaeizadeh, G. and Taylor, C. (eds.) (1997). *Machine learning and statistics: the interface*, John Wiley & Sons, Inc., New York.

Nemhauser, G. (1966). *Introduction to Dynamic Programming*, John Wiley & Sons, Inc., New York.

Neyman, J. and Pearson, E. S. (1933). On the problem of the most efficient test of statistical hypotheses, *Philosophical Transaction of the Royal Society (Series A)* **231**: 286–337.

O'Brien, P. and Fleming, T. (1979). A multiple testing procedure for clinical trials, *Biometrics* **35**: 549–556.

Parmigiani, G. (1992). Minimax, information and ultrapessimism, *Theory and Decision* **33**: 241–252.
URL: *http://www.springerlink.com/content/p10203301581164l/*

Parmigiani, G. (2002). *Modeling in Medical Decision Making*, John Wiley & Sons, Ltd, Chichester.

Parmigiani, G. (2004). Uncertainty and the value of diagnostic information, with application to axillary lymph node dissection in breast cancer, *Statistics in Medicine* **23**(5): 843–855.
URL: *http://dx.doi.org/10.1002/sim.1623*

Parmigiani, G. and Berry, D. A. (1994). Applications of Lindley information measure to the design of clinical experiments, *in* P. R. Freeman and A. F. M. Smith (eds.), *Aspects of Uncertainty. A Tribute to D. V. Lindley*, John Wiley & Sons, Ltd, Chichester, pp. 351–362.

Parmigiani, G., Berry, D., Iversen, Jr., E. S., Müller, P., Schildkraut, J. and Winer, E. (1998). Modeling risk of breast cancer and decisions about genetic testing, *in* C. Gatsonis *et al.* (eds.), *Case Studies in Bayesian Statistics*, Vol. IV, Springer-Verlag, pp. 173–268.

Pencina, M. J., D'Agostino, R. B. and Vasan, R. S. (2008). Evaluating the added predictive ability of a new marker: from area under the ROC curve to reclassification and beyond, *Statistics in Medicine* **27**(2): 157–172; discussion 207–12.
URL: *http://dx.doi.org/10.1002/sim.2929*

Pepe, M. (2003). *The statistical evaluation of medical tests for classification and prediction*, Oxford University Press.

Pfanzagl, J. (1959). A general theory of measurement: application to utility, *Naval Research Logistics Quarterly* **6**: 283–294.

Pham-Gia, T. and Turkkan, N. (1992). Sample size determination in Bayesian analysis, *The Statistician* **41**: 389–397.

Pitman, E. J. G. (1965). Some remarks on statistical inferences, *in* L. M. LeCam and J. Neyman (eds.), *Proceedings of the International Research Seminar, Statistics Lab, University of California, Berkeley*, Springer-Verlag, Berlin.

Pliskin, J. S., Shepard, D. and Weinstein, M. C. (1980). Utility functions for life years and health status: theory, assessment, and application, *Operations Research* **28**: 206–224.

Pocock, S. (1977). Group-sequential methods in the design and analysis of clinical trials, *Biometrika* **64**: 191–199.

Poskitt, D. (1987). Precision, complexity and Bayesian model determination, *Journal of the Royal Statistical Society. Series B.* **49**: 199–208.

Pratt, J. (1962). Must subjective probabilities be realized as relative frequencies? Unpublished seminar paper, Harvard University Graduate School of Business Administration, *Technical report*.

Pratt, J. (1964). Risk aversion in the small and in the large, *Econometrica* **32**: 122–136.

Pratt, J. W., Raiffa, H. and Schlaifer, R. (1964). The foundations of decision under uncertainty: an elementary exposition, *Journal of the American Statistical Association* **59**: 353–375.
URL: *http://www.jstor.org/stable/pdfplus/2282993.pdf*

Raiffa, H. (1970). *Decision analysis: introductory lectures on choices under uncertainty*, Addison-Wesley, Menlo Park, CA.

Raiffa, H. and Schlaifer, R. (1961). *Applied Statistical Decision Theory*, Harvard University Press, Boston, MA.

Ramsey, F. (1926). *The Foundations of Mathematics*, Routledge and Kegan Paul, London, chapter Truth and Probability, pp. 156–211.
URL: *http://socserv2.socsci.mcmaster.ca/ econ/ugcm/3ll3/ramseyfp/ramsess.pdf*

Ramsey, F. P. (1931). *Truth and probability*, John Wiley & Sons, London.

Ramsey, F. P. E. (1991). *Notes on Philosophy, Probability and Mathematics*, Bibliopolis.

Rice, K. M., Lumley, T. and Szpiro, A. A. (2008). Trading bias for precision: decision theory for intervals and sets, *Working Paper 336*, UW Biostatistics.
URL: *http://www.bepress.com/uwbiostat/paper336*

Robbins, H. (1956). An empirical Bayes approach to statistics, *Proceedings of the Third Berkeley Symposium on Mathematical Statistics and Probability, Volume 1*, pp. 157–163.

Robert, C. and Casella, G. (1999). *Monte Carlo Statistical Methods*, Springer-Verlag, New York.

Robert, C. P. (1994). *The Bayesian Choice*, Springer-Verlag.

Sacks, J. (1963). Generalized Bayes solutions in estimation problems, *Annals of Mathematical Statistics* **34**: 751–768.

Sahlin, N. E. (1990). *The Philosophy of F. P. Ramsey*, Cambridge University Press.

Samsa, G., Matchar, D., Goldstein, L., Bonito, A., Duncan, P., Lipscomb, J., Enarson, C., Witter, D., Venus, P., Paul, J. and Weinberger, M. (1988). Utilities for major stroke: results from a survey of preferences among persons at increased risk for stroke, *American Heart Journal* **136**(4 Pt 1): 703–713.

San Martini, A. and Spezzaferri, F. (1984). A predictive model selection criterion, *Journal of the Royal Statistical Society, Series B.* **46**: 296–303.

Savage, L. J. (1951). The theory of statistical decision, *Journal of the American Statistical Association* **46**: 55–67.

Savage, L. J. (1954). *The foundations of statistics*, John Wiley & Sons, Inc., New York.

Savage, L. J. (1971). Elicitation of personal probabilities and expectations, *Journal of the American Statistical Association* **66**: 783–801.

Savage, L. J. (1972). *The foundations of statistics*, Dover.

Savage, L. J. (1981a). A panel discussion of personal probability, *The writings of Leonard Jimmie Savage – A memorial selection*, American Statistical Association, Alexandria, VA, pp. 508–513.

Savage, L. J. (1981b). *The writings of Leonard Jimmie Savage – A memorial selection*, American Statistical Association, Alexandria, VA.

Schervish, M. J. (1995). *Theory of Statistics*, Springer-Verlag.

Schervish, M. J., Seidenfeld, T. and Kadane, J. B. (1990). State-dependent utilities, *Journal of the American Statistical Association* **85**: 840–847.

Schwartz, M. (1978). A mathematical model used to analyze breast cancer screening strategies, *Operations Research* **26**: 937–955.

Seidenfeld, T. (1985). Calibration, coherence, and scoring rule, *Philosophy of Science* **52**: 274–294.
    URL: *http://www.jstor.org/stable/pdfplus/187511.pdf*

Seidenfeld, T. (1988). Decision theory without "independence" or without "ordering". What is the difference? *Economics and Philosophy* **4**: 267–290.

Seidenfeld, T. (2001). Remarks on the theory of conditional probability: some issues of finite versus countable additivity, *Phil-Sci Archive*, p. 92.

Seidenfeld, T., Schervish, M. J. and Kadane, J. B. (1990a). When fair betting odds are not degrees of belief, *PSA: Proceedings of the Biennial Meeting of the Philosophy of Science Association*, pp. 517–524.

Seidenfeld, T., Schervish, M. and Kadane, J. B. (1990b). Decisions without ordering, *in* W. Sieg (ed.), *Acting and Reflecting*, pp. 143–170.

Seidenfeld, T., Schervish, M. J. and Kadane, J. B. (1995). A representation of partially ordered preferences, *Annals of Statistics* **23**: 2168–2217.

Shafer, G. (1986). Savage revisited, *Statistical Science* **1**: 463–485.

Shimony, A. (1955). Coherence and the axioms of confirmation, *Journal of Symbolic Logic* **20**: 1–28.

Shoemaker, P. (1982). The expected utility model: its variants, purposes, evidence and limitations, *Journal of Economic Literature* **20**: 529–563.

Shuster, J. (1993). *Handbook of sample size guidelines for clinical trials*, CRC Press, Boca Raton, FL.

Smith, A. (1983). Bayesian approaches to outliers and robustness, *in* J.-P. Flarens, M. Mouchart, J.-P. Raoult, L. Simar and A.F.M. Smith (eds.), *Specifying Statistical Models*, Springer-Verlag, Berlin, pp. 13–55.

Smith, J. Q. (1987). *Decision Analysis: A Bayesian Approach*, Chapman & Hall.

Sox, H. (1996). The evaluation of diagnostic tests: principles, problems, and new developments, *Annual Reviews in Medicine* **47**: 463–471.

Sox, H. C., Blatt, M. A., Higgins, M. C. and Marton, K. I. (1988). *Medical Decision Making*, Butterworth–Heinemann, Boston, MA.

Spiegelhalter, D., Best, N., Cerlin, B. and Linde, A. (2002). Bayesian measures of complexity and fit, *Journal of the Royal Statistical Society, Series B* **64**: 583–640.

Stangl, D. K. (1995). Modeling and decision making using Bayesian hierarchical models, *Statistics in Medicine* **14**: 2173–2190.

Stein, C. (1955). Inadmissibility of the usual estimator for the mean of a multivariate normal distribution, *Proceedings of the Third Berkeley Symposium on Mathematical Statistics and Probability* **1**: 197–206.

Stein, C. M. (1981). Estimation of the mean of a multivariate normal distribution, *Annals of Statistics* **9**: 1135–1151.

Swartz, R. J., Cox, D. D., Cantor, S. B., Davies, K. and Follen, M. (2006). Inverse decision theory: characterizing losses for a decision rule with applications in cervical cancer screening, *Journal of the American Statistical Association* **101**(473): 1–8.

Tan, S. and Smith, A. (1998). Exploratory thoughts on clinical trials with utilities, *Statistics in Medicine* **17**: 2771–2791.

Torrance, G. (1971). A generalized cost-effectiveness model for the evaluation of health programs, *Doctoral dissertation*, State University of New York at Buffalo.

Torrance, G., Thomas, W. and Sackett, D. (1972). A utility maximization model for evaluation of health care programs, *Health Services Research* **7**: 118–133.

Torrance, G. W. (1986). Measurement of health state utilities for economic appraisal, *Journal of Health Economics* **5**: 1–30.

Tukey, J. W. (1977). *Exploratory Data Analysis*, Addison-Wesley.

Tversky, A. (1974). Assessing uncertainty (followed by discussion) (pkg: P148-191), *Journal of the Royal Statistical Society, Series B, Methodological* **36**: 148–159.

von Neumann, J. and Morgenstern, O. (1944). *Theory of Games and Economic Behavior*, John Wiley & Sons, Inc., New York.

Wakker, P. (1993). Unbounded utility for savageś foundations of statistics, and other models, *Mathematics of Operations Research* **18**: 446–485.

Wald, A. (1939). Contributions to the theory of statistical estimation and testing hypotheses, *Annals of Mathematical Statistics* **10**: 299–326.

Wald, A. (1945). Sequential tests of statistical hypotheses, *Annals of Mathematical Statistics* **16**: 117–186.

Wald, A. (1947a). Foundations of a general theory of sequential decision functions, *Econometrica* **15**: 279–313.

Wald, A. (1947b). *Sequential Analysis*, John Wiley & Sons, Inc., New York.

Wald, A. (1949). Statistical decision functions, *Annals of Mathematical Statistics* **20**: 165–205.

Wald, A. (1950). *Statistical Decision Functions*, John Wiley & Sons, Inc., New York.

Wallis, W. A. (1980). The Statistical Research Group, 1942-1945 (C/R: p331-335), *Journal of the American Statistical Association* **75**: 320–330.

Ware, J. H. (1989). Investigating therapies of potentially great benefit: ECMO (C/R: p306-340), *Statistical Science* **4**: 298–306.
    URL: *http://projecteuclid.org/euclid.ss/1177012384*

Weinstein, M. C., Feinberg, H., Elstein, A. S., Frazier, H. S., Neuhauser, D., Neutra, R. R. and McNeil, B. J. (1980). *Clinical Decision Analysis*, Saunders, Philadelphia, PA.

Weiss, L. (1992). Introduction to Wald (1949) statistical decision functions, *in* S. Kotz and N. Johnson (eds.), *Breakthroughs in Statistics – Foundations and Basic Theory, Volume 1*, Vol. 1 of *Springer Series in Statistics – Perspectives in Statistics*, Springer-Verlag, New York, pp. 335–341.

West, M. (1985). Generalized linear models: scale parameters, outlier accommodation and prior distributions, *Bayesian Statistics 2*, Elsevier/North-Holland, pp. 329–348.

West, M. (1992). Modelling with mixtures, *in* J.O. Berger, J.M. Bernardo, D.V. Lindley and A.F.M. Smith (eds.), *Bayesian Statistics 4*, John Wiley & Sons, Inc., New York, pp. 503–524.

Winkler, R. L. (1969). Scoring rules and the evaluation of probability assessors, *Journal of the American Statistical Association* **64**: 1073–1078.

Winkler, R. L. (1996). Scoring rules and the evaluation of probabilities, *Test* **5**: 1–26.

Yao, T.-J., Begg, C. B. and Livingston, P. O. (1996). Optimal sample size for a series of pilot trials of new agents, *Biometrics* **52**: 992–1001.

Index

Note: The entries in *italics* denote figures

WILEY SERIES IN PROBABILITY AND STATISTICS

Established by WALTER A. SHEWHART and SAMUEL S. WILKS

Editors
David J. Balding, Noel A. C. Cressie, Nicholas I. Fisher, Iain M. Johnstone, J. B. Kadane, Geert Molenberghs, David W. Scott, Adrian F. M. Smith, Sanford Weisberg, Harvey Goldstein

Editors Emeriti
Vic Barnett, J. Stuart Hunter, David G. Kendall, Jozef L. Teugels

The Wiley Series in Probability and Statistics is well established and authoritative. It covers many topics of current research interest in both pure and applied statistics and probability theory. Written by leading statisticians and institutions, the titles span both state-of-the-art developments in the field and classical methods.

Reflecting the wide range of current research in statistics, the series encompasses applied, methodological and theoretical statistics, ranging from applications and new techniques made possible by advances in computerized practice to rigorous treatment of theoretical approaches.

This series provides essential and invaluable reading for all statisticians, whether in academia, industry, government, or research.

*Now available in a lower priced paperback edition in the Wiley Classics Library.

BASILEVSKY · Statistical Factor Analysis and Related Methods: Theory and Applications

BASU and RIGDON · Statistical Methods for the Reliability of Repairable Systems

BATES and WATTS · Nonlinear Regression Analysis and Its Applications

BECHHOFER, SANTNER and GOLDSMAN · Design and Analysis of Experiments for Statistical Selection, Screening and Multiple Comparisons

BELSLEY · Conditioning Diagnostics: Collinearity and Weak Data in Regression

BELSLEY, KUH and WELSCH · Regression Diagnostics: Identifying Influential Data and Sources of Collinearity

BENDAT and PIERSOL · Random Data: Analysis and Measurement Procedures, Third Edition

BERNARDO and SMITH · Bayesian Theory

BERRY, CHALONER and GEWEKE · Bayesian Analysis in Statistics and Econometrics: Essays in Honor of Arnold Zellner

BHAT and MILLER · Elements of Applied Stochastic Processes, Third Edition

BHATTACHARYA and JOHNSON · Statistical Concepts and Methods

BHATTACHARYA and WAYMIRE · Stochastic Processes with Applications

BIEMER, GROVES, LYBERG, MATHIOWETZ and SUDMAN · Measurement Errors in Surveys

BILLINGSLEY · Convergence of Probability Measures, Second Edition

BILLINGSLEY · Probability and Measure, Third Edition

BIRKES and DODGE · Alternative Methods of Regression

BLISCHKE and MURTHY (editors) · Case Studies in Reliability and Maintenance

BLISCHKE and MURTHY · Reliability: Modeling, Prediction and Optimization

BLOOMFIELD · Fourier Analysis of Time Series: An Introduction, Second Edition

BOLLEN · Structural Equations with Latent Variables

BOLLEN and CURRAN · Latent Curve Models: A Structural Equation Perspective

BOROVKOV · Ergodicity and Stability of Stochastic Processes

BOSQ and BLANKE · Inference and Prediction in Large Dimensions

BOULEAU · Numerical Methods for Stochastic Processes

BOX · Bayesian Inference in Statistical Analysis

BOX · R. A. Fisher, the Life of a Scientist

BOX and DRAPER · Empirical Model-Building and Response Surfaces

*BOX and DRAPER · Evolutionary Operation: A Statistical Method for Process Improvement

BOX, HUNTER and HUNTER · Statistics for Experimenters: An Introduction to Design, Data Analysis and Model Building

BOX, HUNTER and HUNTER · Statistics for Experimenters: Design, Innovation and Discovery, Second Edition

BOX and LUCEÑO · Statistical Control by Monitoring and Feedback Adjustment

BRANDIMARTE · Numerical Methods in Finance: A MATLAB-Based Introduction

BROWN and HOLLANDER · Statistics: A Biomedical Introduction

BRUNNER, DOMHOF and LANGER · Nonparametric Analysis of Longitudinal Data in Factorial Experiments

BUCKLEW · Large Deviation Techniques in Decision, Simulation and Estimation

CAIROLI and DALANG · Sequential Stochastic Optimization

*Now available in a lower priced paperback edition in the Wiley Classics Library.

*Now available in a lower priced paperback edition in the Wiley Classics Library.

*Now available in a lower priced paperback edition in the Wiley Classics Library.

*Now available in a lower priced paperback edition in the Wiley Classics Library.

*Now available in a lower priced paperback edition in the Wiley – Interscience Paperback Series.

LEE and WANG · Statistical Methods for Survival Data Analysis, Third Edition
LEPAGE and BILLARD · Exploring the Limits of Bootstrap
LEYLAND and GOLDSTEIN (editors) · Multilevel Modelling of Health Statistics
LIAO · Statistical Group Comparison
LINDVALL · Lectures on the Coupling Method
LINHART and ZUCCHINI · Model Selection
LITTLE and RUBIN · Statistical Analysis with Missing Data, Second Edition
LLOYD · The Statistical Analysis of Categorical Data
LOWEN and TEICH · Fractal-Based Point Processes
MAGNUS and NEUDECKER · Matrix Differential Calculus with Applications in Statistics and Econometrics, Revised Edition
MALLER and ZHOU · Survival Analysis with Long Term Survivors
MALLOWS · Design, Data and Analysis by Some Friends of Cuthbert Daniel
MANN, SCHAFER and SINGPURWALLA · Methods for Statistical Analysis of Reliability and Life Data
MANTON, WOODBURY and TOLLEY · Statistical Applications Using Fuzzy Sets
MARCHETTE · Random Graphs for Statistical Pattern Recognition
MARKOVICH · Nonparametric Analysis of Univariate Heavy-Tailed Data: Research and practice
MARDIA and JUPP · Directional Statistics
MARKOVICH · Nonparametric Analysis of Univariate Heavy-Tailed Data: Research and Practice
MARONNA, MARTIN and YOHAI · Robust Statistics: Theory and Methods
MASON, GUNST and HESS · Statistical Design and Analysis of Experiments with Applications to Engineering and Science, Second Edition
MCCULLOCH and SERLE · Generalized, Linear and Mixed Models
MCFADDEN · Management of Data in Clinical Trials
MCLACHLAN · Discriminant Analysis and Statistical Pattern Recognition
MCLACHLAN, DO and AMBROISE · Analyzing Microarray Gene Expression Data
MCLACHLAN and KRISHNAN · The EM Algorithm and Extensions
MCLACHLAN and PEEL · Finite Mixture Models
MCNEIL · Epidemiological Research Methods
MEEKER and ESCOBAR · Statistical Methods for Reliability Data
MEERSCHAERT and SCHEFFLER · Limit Distributions for Sums of Independent Random Vectors: Heavy Tails in Theory and Practice
MICKEY, DUNN and CLARK · Applied Statistics: Analysis of Variance and Regression, Third Edition
*MILLER · Survival Analysis, Second Edition
MONTGOMERY, PECK and VINING · Introduction to Linear Regression Analysis, Fourth Edition
MORGENTHALER and TUKEY · Configural Polysampling: A Route to Practical Robustness
MUIRHEAD · Aspects of Multivariate Statistical Theory
MULLER and STEWART · Linear Model Theory: Univariate, Multivariate and Mixed Models
MURRAY · X-STAT 2.0 Statistical Experimentation, Design Data Analysis and Nonlinear Optimization

*Now available in a lower priced paperback edition in the Wiley Classics Library.

MURTHY, XIE and JIANG · Weibull Models

MYERS and MONTGOMERY · Response Surface Methodology: Process and Product Optimization Using Designed Experiments, Second Edition

MYERS, MONTGOMERY and VINING · Generalized Linear Models. With Applications in Engineering and the Sciences

†NELSON · Accelerated Testing, Statistical Models, Test Plans and Data Analysis

†NELSON · Applied Life Data Analysis

NEWMAN · Biostatistical Methods in Epidemiology

OCHI · Applied Probability and Stochastic Processes in Engineering and Physical Sciences

OKABE, BOOTS, SUGIHARA and CHIU · Spatial Tesselations: Concepts and Applications of Voronoi Diagrams, Second Edition

OLIVER and SMITH · Influence Diagrams, Belief Nets and Decision Analysis

PALTA · Quantitative Methods in Population Health: Extentions of Ordinary Regression

PANJER · Operational Risks: Modeling Analytics

PANKRATZ · Forecasting with Dynamic Regression Models

PANKRATZ · Forecasting with Univariate Box-Jenkins Models: Concepts and Cases

*PARZEN · Modern Probability Theory and Its Applications

PEÑA, TIAO and TSAY · A Course in Time Series Analysis

PIANTADOSI · Clinical Trials: A Methodologic Perspective

PORT · Theoretical Probability for Applications

POURAHMADI · Foundations of Time Series Analysis and Prediction Theory

PRESS · Bayesian Statistics: Principles, Models and Applications

PRESS · Subjective and Objective Bayesian Statistics, Second Edition

PRESS and TANUR · The Subjectivity of Scientists and the Bayesian Approach

PUKELSHEIM · Optimal Experimental Design

PURI, VILAPLANA and WERTZ · New Perspectives in Theoretical and Applied Statistics

PUTERMAN · Markov Decision Processes: Discrete Stochastic Dynamic Programming

QIU · Image Processing and Jump Regression Analysis

RAO · Linear Statistical Inference and its Applications, Second Edition

RAUSAND and HØYLAND · System Reliability Theory: Models, Statistical Methods and Applications, Second Edition

RENCHER · Linear Models in Statistics

RENCHER · Methods of Multivariate Analysis, Second Edition

RENCHER · Multivariate Statistical Inference with Applications

RIPLEY · Spatial Statistics

RIPLEY · Stochastic Simulation

ROBINSON · Practical Strategies for Experimenting

ROHATGI and SALEH · An Introduction to Probability and Statistics, Second Edition

ROLSKI, SCHMIDLI, SCHMIDT and TEUGELS · Stochastic Processes for Insurance and Finance

ROSENBERGER and LACHIN · Randomization in Clinical Trials: Theory and Practice

ROSS · Introduction to Probability and Statistics for Engineers and Scientists

ROSSI, ALLENBY and MCCULLOCH · Bayesian Statistics and Marketing

ROUSSEEUW and LEROY · Robust Regression and Outline Detection

RUBIN · Multiple Imputation for Nonresponse in Surveys

*Now available in a lower priced paperback edition in the Wiley Classics Library.

†Now available in a lower priced paperback edition in the Wiley – Interscience Paperback Series.

*Now available in a lower priced paperback edition in the Wiley Classics Library.

*Now available in a lower priced paperback edition in the Wiley Classics Library.

Printed and bound by CPI Group (UK) Ltd, Croydon, CR0 4YY

27/10/2024

14580285-0005